Advances in Aquatic Ecology

— Volume 7 —

The Editors

Dr. V.B. Sakhare is Head of Post Graduate Department of Zoology, Yogeshwari Mahavidyalaya, Ambajogai. He has 14 years' experience as an outstanding teacher and researcher. He is recipient of fellowship of Indian Association of Aquatic Biologists, Hyderabad. He has done pioneering work in the field of Reservoir Fisheries and Limnology. Dr. Sakhare has successfully organized *National Conference on Emerging Trends in Fisheries and Aquaculture (ETFA-2012)*, *National Conference on Current Perspectives in Limnology (NCCPL-2009)* and *Regional Workshop* on *Water Quality Assessment (Implications in potability, productivity and pollution control)*.

Dr. Sakhare has been editing an international journal *'Ecology and Fisheries'* (ISSN 0974-6323).He is member of Editorial Advisory Board of *Journal of Science Information* (ISSN 2229-5836) and *E-international Scientific Research Journal (EISRJ)* published by BCTA Inc. Philippines (ISSN 2094-1749). Dr. Sakhare has authored/edited few books such as *'Applied Fisheries'*, *'Reservoir Fisheries and Limnology'*, *'Reservoir Fisheries and Ecology: A Literary survey'*, *'Methodology for Water Analysis'*, *'Aquatic Ecology'*, *'Aquatic Biology and Aquaculture'*, *'Inland Fisheries'*, *' Applied Ecology'*, *'Perspectives in Ecology'*, and *'Advances in Aquatic Ecology (Vols. I–VII)*.

Dr. Sakhare has supervised a research project funded by University Grants Commission, New Delhi and he is a recognized post graduate teacher and research guide of Dr. Babasaheb Ambedkar Marathwada University, Aurangabad, Solapur University, Solapur and J.J.T. University, Rajasthan. One student has completed Ph.D. under his guidance and three students are working for doctoral degree. He has published 30 research articles and reviews in peer reviewed journals and about 70 marathi articles in newspapers and magazines.

Dr. Sakhare has chaired a number of sessions of different seminars/symposia. He has been invited to different colleges/institutes to deliver lectures on different topics in aquatic ecology and reservoir fisheries.

Dr. B. Vasanthkumar is working as Head and Associate Professor in the Department of Zoology, Government Degree Arts and Science College, Karwar. He was awarded with Ph.D. from Gulbarga University. He has published more than 30 research papers and 30 popular articles in different aspects of ecology and environment. He has also published 11 text books for under graduate students of Karnataka University and guided 4 students for M.Phil. Dr. Vasanthkumar is also co-author of a reference books like *'Aquatic Ecosystem and its management'* and *'Applied Ecology'*. Presently he is working with Major Research Project sanctioned by University Grants Commission, New Delhi.

Advances in Aquatic Ecology

— Volume 7 —

Editors

Dr. Vishwas B. Sakhare
Head,
Post Graduate Department of Zoology
Yogeshwari Mahavidyalaya,
Ambajogai – 431 517
Maharashtra
INDIA

&

Dr. B. Vasanthkumar
Head
Department of Zoology
Government Degree Arts and Science College,
Karwar – 581 301
INDIA

2013

Daya Publishing House®
A Division of
Astral International Pvt. Ltd.
New Delhi – 110 002

Published by   :   **Daya Publishing House**®
                            A Division of
                            Astral International Pvt. Ltd. –
                            ISO 9001:2008 Certified Company –
                            4760-61/23, Ansari Road, Darya Ganj
                            New Delhi-110 002
                            Ph. 011-43549197, 23278134
                            E-mail: info@astralint.com
                            Website: www.astralint.com

Laser Typesetting   :   **Classic Computer Services**
                            Delhi - 110 035

Printed at   :   **Salasar Imaging Systems**
                            Delhi - 110 035

PRINTED IN INDIA

Preface

We are delighted to write about the seventh volume of *Advances in Aquatic Ecology*. This volume is the compilation of esteemed articles of internationally acknowledged experts in the field of aquatic ecology with the intention of providing a sufficient depth of the subject to satisfy the need of a level which will be comprehensive and intresting. It is an assemblage of up to date information of rapid advances and developments taking place in the field of aquatic ecology. With its application oriented and interdisciplinary approach, we hope that the students, teachers, researchers, scientists, policy makers and environmental lawyers in India and abroad will find this volume much more useful. The articles in the book have been contributed by eminent scientists/academicians active in the areas of aquatic ecology.

We express our sincere gratitude to Dr. S. T. Khursale, President, Yogeshwari Education Society, Ambajogai who has been source of constant inspiration. We are especially thankful to Dr. R. V. Kirdak, Principal, Yogeshwari Mahavidyalaya, Ambajogai, Prof. Siddaramu, Principal, Government Arts and Science College, Karwar, Prof. K. Vijaykumar, Department of Zoology, Gulbarga University, Gulbarga and Dr. P. K. Joshi, Dnyanopasak Mahavidyalaya, Ambajogai for the encouragement. We thank our publisher Shri Anil Mittal of Daya Publishing House, Delhi for taking pains in bringing out the book.

Our special thanks and appreciation goes to experts and research workers whose contributions have enriched this volume. We are thankful to Dr. Pranab Dutta of West Tripura, Dr. M. V. Radhakrishnan of Annamalai University, Prof. Ravi Shankar Piska of Osmania University, Hyderabad, Dr. P. Neeraja of Sri Venkateshwara University, Tirupati, Dr. A. F. Parmod of Kuvempu University, Shankaraghatta, Dr. Mrinalini Sathe of K. J. Somaiya College of Science and Commerce, Mumbai, Dr. P. Santhi of J.A. College for Women, Periyakulum, Dr. Chhaya Panse of M. D. College, Prel, Mumbai, Dr. S. A. Deshmukh of Kolhapur, Dr. V. S. Hamde of Yogeshwari Mahavidayalaya, Ambajogai, Dr. Medha Tendulkar and Dr. A. S. Kulkarni of R. P. Gogate College of Arts and Science, Ratnagiri, Dr. Krishna Ram of University of Mysore, Mysore, Dr. S. A. Mohite and Suresh Kumar P. S. of College of

Fisheries, Ratnagiri, Dr. V. N. Naik of Karnataka University, Karwar, Dr. Leena Muralidharan of V. K. Krishna Menon College of Commerce and Science, Bhandup (E), Mumbai, Dr. Padmavathi Godavarthy and Y. Sunila Kumari of Osmania Univeristy, Hyderabad, Prof. D. V. Muley and S. A. Manjare of Shivaji University, Kolhapur and Dr. R. P. Mali of Yeshwant Mahavidaylaya, Nanded.

Finally, we will always remain a debtor to all our well-wishers for their blessings, without which this volume would not have come into existence.

Dr. V.B. Sakhare
Dr. B. Vasanthkumar

Contents

Preface   *v*

List of Contributors   *xi*

1. **Performance of Periphyton Based Aquaculture in West Tripura**   1
   Pranab Dutta and Subrato Chowdhury

2. **Effect of Sugarcane Bagasse on Growth Performance of *Clarias batrachus* (Bloch) Fingerlings**   4
   M.V. Radhakrishnan and E. Sugumaran

3. **Studies on the Impact of Fingerlings Stocking on the Production of Major Carps in a Minor Reservoir**   10
   Ravi Shankar Piska, Sreenu Noothi, A. Sunil Kumar and Savalla Murali Krishna

4. **Studies on Effect of Insecticides on Oxygen Consumption of Freshwater Fish, *Lepidocephalichthys guntea* (Ham-Buch)**   17
   M.U. Patil and S.S. Patole

5. **Ambient Ammonia Stress: Its Impact on Ammonia and Urea Levels and Extent of Recovery in Fish *Oreochromis mossambicus***   22
   G. Ravindra Babu and P. Neeraja

6. **Seasonal Variation of Physico-chemical Characteristics of Water in Wetlands of Bhadra Wildlife Sanctuary, Karnataka**   27
   A.F. Pramod and Vijaya Kumara

7. Water Quality of GIP Tank in Ambarnath, Thane District, Maharashtra   40
   Manisha Karpe and Mrinalini Sathe

8. Physico-chemical Characteristics of Coastal Environment of Thootukudi   51
   V. Santhi, V. Sivakumar, R.D. Thilaga and S. Kathiresan

9. Ecological Study of Mankeshwar Beach, Uran, Maharashtra   58
   Chhaya Panse, Vaishali Somani, Nitin Walmiki and Amol Kumbhar

10. Seasonal Variation in Water Quality of Rankala Lake Kolhapur   60
    S.A. Deshmukh and D.S. Patil

11. Physico-chemical and Microbiological Examination of Water from Raviwarpeth
    Lake Ambajogai   62
    V.S. Hamde

12. Effect of Varying Concentrations of Dietary Protein on Growth, Survival and
    Biochemical Aspects of Banana Prawn, *Fenneropenaeus merguiensis* (De Man, 1888)   68
    Medha Tendulkar and A.S. Kulkarni

13. Induced Toxicity of Imidacloprid on Protein Metabolism in Estuarine Clam,
    Katelysia opima (Gmelin)   75
    A.S. Kulkarni and M.V. Tendulkar

14. Population of *Lactobacillus* sp. in Two Freshwater Fishes, *Glossogobius giuris* (Ham)
    and *Labeo rohita* (Ham)   80
    Krishna Ram H., Shivabasavaiah, R. Manjunath and M. Ramachandra Mohan

15. Study of Food and Feeding Habit in *Nemipterus japonicus* along the
    Ratnagiri Coast of Maharashtra   89
    P.S. Suresh Kumar and S.A. Mohite

16. Studies of Monthly Variations in DO, BOD and COD Parameters of
    Gangapur Dam Water at Nashik   97
    R.S. Bhadane

17. Ecological and Protein Estimation Studies in Genus *Cassia* L. from Kolhapur District   104
    Sagar Anant Deshmukh

18. A Checklist of Avifauna in the Mangrove Areas of Aghanashini Estuarine Complex,
    Uttara Kannada, Karnataka, West Coast of India   108
    T. Roshmon and V.N. Nayak

19. Identification of *Vibrio harveyi* in Sea Water Sample by PCR Targeted to *vhh* Gene   119
    Sreenath Pillai and Leena Muralidharan

20. Chronic Effects of Organophosphorus insecticide 'Fenthion' in Melanophore
    Pigments of *Cyprinus carpio* (Linn)   124
    Leena Muralidharan and Sreenath Pillai

21. A Study on Organochlorine Pesticide Accumulation and its Effect on Nutrient Value
    of Edible Fish *Catla catla* (Ham) Sold in Local Market of Mumbai   131
    Leena Muralidharan

22. *In vivo* Studies on the Therapeutic Values of Marine Macroalgae against the
    Fish Pathogen *Aeromonas hydrophila*   135
    M.Parimala Cella and A.P. Lipton

23. Inverse Relationship of Energy Reserves and Body Water Content in
    Anabas testudineus (Bloch) Under Starvation Stress   144
    Padmavathi Godavarthy and Y. Sunila Kumari

24. Physico-chemical Parameters and Zooplankton Population in a Tamdalge Tank,
    Maharashtra   150
    S.A. Manjare and D.V. Muley

25. Physico-chemical Condition and Occurrence of Some Zooplankton in a
    Laxmiwadi Tank, Maharashtra   156
    S.A. Manjare, S.A. Vhanalakar and D.V. Muley

26. Ichthyaofauna of Osmanabad district, Maharashtra   162
    J.S. Mohite, V.B. Sakhare and S.G. Rawate

27. Plankton Diversity in Hangarga Reservoir in Osmanabad District, Maharashtra   166
    V.B. Sakhare, J.S. Mohite and S.G. Rawate

28. Studies on Water Quality Parameters of Siddeshwar Dam, Hingoli District,
    Maharashtra   183
    D.C. Deshmukh and S.O. Bondhare

29. Fishing Crafts and Gears in River Godavari at Nanded, Maharashtra   188
    R.P. Mali and S.O. Bondhare

30. Studies on Groundwater Quality in Villages of Kaij Tehsil in Beed District,
    Maharashtra   191
    V.B. Sakhare and J.S. Mohite

31. Serum Acetylcholinesterase Levels of Freshwater Fishes   195
    Sudhish Chandra

32. Role of Natural Ultraviolet Radiation (UVR) on Freshwater Phytoplankton Growth
on Urban Wetland (Temuco, Chile)   200

Jacqueline Acuña, Patricio De los Ríos-Escalante, Patricio Acevedo and Aracelly Ulloa

33. Role of Natural Ultraviolet Radiation (UVR) and Nutrient Addition on
Phytoplankton Growth on Urban Wetland (Temuco, Chile)   208

Patricio De los Ríos-Escalante, Patricio Acevedo, Jacqueline Acuña, and Aracelly Ulloa

34. Beneficial Effects of Aegle marmelos Leaves on Blood Glucose Levels and
Body Weight Changes in Alloxan- Induced Diabetic Rats   214

Leena Muralidharan

Previous Volumes–Contents   221

Index   237

List of Contributors

Acevedo, Patricio
Universidad de la Frontera, Facultad de Ingeniería, Ciencias y Administración, Departamento de Física.

Acuña, Jacqueline
Universidad de la Frontera, Programa de Doctorado en Ciencias de Recursos Naturales, Casilla 54-D, Temuco, Chile

Babu, G. Ravindra
Department of Zoology, Sri Venkateswara University, Tirupati – 517 502

Bhadane, R.S.
Department of Zoology, L.V.H. Arts, Science and Commerce College, Panchavati, Nashik – 422 003

Bondhare, S.O.
Post Graduate Department of Zoology, Yeshwant Mahavidyalaya, Nanded – 431 602

Cella, M. Parimala
Noorul Islam College of Arts and Science, Kumaracoil – 629 802, Kanyakumari District, Tamil Nadu

Chandra, Sudhish
P.G. Department of Zoology, B.S.N.V. College, Lucknow – 226 001

Chowdhury, Subrato
Divyodaya Krishi Vigyan Kendra, West Tripura Chebri, Tripura (W) – 799 207

Deshmukh, D.C.
Post Graduate Department of Zoology, Yeshwant Mahavidyalaya, Nanded – 431 602

Deshmukh, S.A.
Department of Botany, The New College, Kolhapur – 416 012

Deshmukh, Sagar Anant
Department of Botany, The New College, Kolhapur – 416 012

Dutta, Pranab
Divyodaya Krishi Vigyan Kendra, West Tripura Chebri, Tripura (W) – 799 207

Godavarthy, Padmavathi
Department of Zoology, Osmania University College for Women, Koti, Hyderabad – 500 095

Hamde, V.S.
Department of Microbiology, Yogeshwari Mahavidyalaya, Ambajogai – 431 517

Karpe, Manisha
K.J. Somaiya College of Science and Commerce, Vidyavihar, Mumbai – 400 077

Kathiresan, S.
Ph.D. Research Scholar, SDMRI, Thoothukudi – 628 001

Krishna Ram H.
Department of Studies in Zoology, University of Mysore, Mysore – 570 006

Krishna, Savalla Murali
Department of Zoology, University College of Science, Osmania University, Hyderabad

Kulkarni, A.S.
Department of Zoology, R.P. Gogate College of Arts and Science and R.V. Jogalekar College of Commerce, Ratnagiri – 415 612

Kumar, A. Sunil
Department of Zoology, University College of Science, Osmania University, Hyderabad

Kumara, Vijaya
Department of Wildlife and Management, School of Biosciences, Kuvempu University, Jnana Sahyadri, Shankaraghatta – 577 451

Kumari, Y. Sunila
Department of Zoology, Osmania University College for Women, Koti, Hyderabad – 500 095

Kumbhar, Amol
Department of Zoology, M.D College, Parel, Mumbai – 400 012

Lipton, A.P.
Vizhinjam Research Centre of CMFRI, Vizhinjam, Kerala

Mali, R.P.
Post Graduate Department of Zoology, Yeshwant Mahavidyalaya, Nanded – 431 602

Manjare, S.A.
Department of Zoology, Shivaji University, Kolhapur

Manjunath, R.
Department of Studies in Zoology, University of Mysore, Mysore – 570 006

Mohan, M. Ramachandra
Department of Zoology, Bangalore University, Bangalore – 560 056

Mohite, J.S.
Department of Zoology, Yeshwantrao Chavan Mahavidyalaya, Tuljapur – 413 601

Mohite, S.A.
College of Fisheries, Ratnagiri – 415 629

Muley, D.V.
Department of Zoology, Shivaji University, Kolhapur

Muralidharan, Leena
Department of Zoology, V.K. Krishna Menon College of Commerce and Science, Bhandup (E), Mumbai – 40 0042

Nayak, V.N.
Department Studies in Marine Biology, Karnataka University P.G. Centre, Kodibag, Karwar – 581 303

Neeraja, P.
Department of Zoology, Sri Venkateswara University, Tirupati – 517 502

Noothi, Sreenu
Department of Zoology, University College of Science, Osmania University, Hyderabad

Panse, Chhaya
Department of Zoology, M.D College, Parel, Mumbai – 400 012

Patil, D.S.
Department of Botany, Gopal Krishna Gokhale College, Kolhapur

Patil, M.U.
Department of Zoology, Vimalbai Uttamrao Patil Arts and Science College, Sakri – 424 304

Patole, S.S.
Department of Zoology, S.G. Patil A.S.C. College, Sakri – 424 304

Pillai, Sreenath
Department of Microbiology and Fermentation Technology, Jacob School of Biotechnology and Bio-engineering, Sam Higginbottom Institute of Agriculture, Technology and Sciences (SHIATS), Allahabad – 211 007

Piska, Ravi Shankar
Department of Zoology, University College of Science, Osmania University, Hyderabad

Pramod, A.F.
Department of Wildlife and Management, School of Biosciences, Kuvempu University, Jnana Sahyadri, Shankaraghatta – 577 451

Radhakrishnan, M.V.
Department of Zoology, Annamalai University, Annamalainagar – 608 002

Rawate, S.G.
Post Graduate Department of Zoology, Yogeshwari Mahavidyalaya, Ambajogai – 431 517

Ríos-Escalante, Patricio De los
Universidad Católica de Temuco, Facultad de Recursos Naturales, Escuela de Ciencias Ambientales, Casilla 15-D, Temuco, Chile

Roshmon, T.
Department Studies in Marine Biology, Karnataka University P.G. Centre, Kodibag, Karwar – 581 303

Sakhare, V.B.
Post Graduate Department of Zoology, Yogeshwari Mahavidyalaya, Ambajogai – 431 517

Santhi, V.
P.G and Research Department of Zoology, J.A. College for women (Autonomous), Periyakulam – 625 601

Sathe, Mrinalini
K.J. Somaiya College of Science and Commerce, Vidyavihar, Mumbai – 400 077

Shivabasavaiah
Department of Studies in Zoology, University of Mysore, Mysore – 570 006

Sivakumar, V.
P.G. and Research Department of Zoology V.O. Chidambaram College, Thoothukudi – 628 008

Somani, Vaishali
Department of Zoology, M.D College, Parel, Mumbai – 400 012

Sugumaran, E.
Department of Zoology, Annamalai University, Annamalainagar – 608 002

Suresh Kumar, P.S.
College of Fisheries, Ratnagiri – 415 629

Tendulkar, Medha
Department of Zoology, R.P. Gogate College of Arts and Science and R.V. Jogalekar College of Commerce, Ratnagiri – 415 612

Thilaga, R.D.
P.G. and Research Department of Zoology, St. Mary's College (Autonomous), Thoothukudi – 628 001

Ulloa, Aracelly
Universidad de la Frontera, Programa de Doctorado en Ciencias de Recursos Naturales, Casilla 54-D, Temuco, Chile.

Vhanalakar , S.A.
Department of Zoology, Shivaji University, Kolhapur

Walmiki, Nitin
Department of Zoology, M.D College, Parel, Mumbai – 400 012

Chapter 1

Performance of Periphyton Based Aquaculture in West Tripura

☆ *Pranab Dutta and Subrato Chowdhury*

ABSTRACT

Demonstration of periphyton based aquaculture in five different villages of West district of Tripura was carried out during 2008-09 and 2009-10 with effective result as compared to the check. Highest yield with B:C of 3.50 : 1 was recorded in demonstration as compared to the check as B:C of 1.95:1.

Keywords: Aquaculture, Demonstration, Periphyton.

Introduction

Periphyton is defined as the total assemblage of sessile or attached organisms on any substrate (Weitazal, 1979). It may contain algae, protozoans, bacteria, fungi, rotifers, annelids, insect larvae and crustaceans, all these or some of them (Keshavanath, *et al.*, 2002). Periphyton acts as a delicious food for aquatic browsers and in this culture system autochthonous of food of a pond increased to several fold (Annon, 2006). It is an age old practice based on indigenous knowledge of people aggregating fish. Substrate like bamboo poles, date palm leaves, hizol leaves (*Barringtonia* spp.), tree branches, jute sticks, sugarcane bagasses context palm leaf, turmeric leaf, bamboo trimming, paddy straw, dried water hyacinth etc can be used to promote periphyton growth.

The system of aquaculture have advantages like provide space for additional production of food which can serve as cheap alternative to expansive artificial feed, reduce accumulation of waste in the ecosystem, better mineralization leading to higher nutrient availability, etc. Fishes which have browsing habit are more suitable for this type of culture system.

Munilkumar *et al.* (2006) reported that in India *Labeo rohita, Catla catla, Cirrhinus mrigal, Labeo calbasu, Macrobrachium rosebergii* are some of the species which feed on periphyton. Higher growth in 3 month period with increase production 650 kg/ha in periphyton based aquaculture as compared to normal control (280-320 kg ha^{-1}) (Munilkumar *et al.*, 2006). Similarly a yield of 2305 kg ha^{-1} with 3 months was also reported from ponds with bamboo as substrate without any extra feed. Keshavanath *et al.*, 2002 reported that periphyton grown on bamboo pole feeding alone led to comparable fish yields that were significantly higher (30-75 per cent) that the yields obtained without periphyton or feed. They also reported that the combination of periphyton and feeding resulted in even higher yield increases (54-78 per cent). Moreover the northeastern states are blessed with thick green vegetation throughout the year. The different kind of biodegradable leaves, stems, branches of plant and trees with having utility as substrates in the aquaculture practices.

In Tripura, where poaching is a major problem, substrate are used to prevent casting of nets. Farmers of North eastern region give very less feed as the farmers cannot afford. Keeping all these in view a frontline demonstration on performance of periphyton based aquaculture was undertaken by Divyodaya Krishi Vigyan Kendra during 2008-2009 in farmers field of West Tripura with the aim to develop awareness on the technology and ultimately adopt the same.

Materials and Methods

Bamboo pole was used as substrate during the present demonstration. Six villages *viz*. Sonatala, Jambura, Ganki, and Chebri (KVK Campus) of Khowai sub-division and East and West Isharbil of Teliamura sub-division were selected for the programme. Critical inputs like bamboo pole, lime (500 kg ha^{-1}) and fish seeds (10,000 ha^{-1}) were provided to the farmer. Nearly 20-25 per cent of the surface area of the pond were covered using bamboo poles, palm leaf and bamboo branches. Comparison were made with plant based substrate were tried out for production of periphyton as these are readily viable in localities and its cost effectiveness. Bamboo pole (diameter 2.5-4.5 cm, height 1.5-1.8) @1-2 nos. m^{-2}, similarly palm leaf (diameter 60 cm) @ 1(one) no m^{-2} were used during the demonstration programme

Results and Discussion

Data presented in Table 1.1 showed that using bamboo pole as substrate is profitable in all the demonstrated area. Highest yield of 4375.00 kg ha^{-1} was recorded in village Jambura with B:C ratio of 3.50 : 1, followed by 4337.5 kg ha^{-1} in Ganki (B: C= 3.47: 1) and 3050 kg ha^{-1} in Sonatola (B: C = 1:2.44) as compared to the control pond (1875.00 kg ha^{-1}) with B:C of 1.95 : 1, where the substrate was not used. Comparatively lower yield 2650 kg ha^{-1} and 2375 kg ha^{-1} were recorded in West and East Icharbill area but still found higher than the control pond. Higher yield recorded in substrate used pond might be due to the higher survival of fingerlings and higher individual growth in pond with substrates. The result is in agreement with observation recorded by Munilkumar *et al.*, 2006. Similarly, Keshavanath *et al.* (2002) reported that out the substrates like bamboo poles, PVC pipes and sugarcane bagasse as substrate highest net production was recorded with bamboo substrate. They observed that feeding alone led to comparable fish yields that were significantly higher (30-75 per cent) that the yields obtained without periphyton or feed. The combination of periphyton and feeding resulted in even higher yield increases (54-87 per cent). Significant effect of substrate density on fish survival was also noticed. In a comparative study on effect of substrate type Azim (2001) reported that energy value, and nutritional quality with bamboo substrate is better than jute stick and branches of hizol.

Table 1.1: Performance of Periphyton Based Aquaculture in West Tripura

Name of the Farmer	Village	Substrate Used	Production (kg ha^{-1}yr^{-1})	C:B
Ms. Rita Debnath	Ganki	Bamboo pole	4337.5	1 : 3.47
Sri Rajib Debnath	Jambura	Bamboo pole	4375.0	1 : 3.50
Sri Mantu Debnath	Sonatala	Bamboo pole	3050.0	1 : 2.44
Sri Dipak Sarkar	East Isharbil	Bamboo pole	2375.0	1 : 1.90
Sri Suklal Das	West Isharbil	Bamboo pole	2650.0	1 : 2.12
Control	–	Traditional practices	1875.0	1 : 1.95

Periphyton based aquaculture is a low input sustainable farming operation which is very much suitable in the nort-heast India. The technology particularly simple which can be easily adapted in the local conditions. Such farming practices may be incorporated into rural development programmes where the target groups of beneficiaries are the resource poor farmer. The beneficial effect of periphyton has more than one role in aquaculture as it improves production, water quality and fish health, thus enhancing the efficiency of aquaculture system. Provision of substrate reduces artificial feed and therefore periphyton can be considered for as an alternative to supplementary feeding.

Farmers of the nearby area are now adopting the technology with an adoption percentage of 45-50 per cent.

Keeping the success of the technology more demonstration is taken under action the KVK plan for 2010-2011 so that the technology can be popularized in more villages of the district.

Acknowledgements

The author acknowledges the ICAR, New Delhi and ZPD, ICAR, Zone-III, Barapani, Meghalaya for financial support and SRSK, Kolkata for continuous support and guidance.

References

Anonymous, 2006. *Periphyton Based Aquaculture.* Publ. by College of Fisheries, CAU, Tripura.

Azim, M.E., 2001. The potential of periphyton based aquaculture production systems. *Ph.D. Thesis* Wageningen University, The Netherlands.

Keshavanath, P., Gangadhara, B., Ramesha, T.J. and Priyadarshini, M., 2002. The potential of periphyton based aquaculture. *Fishing Chimes,* 22(6): 19–22.

Muilkumar, S., Dey, A., Paul, A. and Nandeesha, M.C., 2006. Periphyton based aquaculture production system and its ability in North East region, In: *Training Manual on Sustainable Aquaculture Development,* (Ed.) M.C. Nandeesha. College of Fisheries, CAU, Tripura, Lecture, p. 7.

Weitzel, R.L., 1979. Periphyton measurements and applications. In: *Methods and Measurements of Periphyton Communities: A Review,* (Ed.) R.L.Weitzel. American Society for Testing and Material, STP 690, pp. 3–33.

Chapter 2

Effect of Sugarcane Bagasse on Growth Performance of *Clarias batrachus* (Bloch) Fingerlings

☆ *M.V. Radhakrishnan and E. Sugumaran*

ABSTRACT

The present investigation has been conducted to evaluate the effect of sugarcane bagasse as an artificial substrate on growth performance of *Clarias batrachus* (Bloch) figerlings for a period of 120 days. Sugarcane bagasse bundles (Length 50 cm; diameter 4cm) of 5 kg were hung in large cement tanks (5x5x1 m) with 15 cm soil base with well water in 6 of the 9 tanks randomly at the rate of 5 kg each, by suspending the bundles at regular distances from bamboo poles kept across the tanks. Fingerlings of *Clarias batrachus* (av. wt. 3.18 g) were stocked at 40 per tank two weeks after the addition of manure and substrate. No feed was provided to the fish in 3 of the substrate-added tanks (T1), while a pelleted diet was fed to the fish in the remaining 3 substrate-added tanks (T2) and the other 3 tanks without substrate (T3). Individual fish in each tank were weighed at the start and every 14 days to monitor growth response and feed utilization. Water quality parameters (dissolved oxygen, pH and ammonia) were also monitored. Fish growth performance and nutrient utilization were determined for fish Mean weight gain (MWG), Specific growth rate (SGR), Protein efficiency ratio (PER), Feed conversion Ratio (FCR), Protein intake (PI) and fish Survival rates (SR per cent). Fish carcass was also analyzed for crude protein, lipid and ash in all treatments at the end of 120 days. The weight gain of the fish showed a significant response. Specific growth rate showed a high value in bagasse and supplemental feed group (T2) followed by sugarcane bagasse alone used group (T1). The feed intake value is more or less similar in all the three groups. PER and FCR values also showed high in substrate + supplemental feed group. The survival rate value recorded maximum in T1 group. In fish carcass composition, the dry matter showed the order T2>T3>T1. Crude protein and lipid values are more in substrate alone added group (T2), but the ash value is more in T1 group. These results showed that sugarcane bagasse can effectively be use as a substrate for the culture of the catfish *Clarias batrachus*.

Keywords: Sugarcane bagasse, Artificial substrate, Clarias batrachus.

Introduction

In aquaculture feeding is one of the major elements of cost of production and may amount to 50 per cent or more. In most traditional aquaculture practices, herbivorous or omnivorous species have been preferred as they feed on natural food organisms in water, the growth of which can be enhanced through fertilization and water management (Pillay, 2001). But carnivorous species generally need a high protein diet and are therefore considered to be more expensive to produce, even though the costs will depend largely on local availability and price for the required feed stuffs. To compensate for feeding costs, most carnivorous species command higher market prices. Such species generally have greater export markets and therefore attract substantial investments. Species that are hardy and can tolerate unfavourable conditions will have the advantage of better survival in relatively poor environmental conditions that may occur occasionally in culture situations. In the present study *Clarias batrachus* has been selected because the ability to adapt to fresh and brackish waters with very low oxygen content and to grow under generally poor environmental conditions make these fish extremely valuable for small and large scale rural fish farming (Pillay, 2001). Substrate-based farming practices are considered viable low cost technologies as they help in sustainable aquaculture production (Dharmaraj *et al.*, 2002). Sugarcane bagasse is the fibrous residue remaining after sugarcane stalks are crushed to extract their juice, generated in large quantities and is currently used as a renewable resource in the manufacture of pulp and paper products and building materials. This investigation is to find out the effect of substrate (sugarcane bagasse) on growth performance of *Clarias batrachus* fingerlings.

Materials and Methods

Collection of Fish and their Maintenance

Irrespective of sex, healthy fingerlings of *C. batrachus* (2 ± 2 cm length and 3.14 ± 2g weight) were collected locally from a single population and confined to large cement tanks in the laboratory. The experiment was conducted over a period of 120 days in nine 25 m² (5 X5 X1 m) cement tanks with 15-cm soil base following the method of Dharmaraj *et al*. (2002). In all the tanks initially added 0.25 kg of quick lime and 2.5 kg of poultry manure. Water was filled to the tanks from a perennial well and a depth of 90 ± 2 cm was maintained throughout the experimental period. Subsequently, poultry manure was applied at 0.3 kg per tank every 15 days. Sugarcane bagasse, procured locally, was sun dried and bundles were made using nylon rope; they were introduced into 6 of the 9 tanks randomly at the rate of 5 kg each, by suspending the bundles at regular distances from bamboo poles kept across the tanks. After 45 days, once again 1.25 kg of the substrate was supplemented to each of the designated tanks.

Experimental Protocol

Fingerlings of *C. batrachus* (av. wt. 3.18 g) were stocked at 40 per tank (16 000 ha⁻¹) two weeks after the addition of manure and substrate. No feed was provided to the fish in 3 of the substrate-added tanks (T1), while a pelleted diet, formulated according to Varghese *et al.* (1976) (Table 2.1) was fed to the fish in the remaining 3 substrate-added tanks (T2) and the other 3 tanks without substrate (T3) at 5 per cent body weight for the first 30 days and 2 per cent thereafter, in two equal rations daily. Individual fish in each tank were weighed at the start and every 14 days to monitor growth response and feed utilization. Water quality parameters which include dissolved oxygen, pH and ammonia were kept within the range of 6.7-6.9 (mg/l), 7.2-7.8 and 0.16-0.18 (mg/l), respectively and were considered favourable in fish culture tanks according to Boyd (1990). Fish growth performance and nutrient utilization were determined according to the methods of Jobling (1983) for fish Mean weight gain (MWG), Specific growth rate (SGR), Protein efficiency ratio (PER), Feed conversion ratio (FCR), Protein intake (PI) and fish Survival rates (SR per cent). Fish carcass was analyzed for crude protein, lipid and ash using the methods of AOAC (1990) in all treatments at the end of 120 days.

Table 2.1: Ingredient Proportion and Proximate Composition of Feed (Wet weight basis)

Ingredient	Per cent	Proximate Composition	Per cent
Fish meal	25	Moisture	7.29 ± 0.13
Rice bran	40	Crude protein	28.17 ± 0.66
Groundnut oil cake	25	Crude fat	3.15 ± 0.08
Tapioca flour	10	Crude fibre	15. 90 ± 0.54
		Ash	13.80 ± 0.62
		NFE	31.69
		Energy content	12.50 (KJ^{-1})

* Average of three values ± S.E.

Statistical Analysis

The experiment consisted of a completely randomized design with three replicates for each three dietary treatments. Statistical analysis of the data included the one-way analysis of variance (ANOVA) using the SPSS version 10.0 for windows on PC (Statistical Graphics Corp, US). Significant mean differences were separated at 5 per cent using the methods of Steel *et al.* (1997) whereas appropriate and values are expressed as means ± SE.

Table 2.2: Growth Response of *Clarias batrachus* Fingerlings

Parameters	T1	T2	T3
Initial Weight (g)	3.18 ± 0.21	3.20 ± 0.22	3.18 ± 0.21
Weight gain (g)	20.37 ± 0.50*	24.17±0.09*	21.34 ± 0.19*
Feed intake (g)	25.23 ± 1.09	25.24± 0.00	25.23 ± 1.10
SGR (per cent/day)	1. 98±0.05	2.00 ± 0.02	1. 74 ±0.00
PER	0.72 ± 0.10	0.80 ± 0.24	0.78 ± 0.20
FCR	1.27 ± 0.01	1.37 ± 0.03	1.32 ± 0.05
Survival Rate (per cent)	99.00	98.00	98.50

* $p< 0.05$

Results and Discussion

The growth response of *C. batrachus* using sugarcane bagasse as a substrate for 120 days is given in Table 2.2. The weight gain of the fish after 120 days showed a significant response. Specific growth rate showed a high value in bagasse and supplemental feed group (T2) followed by sugarcane bagasse alone used group (T1). The feed intake value is more or less similar in all the three groups. PER and FCR values also showed high in T2 group. The survival rate value recorded maximum in T1 group (Table 2.2). In fish carcass composition, the dry matter showed the order T2>T3>T1. Crude protein and lipid values are more in T2 group, but the ash value is more in T1 group. More details are given in Table 2.3.

Table 2.3: Fish Carcass Composition after 120 Days

Parameters (per cent)	T1	T2	T3
Dry matter	82.16 ± 2.14	87.52±1.55	82.19 ± 2.10
Crude protein	15.18 ± 3.22	15.05 ± 2.20	17.31 ± 2.00
Crude lipid	12.80 ±1.10	11.84 ± 1.72	15.80 ± 2.70
Ash	0.56 ± 0.11	0.48 ± 0.16	0.45 ± 0.02

For optimum growth of fish and overall productivity of fish culture a variety of nutrients carriers have been subject of many studies, and received much attention (Brady, 1991). The amount of different nutrients present in any natural aquatic environment is generally low and so unproductive, while the quantity required for the growth of fish and fish food organisms (phytoplankton, zooplankton and bottom fauna/flora) is comparatively large (Kumaraiah *et al.*, 1997). When too much nutrient carriers (organic and inorganic) are used for fish culture, a substantial amount are lost through several ways and may become pollutants. Therefore, a large application of nutrients carriers may become hazards to fish (Kumaraiah *et al.*, 1997). Although inorganic fertilizers and organic manure contain various essential elements, all of them are not necessary for fish growth, so need to be conserved and carefully managed (Brady, 1991). Organic substances such as plant materials, food scraps and paper products can be recycled using biological process. The intention of biological processing is to control and accelerate the natural process of decomposition of organic matter. Development of viable low-cost technologies and their application to current farming practices would help in enhancing aquaculture production (Dharmaraj *et al.*, 2002). Substrate based aquaculture is one such technology that has generated a lot of interest in recent years ((Wahab *et al.*, 1999, Tidwell *et al.*, 2000, Azim *et al.*, 2001, Keshavanath *et al.*, 2001). By providing organic matter and suitable substrates, heterotrophic food production can be increased several folds which in turn would support fish production. Substrates provide the site for epiphytic microbial production, consequently eaten by fish-food organisms and fish. Fish harvest microorganisms directly in significant quantities, either from microbial biofilm on detritus or from naturally occurring flocks in water column.

Our earlier study revealed that sugarcane bagasse not only affects water quality parameters but help to increase the growth of zooplankton (Radhakrishnan and Sugumaran, 2010). Hence, it can effectively be used as a substrate for the growth of planktons in aquaculture ponds. Provision of substrate would therefore, be useful for the growth of microbial biofilm (Radhakrishnan and Sugumaran, 2010a). Apart from forming food for fish, biofilm improves water quality by lowering ammonia concentration (Langis *et al.*, 1998, Ramesh *et al.*, 1999). The growth performance of fish was the best in substrate + feed treatment, being significantly higher than in substrate-alone or feed-alone treatments; growth under T2 treatment was higher over T1 and T3 treatments respectively. Fish growth in feed alone treatment was higher than in substrate-alone treatment. In feed-alone treatment (T3) fish achieved a 5 fold weight increment over the experimental period. This superior performance as compared to T1 treatment could be attributed to the nutritional quality of the diet employed. The significantly better growth of fish in the combination treatment indicates that the species can efficiently utilize natural as well as artificial diets when provided together.

Acknowledgements

The authors are thankful to University Grants Commission, New Delhi, India for financial assistance and Dr. M. Sabesan, Professor and Head for the encouragement.

References

AOAC (Association of Official Analytical Chemists), 1990. *Official Method of Analysis*, 17ᵗʰ Edition, Washington D.C., U.S.A., p. 1298.

Azim, M.E., Wahab, M.A., Van Dam, A.A., Beveridge, M.C.M. and Verdegem, M.C.J., 2001. The potential of periphyton-based culture of two Indian major carps, rohu *Labeo rohita* (Hamilton) and gonia *Labeo gonius* (Linnaeus). *Aquacult. Res.*, 32: 209–216.

Boyd, C.E., 1990. *Water Quality in Ponds for Aquaculture*. Alabama Agricultural Experimental Station, Aurburn University, Alabama, pp. 483.

Brady, N.C., 1991. *The Nature and Properties of Soil*. McMillan Company, New York.

Dharmaraj, M., Manissery, J.K. and Keshavanath, P., 2002. Effects of a biodegradable substrate, sugarcane bagasse and supplemental feed on growth and production of fringe-lipped peninsula carp, *Labeo fimbriatus* (Bloch). *Acta Ichthyol. Piscat.*, 32(2): 137–144.

Jobling, M., 1983. A short review and critique of methodology used in fish growth and nutrition studies. *J. Fish. Biol.*, 23: 686–703.

Keshavanath, P., Gangadhar, B., Ramesh, T.J., Van Rooij, J.M., Beveridge, M.C.M., Baird, D.J., Verdegem, M.C.J. and van Dam, A.A., 2001. Use of artificial substrates to enhance production of freshwater herbivorous fish in pond culture. *Aquacult. Res.*, 32: 189–197.

Kumaraiah, P., Chakrabarthy, N.M. and Radhavan, S.L., 1997. Integrated aquaculture with agricultural (animal and plant crop) components including processing of organic waste and their utilization. *Fish. Chim.*, 17: 19–20.

Langis, R., Proulex, D., de la Noue, J. and Couture, P., 1998. Effect of bacterial biofilms on intensive *Daphnia* culture. *Aquacult. Engg.*, 7: 21–38.

Pillay, T.V.R., 2001. *Aquaculture: Principles and Practices*. Blackwell Science Ltd., Oxford.

Radhakrishnan, M.V. and Sugumaran, E., 2010. Utilization of sugarcane bagasse as a substrate for plankton productivity. *Recent Research in Science and Technology*, 2(3): 19–22.

Radhakrishnan, M.V. and Sugumaran, E., 2010a. Efficacy of sugarcane bagasse to produce bacterial biofilm in water. *J. Ecobiotechnol.*, 2(2): 41–44.

Ramesh, M.R., Shankar, K.M., Mohan, C.V. and Varghese, T.J., 1999. Comparison of three plant substrates for enhancing growth through bacterial biofilm. *Aquacult. Engg.*, 19: 119–131.

Steel, R.G., Torrie, J.H. and Dickey, D.A., 1997. *Principles and Procedures of Statistics: A Biometric Approach*, 3rd Edn. McGraw Hill Companies Inc., New York, USA, p. 121.

Tidwell, J.H., Coyle, S.D., Van Arnum, A. and Weibel, C. 2000. Production response of freshwater prawn, *Macrobrachium rosenbergii* to increasing amounts of artificial substrate in ponds. *J. World Aquacult. Soc*, 31: 452–458.

Varghese, T.J., Devaraj, K.V., Shantaram, B. and Shetty, H.P.C., 1976. Growth response of common carp, *Cyprinus carpio* var. communis to protein-rich pelleted feed. In: *Symposium on Development and Utilization of Inland Fishery Resources*, Colombo (Sri Lanka). FAO Regional Office for Asia and the Far East, Bangkok, Thailand, p. 408–416.

Wahab, M.A., Azim, M.E., Ali, M.H., Beveridge, M.C.M. and Khan, S., 1999. The potential of periphyton-based culture of the native major carp calbaush, *Labeo calbasu* (Hamilton). *Aquacult. Res.*, 30: 409–419.

Chapter 3

Studies on the Impact of Fingerlings Stocking on the Production of Major Carps in a Minor Reservoir

☆ *Ravi Shankar Piska, Sreenu Noothi, A. Sunil Kumar*
and Savalla Murali Krishna

ABSTRACT

Present study deals with impact of fingerlings stocking on the major carp production in a minor reservoir Nadargul Ranga Reddy district. The water spread area is 15.2/ha and water is useful for agriculture and fisheries activities. Fingerlings with 75-80mm size were stocked@2000/ha. Catla, rohu, mrigal, common carp and grass carp fingerlings were stocked @500/ha, 500/ha, 400/ha, 400/ha and 200/ha respectively. The fishes were harvested after one year. Survival rate was found to be 39.2 per cent, 36.5 per cent, 30.5 per cent, 39.5 per cent, and 31 per cent in catla, rohu, mrigal, common carp and grass carp respectively. Major carps yield was found to be 662kg/ha/yr, which is more than Indian average reservoir fish production (29.07kg/ha/yr).Catla attained maximum growth, followed by common carp, grass carp, rohu and mrigal and all the carps attained more than 1kg in one year and catla (29.25 per cent) yield was maximum, followed by rohu (24.24 per cent), common carp (23.04 per cent), mrigal (14.59 per cent) and grass carp (09.09 per cent).

Keywords: Fingerlings, Stocking density, Minor reservoirs, Major carps, Fish production.

Introduction

Reservoir constitutes the single largest inland fishery resource, both in terms of size and production potential. These manmade ecosystems offer enough scope for stock manipulation through ecological adoption, increasing production with relatively low capital investment. Reservoir fisheries development is labor productive and ensure employment for weaker section of our society. These water bodies, especially small and minor reservoirs, have immense potential for fish husbandry through extensive aquaculture technologies stocking cum capture.

The total surface area or reservoir in India is 3.15m.ha and Indian average reservoir fish production is 29.7kg/ha/yr, which is very low. Seed is prerequisite for fish culture. Major carp seed can be procured either from natural resources like rivers, or from hatcheries. Major carp seed is in the form of hatchling spawn, fry and fingerlings. Both fry and fingerlings are useful for stocking natural water bodies like reservoirs, lakes and ponds etc. Seed stocking is one of the important management measures in the rearing systems, including extensive systems like reservoirs. Not only stocking densities but also stocking sizes play a pivotal role in fish rearing. Stocking of either fry or fingerlings play an important role in fish growth, survival rates and production. The present study was undertaken to observe the impact of stocking of fingerlings of major carp.

Materials and Methods

The present study was conducted in a minor reservoir Nadergul in Ranga Reddy District, Andhra Pradesh. The water spread area is 15.2 hectares and water is useful for agriculture and fisheries. The study was carried for one year period during 2009-10.

Various parameters were studied during the present work like stocking densities, growth, survival rates and fish yield.The major carps such as catla (*Catla catla*), rohu (*Labeo rohita*), mrigal (*Cirrhina mrigala*), common carp (*Cyprinus carpio*) and grass carp (*Ctenopharyngodon idella*) fingerlings(75-80mm) were stocked in the reservoir. These carp fingerlings were stocked in July 2009 and harvested in may 2010. Fingerlings were introduced in the reservoir @2000/ha. Yearly carp production was analyzed.

Results and Discussion

Stocking Density

In the present study the stocking density was maintained has 2000 fingerlings/ha. Catla 500/ha(25 per cent), Rohu 500/ha(25 per cent), Mrigal 400/ha(20 per cent), Common carp 400/ha (20 per cent) and Grass carp 200/ha(10 per cent) fingerlings were stocked in the reservoir.

Table 3.1: Stocking Densities, Growth Rate, Survival Rates and Production of Major Carps in the Minor Reservoir

Major Carps	Stocking Density		Maximum Growth Rate (g)	Survival Rate		Production	
	No.	per cent		No	per cent	Kg/ha/yr	per cent
Catla catla (Catla)	500	25	1380	194/ha	39.20	193.65	29.25
Labeo rohita (Rohu)	500	25	1240	182/ha	36.50	160.52	24.24
Cirrhina mrigala (Mrigal)	400	20	1020	122/ha	30.50	96.62	14.59
Cyprinus carpio (Common carp)	400	20	1310	156/ha	39.50	152.57	23.04
Ctenopharyngodon idella (Grass carp)	200	10	1280	62/ha	31	60.02	09.09

Total major carp production: 662kg/ha/yr

Indian Institute of Management (IIM, Ahmedabad) conducted survey during 1983 and recommended the stocking rates of 1000 fingerlings/ha in minor reservoirs (Srivastava, 1985).

Government of India suggested that on an average all the reservoirs in India were under – stocking range of 2-150 fingerlings per hectare as against the required density of 500 fingerlings per hectare. In China, the stocking was very high varying from 1200-3000 fish seed per hectare per year (Mohanty, 1984). Mohanty (1984) suggested 1000 seed/ha in reservoirs of Orissa. Das *et al.* (1984) reported the stocking rate of 200 fingerlings/ha in few reservoirs like Mandira, Hadgarh, Talsara etc. In most of the reservoirs the stocking rate was low when compared to the proposed stocking rate. In the present study the stocking rate was higher than that of the above studies.

Piska (2003) suggested the stocking density of 2000 seed in reservoirs and tanks and 5000 seed ha in the ponds in Andhra Pradesh. He also stated that the stocking density of 2000/ha is slightly high, which in due to compensate the mortality of sensitive fingerlings in the open waters. He also stated that 31, 09, 580 fingerlings are required to stock the reservoirs and 72, 49, 540 are required to stock the tanks in Andhra Pradesh. Mathew and Mohan (1990) reported the stocking rate was 841 fingerlings/ha in Kerala and they never considered the stocking rate was low. They also stated that a stocking density of 1000-5000 advanced fry/ha was planned for the reservoirs in Kerala. They stated that the major carp stocking of 250/ha was desired for reservoirs without predatory fishes and 600/ha for reservoirs with abundance of cat fishes.

In the present study, 75-80mm size fingerlings were introduced into the reservoir. Different authors tried with fingerlings (Srinivastava, 1985; Mathew and Mohan, 1990) or advanced fingerlings (Ahmed and Singh, 1992). Piska and Rao (2005) conducted an experiment with the same stocking sizes of major carp seed in a minor reservoir, Bibinagar. Some of the reservoirs also stocked with fingerlings (Mathew and Mohan, 1990). Srivastava (1985) recommended 12.5 cm fingerlings for stocking in reservoirs.

Many authors reported that the reservoirs were stocked with fry, and found low fish production (Devi, 1997; Chary, 2003; Srinivas, 2005; and Ansar, 2010). This was due to delicate nature of fry which were perished due to the environmental changes and predation. The predators which were found in the reservoir attack easily on fry and kill them. Due to the above reasons, fingerlings were used has the stocking material in the present study. The mortality due to the rate was more in case of fry when compared to fingerlings. Mathew and Mohan (1990) reported that the fish seed mortality was also due to predatory fishes in the reservoir. They also suggested high stocking of fish seed in the reservoirs with more predators.

Growth

All carps except mrigal crossed 100g during monsoon period. Catla and grass carp grew to maximum to 120g. During post monsoon period catla (840g), grass carp (820g) and common carp (800g) reached 800g. All carps crossed 1kg at the end of the year catla grew to maximum (1380g), followed by common carp (1310g), grass carp (1280g), rohu (1240g) and mrigal (1020g).

Ansar (2010) reported that all major carps crossed 1kg mark in the first year in Jamulamma reservoir, Gadwal. Venugopal *et al.* (1998) reported the growth of common carp was more than 600g and catla - 400g and rohu - 300g in seven months rearing in a seasons rain-fed tank. Piska (2000) reported that rohu grew to 401.35g, catla to 407.32g and common carp to 653.46g in six months in a seasonal rain-fed tank without supplementary feeds and inorganic fertilizers. The growth rates in present study were less when compared to above studies. The less growth in the present study due to the presence of high population of Tilapia and pollution load in the tank.

Survival Rates

In the present study, out of 2000/ha 718/ha (35.90 per cent) fishes were recovered during harvesting. Maximum survival rate was observed in catla (194/ha, 39.20 per cent), and followed by rohu(182/ha, 36.40 per cent), mrigal(122/ha, 30.50 per cent), common carp(156/ha, 30.50 per cent) and grass carp(62/ha, 31 per cent). Piska and Rao (2005) reported the range of survival rates as 9.85-51.10 per cent in Bibinagar.

Piska and Rao (2005) reported, 684 with 75-80 mm, in Bibinagar, Nalgonda. The catla was dominated with 193.65kg/ha/yr or 2943.45kg with the percentage of 29.25 per cent, catla was followed by rohu with 160.52kg/ha/yr or 2439.96kg with the percentage of 24.25 per cent, Mrigal with 96.62kg/ha/yr or 1468.68kg with the percentage of 14.60 per cent,common carp with 152.57kg/ha/yr or 2319.03kg with percentage of 23.05 per cent. and grass carp with 60.02kg/ha/yr or 912.24kg,with the percentage of 9.09 per cent.

Major Carp Production

The total major carp production was found to be 662 kg/ha/yr (92.30 per cent). Among different major carps, catla production was maximum with 193.65 kg/ha/yr(29.25),followed by Rohu 160.52 kg/ha/yr(24.24 per cent), Mrigal 96.62 kg/ha/yr(14.59 per cent), Common carp 152.57 kg/ha/yr(23.04 per cent) and grass carp 60.02 kg/ha/yr(9.09 per cent).

Piska and Rao (2005) opined that the major carp production increased with increase of stocking sizes of carp seed. They reported that the major carp production was 632.91 with the stocking size of 75-80 in Bibinagar. The production levels in the present study were more when compared to that of Piska and Rao (2005) study. This indicates that the productivity of the present study was more due to nutrient-rich waters. Most of the authors tried with fry to produce major carps with different stocking densities. Devi (1997), Chary (2003) and Rao (2004) used fry to improve the major carp production in the reservoirs of in and around Hyderabad. Srinivas (2005) tried with advanced fry in Edulabad reservoir.

Das *et al.* (1984) reported the dominance of catla (84 per cent), followed by mrigal (14 per cent) and rohu (2 per cent) in Pitamahal reservoir. They reported the total dominance of catla, around 98 per cent and followed by common carp (1 per cent) and rohu (0.50 per cent) and calabasu (0.50 per cent) in Sanamachkandana reservoir. The percentage of catla, rohu and mrigal was estimated 21, 33 and 46 respectively. They also reported that in Hadgarh reservoir catla constitute about 80 per cent of the catch followed by mrigal and calabasu. Rohu was scarce in the catch. This was no common carp in the catches.

Jhingran and Sugunan (1990) reported that the major carps contributed 90.54 – 94.76 per cent in Tillaiya and 57.44 – 76.61 per cent in Konar reservoirs. Catla showed a reduction only in Tillaiya. In Konar, a general stock reduction was noted particularly that of mrigal. He also observed the dominance of mrigal (21.1-71.4 per cent) among overall catches of Indian major carps in Ranapratap Sagar reservoir in Rajasthan.

Selvaraj and Murugeshan (1990) reported that the contribution of major carps was 93.09 per cent in Aliyar reservoir, Tamil Nadu. Devi (1997) reported that the contribution of major carps was 91.42 per cent in Ibrahimbagh and 96.40 per cent in Shathamraj reservoirs of Andhra Pradesh during 1993-1995. The major carp composition was 31.30-33.29 per cent catla, 24.65 per cent -27.24 per cent. Rohu, 18.37 per cent -20.68 per cent mrigal, 21.65 per cent -15.18 per cent common carp and grass carp was 4.03 per cent -3.60 per cent in Ibrahimbagh during 1993-95. In Shathamrai reservoir 41.62-33.06 per cent catla, 20.70-26.30 per cent rohu, 16.50-19.20 per cent mrigal, 19.62-20.44 per cent common carp and 1.0-1.56 per cent grass carp was reported during 1993-95.

L. calbasu too used to have a significant presence (3 tonnes) in the 1997-98 catches (Singh, 2001). In the Markonahalli reservoir, Karnataka fish yield which was low of 5.6kg/ha in 1990-91 enhanced nearly 13 folds to 74.8kg/ha in 1993-94 due to stocking of major carps and increased fishing efforts. Major carps consisting rohu (41.5 per cent) and catla (16.1 per cent) accounted for more than 50 per cent of the catch (Rao _et al.,_ 2002). According to Patel _et al._ (2002) the fish production in Ukai reservoir was 159.0kg/ha and is capable of producing 220 kg/ha on the basis of primary productivity. This production is to be rated as high, considering the average natural production of 11.43kg/ha and potential production of 49.99kg/ha from the category of large reservoirs.

The productivity of the reservoir was much higher when compared to other minor reservoirs of India, medium and large reservoirs of India. The present figures were much higher than small reservoirs of India-49.9kg/ha/yr (Piska, 2000). Devi (1997) and Piska (2000) recorded the productivity of 445kg/ha/yr and 528kg/ha/yr during 1993-95 in Ibrahimbagh and Shanthamrai reservoirs of Rangareddy district, Andhra Pradesh. The present productivities were higher than other minor reservoirs like Baghla – 106kg/ha/yr, Bachra – 139kg/ha/yr and Gularia – 100kg/ha/yr which were managed by scientific methods as described by Jhingran and Sugunan (1990).

The present productivity of fish was much higher than average Indian large reservoirs, which were observed by Srivastava (1985)- Pong dam 4.1 to 25.08kg/ha/yr, Rihand – 3.7 to 14.24kg/ha/yr, Tenughat – 0.53 to 1.471kg/ha/yr, Kangsabati – 0.55 to 1.10kg/ha/yr, Kodana 6kg/ha/yr. Gandhisagar 0.52 to 13.3kg/ha/yr, Hirakud – 10.5kg/ha/yr, Santhamur 3.5 to 11kg/ha/yr, Tungabhadra 5.54kg/ha/yr. Pilit 08-35.30kg/ha/yr and Shardarsagar 42 to 56kg/ha/yr. The fish production of 7kg/ha/yr in Nizamsagar, 107kg/ha/yr in Kolleru, 8kg/ha/yr in Bhadha and 6kg/ha/yr in Panam reservoirs. According to Srivastava, (1985) the average reservoir annual fish yield was estimated to increase about 60kg/ha/yr.

The present production was many times more than the average fish production in Indian reservoirs, 29.70kg/ha/yr (Dehadrai, 2001). Mahapatra (2003) recorded only 15.6kg/ha/yr in Hirakud reservoirs and 5-10kg/ha/yr in other major reservoirs in Orissa and concluded that there was scope for increase the yield rate to 100 kg/ha/yr by proper management. Sreenivasan (2001) estimated the production potential of Indian reservoirs at 100kg/ha/yr. Even according to a conservative administrative estimate the potential yield of Indian reservoirs is around 50kg/ha/yr. Bihar holds the record for the lowest fish yield from reservoirs, 0.54kg/ha/yr. In 1997-98, the TNFDC operated 11,088 ha of reservoirs, producing 426.73mt of fish (38.5kg/ha) (Sreenivasan, 2001).

According to Dwivedi _et al._ (2000) fish production of 133.5kg/ha/yr was achieved against an average potential fish yield of 166.5kg/ha/yr was harvested (80.5 per cent) from the Naktara reservoir Madhya Pradesh, major carps dominated the catch, particularly due to continuous stocking and due to absence of large predatory fish.

The number of fingerlings required for the production of 1 kg fish was 3.02. Among major carps, catla required less number of seed to produce 1kg, catla with 5.16 seed followed by common carp with 6.21 seed, rohu with 7.10 seed, grass carp with 7.61 seed and mrigal 14.07 seed to produce 1kg of fish, 2.58, 3.12, 4.14, 2.62 and 3.33 fingerlings required to produce 1kg of catla, rohu, mrigal, common carp and grass carp respectively.

Piska and Rao (2005) also reported that the number of seed required to produce 1kg of major carp was decreased gradually from fry to advanced fingerlings in Bibinagar. The total number of fry required to produce 1kg of carps were 13.89, whereas advanced fry 8.64, fingerlings 3.16 and advanced fingerlings 1.99 with overall figure of 3.97 in Bibinagar. They also reported more number of fry required and less number of advanced fingerlings required to produce 1kg carps.

Conclusion

The present study indicate the stocking of fingerlings yielded better result, hence recommend the fingerlings as stoking material in reservoirs to get maximum fish production instead of fry, which were stocked most commonly in Indian reservoirs.

References

Ansari, 2010. Studies on fisheries of Jamulamma reservior, Gadwal, Mahaboobnagar Dist., Andhra Pradesh. *Ph.D. Thesis*, Osmania University.

Chary, K.D., 2003. Present status, management and economics of fisheries of a minor reservoir, Durgamcheruvu of Ranga Reddy district. *Ph.D. Thesis*, Osmania University, Hyderabad.

Das, R.K., Misra, D.C., Murthy, T.S., Pradhan, N.K. and Mitra, S., 1984. Managing reservoir fisheries, certain case studies. *Proc. Freshwater Fish–Rural Devel. Sec.*, 1: 57–63.

Dehadrai, 2001. Reservoir fisheries in India. *Proc. Nat. Sem. Riverine and Reservoir Fisher.*, p. 97–104.

Devi, B.S., 1997. Present status, potentialities, management and economics of fisheries of two minor reservoir of Hyderabad. *Ph.D. Thesis*, Osmania University.

Dwivedi, R.K., Khan, M.A., Singh., H.P., Singh, D.N. and Tyagi, R.K., 2000. Production dynamics and fisheries development in Naktara reservoir, Madhya Pradesh, India. *J. Inland Fish Soc., India*, 32(2): 81–86.

Jhingran, A.G. and Sugunan, V.V., 1990. General guidelines and planning criteria for small reservoir fisheries management. *Proc. Nat. Workshop Reservoir Fish*, p. 1–8.

Mahapatra, D.K., 2003. Present status of fisheries of Hirakud reservoir, Orissa. *Fishing Chimes*, 22(10 and 11): 76–79.

Mathew, P.M. and Mohan, M.U., 1990. Reservoir fisheries in Kerala: Status and scope for development. *Proc. Nat. Workshop Reservoir Fish.*, p. 111–117.

Mohanty, S.K., 1984. An approach for the development of reservoir fisheries in Orissa. *Proc. Freshwater Fish. Rural Devel. Sec.*, 1: 40–46.

Patel, M.I., Thaker, D.B. and Shukla, N.M., 2002. Fisheries of Vallabhasagar reservoir Ukai, Gujarat. *Fishing Chimes*, 22(6) : 27–31.

Piska, R.S., 2000. *Concepts of Aquaculture*. Lahari Publications, Hyderabad.

Piska, R.S. and Rao, A.M., 2005. Impact of juvenile stocking size on the major carp production in a minor reservoir, Bibinagar, India. *Rev. Fish Biol. Fish.*, 15: 167–173.

Rao, A.M., 2004. Studies on limnology and fisheries in a *Tilapia* – dominated perenial tank, Julur, Nalgonda district, Andhra Pradesh, *Ph.D. Thesis*, Osmania University, Hyderabad.

Rao, D.S.K., Ramakrishnaiah, M., Karthikeyan, M. and Sukumaran, P.K., 2002. Limnology and fish yield enhancement through stocking in Markonahalli reservoir. *Nat. Symp. Fish. Enhanc. Inland Waters, IFSI*, p. 17.

Selvaraj, C. and Murugesan, V.K. 1990. Management techniques adopted for achieving a record fish yield from Aliyar reservoir, Tamil Nadu. *Proc. Nat. Workshop. Reservoir Fish.*, p. 86–96.

Sreenivasan, A. 2001. Stagnating reservoir fish production of India. *Fishing Chimes*, 21(1): 39–43.

Srinivas, Ch., 2005. Impact of river Musi pollution on the fisheries of Edulabad reservoir, Ranga Reddy District, Andhra Pradesh, India. *Ph.D. Thesis*, Osmania University, Hyderabad.

Srivastava, A., 1985. *Inland Fish Marketing in India*. Indian Institute of Management, Ahmedabad.

Venugopal. G., Hingorani, H.G.,Venkateshwaran, K. and George, J.P., 1998. Culture of carps in a seasonal tank. *Comp. Physiol. Eco.*, 13(1): 20–23.

Chapter 4

Studies on Effect of Insecticides on Oxygen Consumption of Freshwater Fish, *Lepidocephalichthys guntea* (Ham-Buch)

☆ *M.U. Patil and S.S. Patole*

ABSTRACT

The toxic impact of two insecticides *viz.*, Synthetic pyrethroids, cypermethrin and Organophosphorus compound, malathion at their sublethal levels were evaluated for oxygen consumption in freshwater fish, *Lepidocephalichthys guntea* for 96 hours. After determining the LC_{50} concentrations of both insecticides individually, ¼ and ¾ of 96 h LC_{50} values were taken as sublethal concentration for the studies on oxygen consumption. In both concentrations, which were evaluated, significant reduction in the rate of oxygen consumption over the control group of fish was reported. In both ¼ and ¾ concentrations indicated that ¾ concentrations had profound effect than ¼ concentrations and cypermethrin insecticide was found to be more deleterious than Malathion. During experimentation, severe respiratory distress, rapid opercular movements leading to decrease in oxygen uptake efficiency were observed in toxicated group of fishes.

Keywords: Insecticides, Cypermethrin, Malathion, Oxygen consumption, Sublethal.

Introduction

The wide applicability of insecticides provide many occasion for their entry into an aquatic environment as surface run off or through the direct application to the crop production, program and public health operation for pest and mosquito control. The indiscriminate use of pesticides and industrial effluents increase water pollution in developing countries (Alderic, 1967). The use of organophosphorus and synthetic pyrethroid group of pesticides on crop field is highly toxic to the aquatic organism including fish (Verma *et al.*, 1979, Susan and Sobha, 2010).

In accordance with general practice the metabolic rate of poikilotherms animals is ordinarily measured in terms of O_2 consumption, which is highly complex process and is subjected to the influence of various intrinsic and extrinsic factors. It is the well known fact that the rate of O_2 consumption is used as important tool for understanding the physiological state of metabolic activity of an organism (Tilak *et al.*, 2005). In response to the effect of various pesticides, changes in respiratory metabolism have been documented in fishes. These pesticides affect entire body of fish. The gills are supposed to be first site of attack. They get affected and ultimately oxygen consumption rate is reduced (Mahajan and Patole, 2003).

Oxygen is necessary to provide energy for life and its availability imports limit on distribution and survival of animal. The rate of O_2 consumption of a fish is considered as a relation and reflection of total metabolism and an opercular movement is taken as an index model to verify respiratory activity of animal under environmental stress (Vijayalakshmi and Tilak, 1996; Beat Gassner *et al.*, 1997; Susan and Sobha, 2010). Both insecticides *i.e.* cypermethrin and malathion are widely used in large proportion to combat the agricultural pest.Hence in present investigation they are used to study the oxygen consumption rate in freshwater fish, *Lepidocephalichthys guntea*.

Materials and Methods

Live specimen weighing 1.0 ± 0.3 g and 4.0 ± 0.5 cm length were collected from Panzara River. The collected fishes were acclimatized to laboratory condition for 15 days. Before experimentation glass aquaria is containing 50 litres dechlorinated tap water with continuous aeration under normal photoperiod. There were fed with fish diet, the water was replaced in every alternate day; if mortality occurs during this period dead fishes were removed immediately because such dead animals may deplete dissolved oxygen with resultant effect on other fish. (Jain, 2000). Only healthy and acclimatized fishes (10 number) weighing between 10.0 ± 2.0 g were taken for experimental purpose. The experimental doses of sub lethal concentration are arranged in five groups with control for the period of 96 h. After exposure period from their five groups survival animal are kept for experiment of O_2 consumption.

The respiratory chamber is prepared by using 1.5 litre capacity reagent bottles to keep fishes for estimation of O_2 consumption. The initial amount of O_2 present in tap water was calculated. The test fish *Lepidocephalichthys guntea* were transferred in treatment free water of respiratory chamber and kept there for 1 h, five healthy treated fishes were kept in respirometer containing 1000 ml freshwater. Liquid paraffin was added to avoid entry of atmospheric O_2 in the chamber and cork was tightly closed with plaster of Paris. After 1 h the fish were removed from respiratory chamber and O_2 content was estimated by Winkler's iodometric method as described by (Golterman and Clymo, 1969) 250 ml, water sample was taken out in another bottle to estimate O_2 content. The difference in the rate of O_2 consumption between control and the experimental fishes denotes the effect of the toxicant on O_2 consumption. Likewise experiment was conducted in sub lethal concentration of cypermethrin and malathion. The obtained data was compared with control. The O_2 consumption was calculated by employing the following formula and expressed in mg of O_2 consumed/g/body wt. of fish/h.

$$O_2 \text{ mg/L} = \frac{O_2 \text{ consumed by animal in 1 h}}{\text{Weight of animal in g}} \times 100$$

Results and Discussion

Oxygen consumption by test fish exposed to sublethal dose concentrations of Cypermethrin and Malathion is shown in Table 4.1. It indicates, significantly decreased in oxygen consumption were recorded in all treated groups. The action of toxicant on the experimental fish is found to be oxygen consumption rate decrease in order. ¾ c > ¼ c > ¾ m > ¼ m.

Table 4.1: Oxygen Consumption of *L. guntea* Exposed to Cypermethrin and Malathion

Name of Pesticide	96hLC$_{50}$values (ppm)	Sub Lethal Doses (ppm)		O$_2$ Consumption mg/g/h	Per cent Variation
		¼	¾		
Malathion	11	2.75	–	156. 24 ± 0. 205	(-23. 53) *
		–	8.25	122. 30 ± 0. 153	(-42. 21) **
Cypermethrin	5.2	1.3	–	95. 230 ± 0. 320	(-53. 28) **
		–	3.9	74. 695 ± 0. 223	(-63.28) ***
Control	–	–	–	204. 282 ± 0. 256	

All values expressed in mg/100g wet weight tissues and mean ± S.D. of six observations.

* Significant value *P < 0.05, ** P< 0.01, *** P< 0.001. NS= Non significant value is P > 0.05.

Values in parenthesis indicate percentage change over control (taken as 100 per cent).

The water characteristics analyzed at beginning of experiment (APHA, 1998) there were pH = 7.1 ± 1, Temperature 28 ± 1°C and dissolved oxygen 5.5 mg/l. The rate O$_2$ concentration in fish has been considered to be indication of metabolism it is influenced by many factors like temperature, species specification, size, crowding and types of pesticides. Incase of aquatic animal there is no escape from toxicant and maximum part that continuously expose to pesticides, the respiratory surface *i.e.*, gill surface.

In present study it is observed that, as the concentration of pesticide increased the rate of O$_2$ consumption declined gradually. The declined rate of O$_2$ consumption might be the result of failure of compensating mechanism to achieve a new steady state of metabolism causing to pollutant stress. Thus, decrease might also be due to penetration of pollutants and its action alters metabolic cycle at sub cellular level. Natrajan (1978) suggested that decreased O$_2$ consumption at maximum activity affected swimming performance as well as damage of the tissue the vital system performance. Further he reported that the degree in O$_2$ consumption is brought about by surviving of links between the oxidative and phosphorelative processes.

From the result, it indicate that, the organophosphorus and synthetic pyrethroid group of pesticides might have an immediate action on the central nervous system, which is involved in controlling most of the vital activities in fishes. Earlier it was reported that the toxic effects were pronounced at very early hours of exposure to organophosphorus and synthetic pyrethroids observed and depletion O$_2$ consumption due to their toxic actions on the nerve enzyme system, which interference in the regulation of metabolism (Jaysurya *et al.*, 1991; Roy and Dattamunshi, 1988). Another reason is the insecticides acts as potent inhibitors of mitochondria and exhibited sigmoid inhibition kinetics (Beat Gassner *et al.*, 1997). Asifa and Vasantha (2001) stated that aquatic animals have to pass a large quantity of water over their respiratory surfaces and are subjected to a relatively greater risk to exposure to the toxic substances, as well as reduction in oxygen consumption and effect of energy synthesis.

Susan and Sobha (2010) revealed that the depletion in oxygen consumption with severe respiratory distress and other behavioral changes. Decreased rate O_2 consumption of *Lepidocephalichthys guntea* observed in the present study may be because of similar toxic action of cypermethrin and malathion as reported in fishes by earlier researchers.

High activity of gills and continuous exposure to toxicants causes severe damage to gills surface and reduces to O_2 uptake capacity of respiratory organs. Few earlier workers have suggested that the toxic substances enter in to the blood stream of fish through the gills and turns spread over vital organs and inhibit the respiratory reaction in the mitochondria. Decreased rate of O_2 consumption may be due to the either reduction in the R. B. Cs or acts as haemolytic agents. The toxic substances interfere with respiration causing alternation in a structural architecture of gills. This inhibits the enzyme system of mitochondria which result in to reduction of rate of O_2 consumption (Pandey *et al.*, 1976). The rate of O_2 consumption decreases on exposure to sub lethal concentration of cypermethrin and malathion at 96 h exposure, when compared with control fish. The reason for depletion in O_2 consumption is alternating in the gills architecture and permeable property may be responsible for the observed depletion in O_2 uptake and the pesticides alter the ionic content of the biomass.

References

Alderic, D.E., 1967. Detection and measurement of water pollution biological assay. Canada Deptt. Fisheries. *Canadian Fisheries* 9: 33–39.

APHA/AWWA/WEF, 1998. *Standard Methods for the Examination of Water and Wastewater*, 20th Edn, American Public Health Association, New York, USA.

Asifa, P. and Vasantha, N., 2001. Effect of ammonia on respiratory activity of an air breathing fish, *Clarias batrachus. Nat. Conference. Hyderabad Abs.*, p. 59.

Gassner, Beat, Wuthrich, Andreas, Scholtysik, Gunter and Marc Soliz, 1997. The pyrethroids permethrin and cyhalothrin are potent inhibitors of the mitochondrial complex–I. *Jour. Pharma and Expt. Therapeu.*, 28(2): 855–860.

Golterman, H. and Clymo, C., 1969. *Methods for the Chemical Analysis of Freshwater*. Blackwell Scientific Publication, pp. 116.

Jain, G.K., 2000. *Pesticides and Fish*. Aditya Publishers, Bina, M.P.

Jayasuriya, S., Subramanian, M.A. and Varadaraj, G., 1991. Effect of detergent on the oxygen consumption of the cat fish *Mystus vittatus J. Ecobiol.*, 3: 217–220.

Mahajan, R.T. and Patole, S.S., 2003. Effect of plants extracts on oxygen consumption rate in fish *Nemacheilus evezardi. J. Freshwater Biol.*, 15(4): 109–113.

Natrajan, G.M., 1978. Observation on the oxygen consumption of Indian air breathing fishes, oxygen consumption in the climbing perch *Anabas scandens (Curvier). Comp. Physiol. Ecol.*, 3: 246–248.

Pandey, B.N., Chanchal, A.K. and Singh, M.P., 1976. Effect of malathion on the oxygen consumption and blood of *Channa punctatus (BI). Ind. J. Zootamy*, 97: 95–100.

Roy, P.K. and Dattamunshi, J.S., 1988. Oxygen consumption and ventilation rate of a major carp. *Cyprinus mrigala* in fresh and Malathion treated water. *J. Environ. Biol.*, 9: 5–13.

Susan, Anita and Sobha, K., 2010. A study of acute toxicity, oxygen consumption and behavioural changes in the three major carps, *Labeo rohita, Catla catla* and *Cirrhina mrigala* exposed to Fenvalerate. *Biores. Bulletin*, 1: 33–40.

Tilak, K.S., Vardhan, K.S. and Kumar, B.S., 2005. The effect of ammonia, nitrate on the oxygen consumption of the *Ctenopharyngodon idella (Valenciennes). J. Aqua. Biol.,* 20(1): 117–122.

Verma, S.R., Bansal, S.K., Gupta, A.K., Pal, N., Tyagi, A.K., Bhatnagar, M.C., Kumar, K. and Dalela, R.C., 1979. Acute toxicity of twenty three pesticides to a freshwater teleost *Saccobranchus fossilis. Proc. Symp. Environ. Biol.,* p. 481–497.

Vijayalakshmi, S. and Tilak, K.S., 1996. Effect of pesticides on the gill morphology of *Labeo rohita. J. Ecotoxicol. Environ. Monit.,* 6(1): 59–64.

Chapter 5

Ambient Ammonia Stress: Its Impact on Ammonia and Urea Levels and Extent of Recovery in Fish *Oreochromis mossambicus*

☆ *G. Ravindra Babu and P. Neeraja*

ABSTRACT

Ammonia is an important component of fertilizers. The excessive pesticides and other industrial effluents on retention in soil and water are reported to result in production of ammonia during their degradative processes. Ammonia is also a byproduct of fish metabolism. But the increase in ammonia concentration in ambient medium is toxic to fish. The present investigation is to understand the response of the animal to low concentration of ammonia and its recovery potential on removal of ammonia stress. Fish *Oreochromis mossambicus* is taken for the present study. Animals weighing 12 g and 8 cm long are exposed to 3.26 ppm of ammonia solution for seven and fourteen days. In order to understand the extent of recovery after ammonia stress, they were kept in ammonia free water for 7 and 14 days. Ammonia and urea levels were estimated in the liver, brain, kidney and gill tissues of the animal. There was increment of ammonia and urea levels in all the tissues in 7 days and 14 days ammonia exposed experimental. The increment was more in 14 days than 7 days exposure. The increment in brain tissue was highest followed by kidney, liver and gill in both the exposed experimental. There was increment in ammonia and urea levels even in 7 and 14 days recovery experimental also, but the increment was less than the exposed ones. The impact of recovery was more in 14 day than 7 days recovery experimental. The above changes will be discussed taking the impact and recovery capacity of the fish.

Keywords: Ammonia stress, Impact, Oreochromis mossambicus.

Introduction

Ammonia is the chief nitrogenous excretory product of teleostean fishes. It is also a common product of freshwater bodies contaminated with the runoff water containing nitrogenous fertilizers (ammonium salts and urea) from the near by agricultural fields and other pollutants from industries and pesticides (Singh *et al.*, 1991), (Van den Heuvel *et al.*, 2008). Metabolism for ammonia in fish is met as in other animals, more or less incidentally to its formation as an excretory product to its conversion into urea (Moraes and Polez, 2004). But the increase in ammonia concentration in ambient medium is toxic to fish. The toxicity and ammonia stress lead to number of metabolic changes in the fish (Walsh *et al.*, 2004, Santos *et al.*, 2010). Recovery studies have been scantily reported. Recovery trend in fish exposed to malathion was reported by Kausal and Ansari, 1986. Hence, the present investigation is to understand the response of the animal to low concentration of ammonia and its recovery potential on removal of ammonia stress.

Materials and Methods

Fish, *Oreochromis mossambicus* weighing about 12±2 gms and 8±2 cm long are selected and maintained in the laboratory. Temperature and pH were maintained throughout experimentation. Toxicity tests were conducted using ammonia solution. LC 50 was determined using Finney's method and it is 16.3 ppm for 48 hours. To understand the impact of low concentration, 3.26 ppm or 1/5th of LC 50 was selected as experimental concentration. Fish were exposed to this concentration for seven and fourteen days. After 7 and 14 days of exposure to test chemical, the fishes were transferred to normal tap water and kept for 7 and 14 days to allow them to recover. Steps were taken to maintain the experimental concentration constant throughout the experimentation. Ammonia and Urea contents were measured using Bergemeyer (1965) and Natelson method (1971).The experimental values were subjected to Student's 't' test and one way ANOVA using Tukey–Kramer multiple comparison test in Graph pad InStat-3 package.

Results and Discussion

Ammonia levels showed an increment in all the tissues of both 7 and 14 days exposure and post exposure (recovery) also (Table 5.1). The increment was more in exposure than recovery or post exposure to ammonia free medium. The order of increment in the tissues is Brain>Kidney>Liver>Gills in both 7 and 14 days of exposure. In the recovery experimentals it was Kidney>Brain>Liver>Gills in 7 day periods while it was Brain>Kidney>Gills>Liver in 14 days recovery. There was decrement in ammonia levels in all tissues in recovery experimentals compared to exposed experimentals.

Increased levels of ammonia on ammonia exposure might be due either increase production of ammonia by tissues due to stress or inability to remove the ammonia and hence retention of ammonia in the tissues. These increased levels of ammonia probably might be due to metabolic lethargy of the animal to eliminate it or detoxify into other products like urea or glutamine (Erdogan *et al.*, 2005). And this seems to continue in prolonged exposure for 14 days. Hence, the increment was more in 14 days exposure than 7 days exposure. The values were significant by ANOVA analysis between 7 and 14 day exposure for all the tissues except liver. This suggests that liver tissue is able to tolerate prolonged exposure and hence, the increment was also very less.

When the animals were shifted to ammonia free water to understand whether it can recover from the ammonia stress, the increment in ammonia levels was less than exposed experimentals suggesting there was recovery by the animal and it was able to convert into other products. Hence, the increment was less. This contention is supported by the results of statistical treatment through ANOVA and using Tukey Kramer multiple comparisons tests.

Table 5.1: Change in Ammonia Content in Different Tissues of the
Fish, *Oreochromis mossambicus* on Ambient Ammonia Exposure and Recovery
for 7 and 14 Days (Units are micromoles/g wet wt. of tissue)

Tissue	7 Days			14 Days		
	Control	Experimental	Recovery	Control	Experimental	Recovery
Brain	1.472	2.412	1.914	1.468	2.742	1.542
SD±	0.127	0.061	0.182	0.116	0.058	0.068
Per cent change	–	+63.85	+30.02	–	+86.78	+5.04
			–20.64*			–43.76*
Liver	2.648	2.942	2.792	2.641	3.121	2.694
SD±	0.158	0.047	0.154	0.162	0.027	0.028
Per cent change	–	+11.10	+5.43	–	+21.8	+2.00
			–5.09*			–13.68*
Gills	2.481	2.756	2.592	2.461	2.948	2.524
SD±	0.085	0.047	0.074	0.062	0.072	0.082
Per cent change	–	+11.08	+4.47	–	+19.78	+2.55
			–5.95*			–14.38*
Kidney	0.219	0.349	0.318	0.217	0.394	0.254
SD±	0.02	0.01	0.04	0.03	0.04	0.02
Per cent change	–	+59.36	+45.20	–	+81.56	+17.05
			–8.88*			–35.53*

Each value is mean and SD of six observations. All values are significant at $P<0.01$ level.

*: Per cent change of recovery over experimental.

The urea levels were also found to increase in all the experimentals similar to ammonia levels (Table 5.2). But the percent increment was less than percent increment of ammonia. The order of increment in the tissues is Kidney>Gills>Brain>Liver in both 7 and 14 days of exposure and 7day recovery experimentals, while it was Brain>Kidney>Gills>Liver in 14 days recovery. There was decrement in urea levels in all tissues in recovery experimentals compared to exposed experimentals. The main detoxification mechanism of ammonia is through urea conversion and these two are the main excretory products. This conversion seems to operate well in the present investigation. The results of the ANOVA also support this contention as most of the comparisons were significant at $P<0.001$. Liver tissue gave highly significant values in all comparisons supporting that the liver tissue, a centre for ureogenesis, and has been active in chronic ammonia stress.

Increased urea levels in gills of ambient ammonia exposed fish suggest increased vascular mobilization of urea from other tissues for its disposal into the medium (Walsh *et al.,* 2001). The increased level of urea in kidney tissue also supports the present possibility.

**Table 5.2: Change in Urea Content in Different Tissues of the
Fish, *Oreochromis mossambicus* on Ambient Ammonia Exposure and Recovery
for 7 and 14 days (Units are mg/g wet wt. of tissue)**

Tissue	7 Days			14 Days		
	Control	Experimental	Recovery	Control	Experimental	Recovery
Brain	1.572	1.914	1.714	1.568	2.112	1.614
SD±	0.028	0.041	0.042	0.026	0.028	0.019
Per cent change	–	+21.75	+9.03	–	+34.09	+2.93
			−10.44*			−23.57*
Liver	3.872	4.286	4.107	3.864	4.782	4.092
SD±	0.042	0.061	0.072	0.028	0.019	0.021
Per cent change	–	+10.69	+6.06	–	+23.75	+15.9
			−4.17*			−14.42*
Gills	2.384	3.124	2.784	2.378	3.748	2.456
SD±	0.034	0.14	0.036	0.032	0.027	0.019
Per cent change	–	+31.04	+16.77	–	+57.61	+3.28
			−10.88*			−34.47*
Kidney	0.972	1.341	1.251	0.974	1.562	1.164
SD±	0.017	0.002	0.004	0.016	0.012	0.006
Per cent change	–	+37.96	+28.78	–	+60.36	+19.51
			−6.71*			−25.48*

Each value is mean and SD of six observations. All values are significant at P<0.01 level.

*: Per cent change of recovery over experimental.

The increment in urea in recovery is less than experimentals suggesting that there is no requirement for urea conversion and removal as ammonia levels were less than exposed experimentals.

Urea plays an important role in maintaining osmotic balance in hypertonic environment. Since *O. mossambicus* is a euryhaline fish, it is quite reasonable to assume that elevated urea levels might have a role in maintaining osmotic balance under ammonia stress.

The present investigation is a beginning to understand the basic mechanism of the main excretory products. Further work on urea cycle enzymes is necessary to substantiate the present investigation.

References

Bergmeyer, H.V., 1965. In: *Methods of Enzymatic Analysis*. Academic Press, New York, pp. 1802.

Erdogan, O., Hisar, O., Korughu, G. and Ciltas, A.K., 2005. Sublethal ammonia and urea concentration inhibit rainbow trout (*Oncorhynchus mykiss*) erythrocyte glucose 6-phosphate dehydrogenase. *Comp. Biochemis. and Physiol., C*, 141(2): 145–150.

Kausal, K. and Ansari, B.A., 1986. Malathion toxicity: Effect on the liver of the fish *Brachydanio rerio* (cyprinidal). *Ecotoxico. Environ. Safety*, 12(3): 199–205.

Moraes, G. and Polez, V.L.P., 2004. Ureotelism is inducible in the Neotropical freshwater *Hoplias malabaricus* (Teleostei, Erythrinidae). *Brazilian Journal of Biology*, 64(2): 265–271.

Natelson, S., 1971. In: *Techniques of Clinical Chemistry*. Thomas, C.C., Springfield, Illinois, pp. 261.

Santhos, G.A., Schrama, J.W., Mamuay, R.E.P., Rombout, J.H.W.M. and Verreth, J.A., 2010. Chronic stress impairs performance, energy metabolism and welfare indicators in European seabass (*Dicentrarchus labrax*): The combined effecs of fish crowding and water quality deterioration. *Aquaculture*, 299(1–4): 73–80.

Singh, R.K., Sethi, N. and Trivedi, C.P., 1991. Risk assessment and management of pesticides. *J. Environ. Res.* 1(2): 7–15.

Van den Heuvel, M.R., Landman, M.J., Finley, M.A. and West, D.W., 2008. Altered physiology of rainbow trout in response to modified energy intake combined with pulp and paper effluent exposure. *Ecotoxicol. and Environmental Safety*, 69(2): 187–198.

Walsh, P.J., Grosell, M., Goss, G.G., Bergnan, H.L., Bergman, A.N., Wilson, P., Lavent, P., Alper, S.L., Smith, C.P., Kamunde, C. and Wood, C.M., 2001. Physiological and molecular characterization of urea transport by the gills of the lake Magad Tilapia (*Alcolapsia grahami*). *Journal of Experimental Biology*, 204(3): 509–520.

Walsh, P.J., Wei, Z., Wood, C.M., Loong, A.M., Hiong, K.C., Lee, Song,W.P., Chew, S.F. and Ip, Y.K., 2004. Nitrogen metabolism and excretion in *Allenbatrachus grunniens* (L): Effects of variable salinity, confinement, high pH and ammonia loading. *Journal of Fish Biology*, 65(5): 1392–1411.

Chapter 6

Seasonal Variation of Physico-chemical Characteristics of Water in Wetlands of Bhadra Wildlife Sanctuary, Karnataka

☆ *A.F. Pramod and Vijaya Kumara*

ABSTRACT

A systematic study has been carried out to evaluate physico-chemical characteristics of the selected wetlands of the Bhadra Wildlife sanctuary from May 2009 to April 2010. Three major wetlands which come across the temperate zone of the sanctuary *viz.*, Kalhalla Kere, Kesrhalla Kere and Hadhirayana halla have been selected. Eighteen physico-chemical water quality parameters have been analyzed for pre, post and monsoon seasons. The obtained data analyzed as per BIS and WHO standards. The physico-chemical characters of water were well with in the permissible limit and the data fluctuated as the season changes. The results were provided information for wetland water quality management in the region.

Keywords: Wetlands, Physico-chemical characteristics, Water.

Introduction

The physico-chemical parameters are very important in study of any environment, especially aquatic environment. Apart from the general interest in understanding the conditions of water and its impact on the aquatic biota, observation on the short term changes on the physico-chemical parameter may also have practical implication in pollution studies (Singabal, 1973).Wetlands are often referred to as "biological supermarkets" for the extensive food chain and rich biodiversity they support (Mitsch and Gosselink, 1993). Wetlands are one of the most important ecosystems, which have multiple utilities and covers 58.2 million hectares in India, out of which 40.9 million hectares are under rice cultivation (Anon, 2007). Water is most important chemical compound for the perpetuation of life on this planet. It is not only essential for lives but also important chemical compound from engineering point of view. Nearly 2/3 portion of this planet is occupied by water. It is present in three physical forms e. g. solid, liquid and gaseous. It has many unique properties. It is the compound which becomes rarer on solidification. It finds extensive use in the field of agriculture, hydro electric power generation and air conditioning (Prasad *et al.,* 2009). Water is the one of most important compound to ecosystem. Good quality of water described by its physical, chemical and microbial characteristics. But some correlation were possible among these parameters and the significant one would be useful the indicate quality of water (Kamble *et al.,* 2009). Contamination of water bodies might lead to a change in their tropic status and render them unsuitable for aquaculture. Several physicochemical or biological factors could act as stressors and adversely affect growth and reproduction (Iwama *et al.,* 2000).

Materials and Methods

The Bhadra Wildlife Sanctuary of Karnataka lies in the tropical forests of the Western Ghats in Chikmagalur district of Karnataka covering an area of 492.46 sq. km. Temperature varies from 10° C in winter maximum 32° C in summer. Here we made an effort to cover one range of Bhadra Wildlife Sanctuary that is Tanigebile range. Tanigebile section lies from 13°22' to 13°47' N latitude, 75°29' to 75°45' E longitude. The total study area comprising of 3 station *i.e.,* Kalhalla Kere, Kesrhalla kere, Hadhirayana halla (Table 6.1). The study was carried out for a period of 12 months from May, 2009 to April, 2010. Water sample was collected in the morning between 7 am and 11 am. The exact sample locations were fixed by Global Positioning System (GPS). Water samples were collected by 2 litres blue polythene bottle, physico-chemical parameters like pH, air temperature, water temperature of the sample was determined on the spot, dissolved oxygen was also fixed on the spot, electrical conductivity, total dissolved solid, biochemical oxygen demand, chemical oxygen demand, alkalinity, acidity, free carbon dioxide, chloride, calcium hardness, magnesium hardness, total hardness, phosphate, sulphate and iron are analyzed in the laboratory by following standard methods as prescribed by APHA, 2005. Statistical analysis was done by using PAST software.

Table 6.1: Sampling Locations with Codes

Code	Sampling Location	Latitude	Longitude	Elevation
TBL1	Kalhalla Kere	13°,35',524"	75°,42',789"	2380±22
TBL2	Kesrhalla Kere	13°,35',619"	75°,43',733"	2502±14
TBL3	Hadhirayana halla	13°,35',761"	75°,44',406"	2520±18

Note: Elevations are expressed in meters.

Results and Discussion

The present investigation of physico-chemical parameters of water samples in Bhadra wildlife sanctuary had shown much fluctuation. No work has been carried out in this sanctuary. pH of water is as measure of hydrogen ion concentration in water and indicates how much water is acid or alkaline in the present study reveals that during monsoon the pH was high 7.14±0.21 and less during the pre monsoon 6.82±0.10 and pH desired limit was also observed by Kulkarni *et al.* (2009). Gupta and Gupta (2006) stated that intense photosynthetic activities of phytoplankton will reduce the free carbon dioxide content resulting in increased pH values.

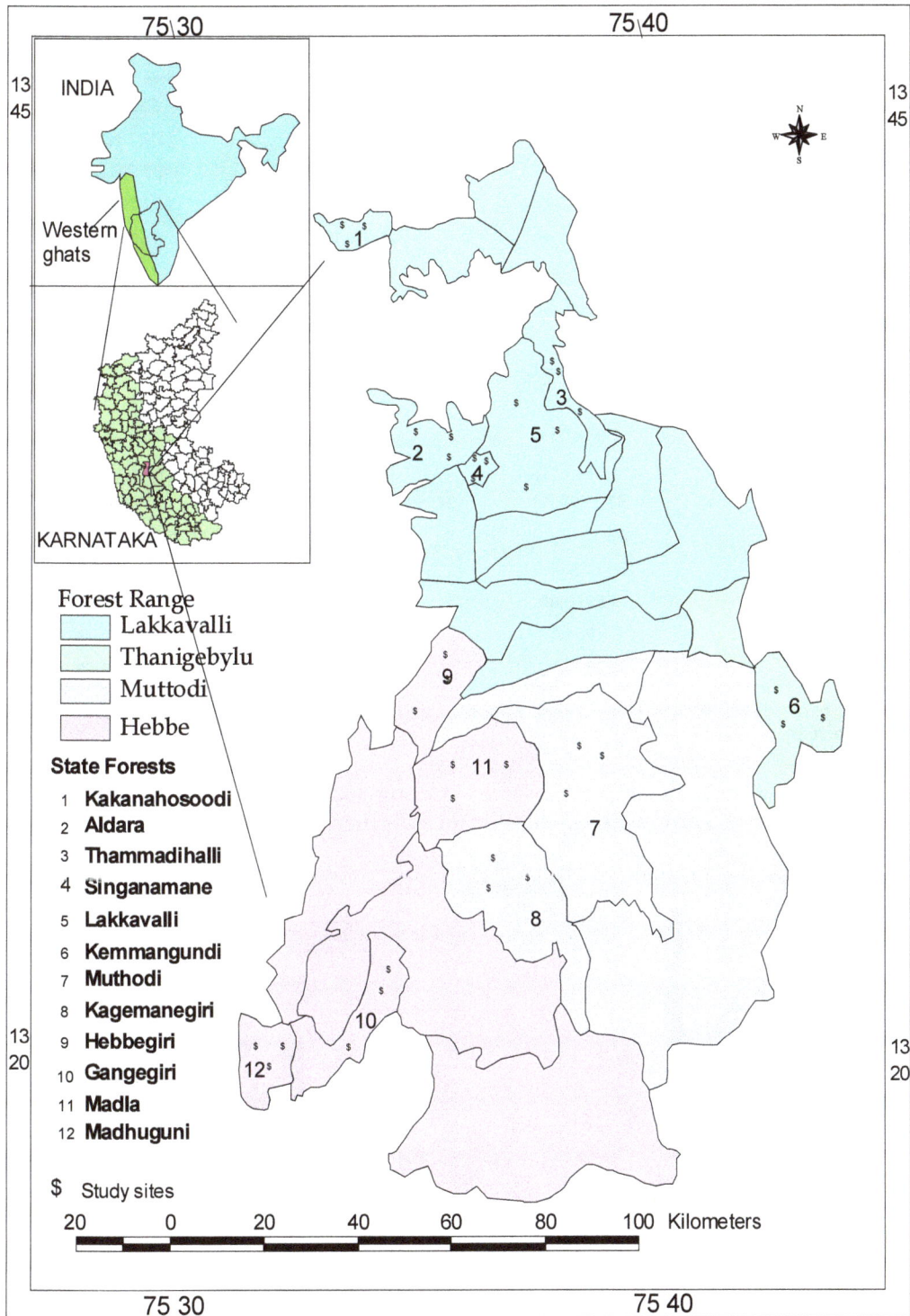

Figure 6.1: Map Showing the Bhadra Wildlife Sanctuary with Tanigaebile Range

Table 6.2: Average Values of Physico-chemical Parameters of the Water

Parameter	TBL1	TBL2	TBL3
pH	7.63±0.51	7.75±0.32	7.67±0.54
WT	21.50±1.86	22.35±2.81	23.13±2.62
AT	24.50±2.54	25.40±2.58	25.25±2.45
EC	149.30±50.74	134.60±41.99	102.25±17.83
TDS	48.88±5.46	47.00±13.89	51.01±5.36
DO	8.12±0.27	7.84±0.42	7.72±0.80
BOD	0.60±0.33	0.57±0.33	0.66±0.29
COD	5.07±1.90	4.14±1.13	5.07±1.88
ALK	42.68±5.61	42.28±3.95	45.81±6.86
Aci	5.74±1.21	5.99±1.34	5.25±1.51
Free CO_2	1.96±1.14	1.90±1.26	2.43±1.31
Cl	60.54±10.83	51.60±9.46	43.68±8.02
Ca	57.77±9.62	57.41±5.99	37.80±9.42
Mg	23.78±5.60	22.21±5.65	19.71±3.55
TH	130.83±24.01	108.33±9.54	63.68±6.94
PO_4	0.85±0.42	0.78±0.30	0.73±0.24
SO_4	8.72±1.73	9.31±1.69	8.16±2.74
Fe	0.86±0.26	0.92±0.24	0.84±0.23

Note: All the parameters are expressed in mg/L except air and water temperature (°C), pH, electrical conductivity (mmhos/cm).

The water and air temperature were observed by using field mercuric thermometer and expressed in degree celsius. The water and air temperature were the highest during the pre monsoon 21.87±1.21°C

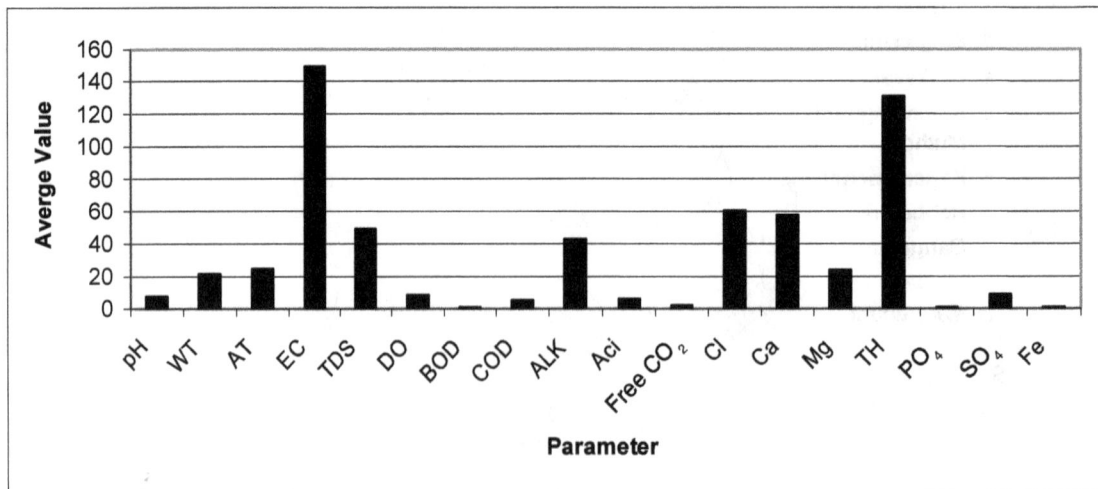

Figure 6.2: Average Value of Physico-chemical Characteristics TBL 1 during 2009-2010

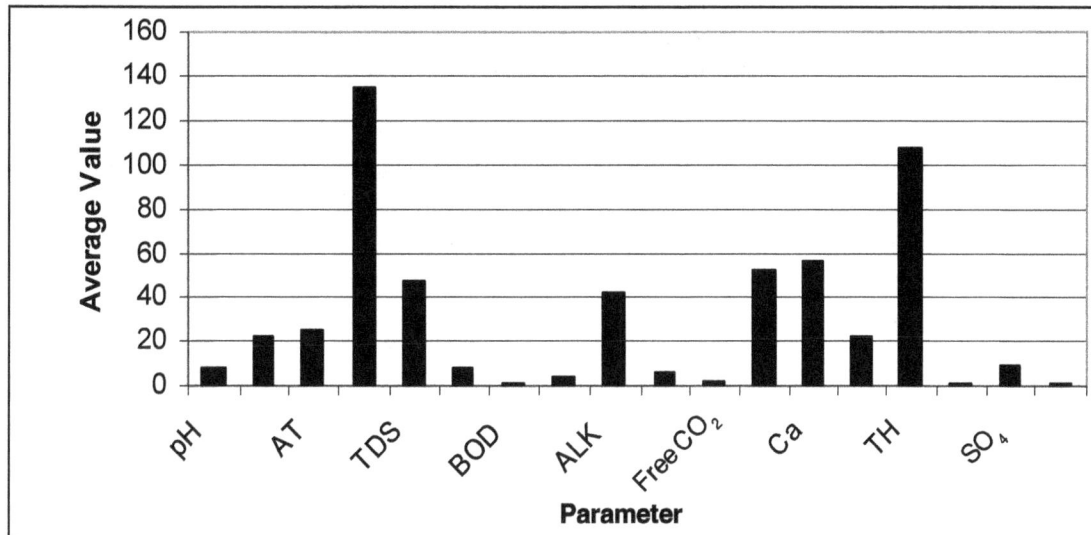

Figure 6.3: Average Value of Physico-chemical Characteristics TBL 2 during 2009-2010

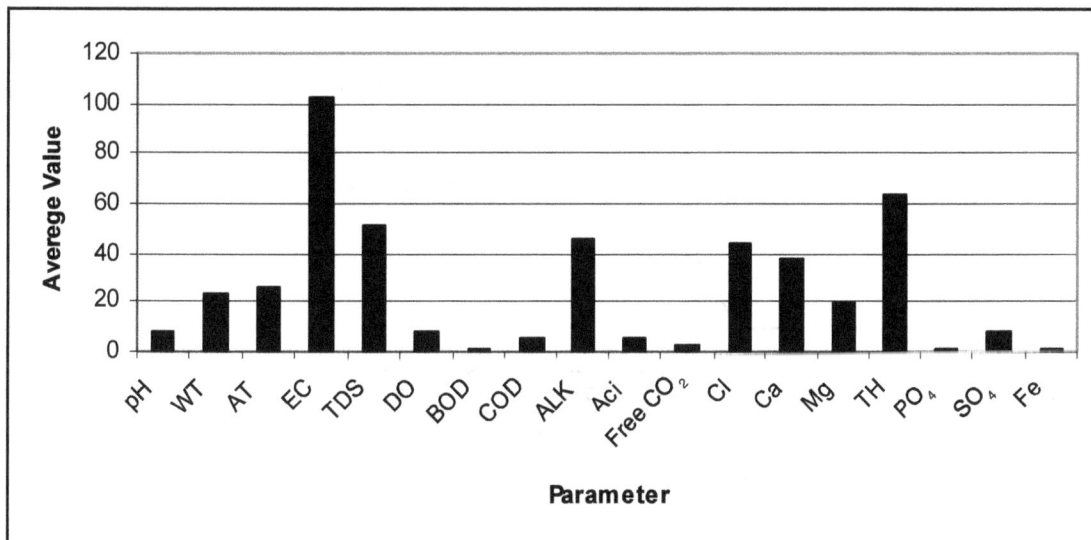

Figure 6.4: Average Value of Physico-chemical Characteristics TBL 3 during 2009-2010

and 23.07±0.26°C and water temperature the lowest at post monsoon season 20.51±0.75°C, air temperature decrease at 20.17±0.47°C during monsoon season respectively (Table 6.3). The electrical conductivity was ranging from 140.33±30.82 to113.33±14.72. It was recorded, the highest during pre monsoon because of recharged of rain water and also by streamlets and gradually decrease in post monsoon season Table 6.4 and similar observation was made by Sulabha and Prakasam (2006).

In the present investigation revealed that Total Dissolved Solid of water maximum during the post monsoon season 47.81±6.30 as compare to pre and post monsoons Table 6.4 and similar finding were observed by Rajurkar *et al.* (2003). There were slight variations in the DO during monsoon 8.18±26 mg/L and during pre monsoon and post monsoon, it ranging from 7.21±0.13 mg/L and 7.05±0.17 mg/L respectively Table 6.4. Similar observation was done by Sahu *et al.* (2000). In post monsoon season a maximum value of BOD was 0.67±0.01 mg/L and very less during monsoon season 0.41±0.12 mg/L Table 6.5. Fluctuations in the values may be due to the increase or decrease in temperature which results in decrease in microbial activity and algal bloom. These observations were supports the above findings Sachidanandamurthy and Yajurvedi (2004).

Table 6.3: Seasonal Mean Values of the Physico-chemical Characters of Water

Sampling Stations	pH			WT			AT		
	Monsoon	Post Monsoon	Pre Monsoon	Monsoon	Post Monsoon	Pre Monsoon	Monsoon	Post Monsoon	Pre Monsoon
TBL1	7.50	6.92	6.68	20.75	19.77	23.50	20.50	19.50	22.82
TBL2	7.63	7.03	6.90	22.25	20.21	21.50	19.50	21.00	23.43
TBL3	7.38	7.16	6.90	22.25	21.54	20.60	20.50	21.00	22.95
Mean	7.14	7.04	6.82	21.75	20.51	21.87	20.17	20.50	23.07
SD	0.21	0.10	0.10	0.71	0.75	1.21	0.47	0.71	0.26

Table 6.4: Seasonal Mean Values of the Physico-chemical Characters of Water

Sampling Stations	EC			TDS			DO		
	Monsoon	Post Monsoon	Pre Monsoon	Monsoon	Post Monsoon	Pre Monsoon	Monsoon	Post Monsoon	Pre Monsoon
TBL1	162.50	132.59	161.51	48.45	44.00	42.21	8.23	7.26	7.18
TBL2	161.75	110.53	133.87	39.79	45.89	46.19	7.85	7.05	7.07
TBL3	96.75	96.87	97.28	55.19	45.39	47.29	8.48	6.84	7.38
Mean	140.33	113.33	130.89	47.81	45.09	45.23	8.18	7.05	7.21
SD	30.82	14.72	26.31	6.30	0.80	2.18	0.26	0.17	0.13

Table 6.5: Seasonal Mean Values of the Physico-chemical Characters of Water

Sampling Stations	BOD			COD			ALK		
	Monsoon	Post Monsoon	Pre Monsoon	Monsoon	Post Monsoon	Pre Monsoon	Monsoon	Post Monsoon	Pre Monsoon
TBL1	0.39	0.67	0.58	5.43	4.28	3.75	41.00	39.55	38.55
TBL2	0.28	0.69	0.51	4.85	3.34	3.53	41.89	38.20	37.84
TBL3	0.56	0.66	0.61	5.80	4.42	4.35	47.27	41.58	42.23
Mean	0.41	0.67	0.57	5.36	4.02	3.88	43.39	39.78	39.54
SD	0.12	0.01	0.04	0.39	0.48	0.35	2.77	1.39	1.92

Chemical oxygen demand was found a minimum in pre monsoon season 3.88±0.35 mg/L and maximum during monsoon 5.38±0.39 mg/L and during post monsoon it was found to be 4.02±0.48 mg/L Table 6.5. The above finding agrees with Kulasherstha and Sharma (2006). Alkalinity was observed during monsoon season was recorded high 43.39±2.77 mg/L followed by the post and pre monsoon 39.78±1.39 mg/L and 39.54±1.92 mg/L (Table 6.5). Acidity of the water was found to be more during the monsoon season 5.78±0.26 mg/L and during pre monsoon it was found to be 5.26±0.19 mg/L and in post monsoon season it found that 5.11±0.24 mg/L Table 6.6. In season wise data revealed that seasonal changes in Bhadra wildlife sanctuary of Free CO_2 of the water were recorded significantly high in monsoon season 3.02±0.44 mg/L and less during the post monsoon season 1.59±0.18 mg/L, 2.08±0.19 mg/L in pre monsoon season. The less value of CO_2 during rainy and winter season might be due its utilization photosynthesis activity or it was being inhabited by presence of appreciable amount of carbonate in water Koorosh Jalilzadeh *et al.* (2009) Table 6.6. Chloride contents of water during the monsoon was found to be more 50.27±13.12 mg/L and 47.65±3.37 mg/L and 46.70±6.29 mg/L during post monsoon and pre monsoon Table 6.6.

Table 6.6: Seasonal Mean Values of the Physico-chemical Characters of Water

Sampling Stations	Acidity			Free CO_2			Cl		
	Monsoon	Post Monsoon	Pre Monsoon	Monsoon	Post Monsoon	Pre Monsoon	Monsoon	Post Monsoon	Pre Monsoon
TBL1	5.78	5.16	5.26	2.78	1.45	1.89	67.96	50.83	54.01
TBL2	6.10	5.37	5.49	2.63	1.47	2.00	46.26	49.13	47.44
TBL3	5.48	4.79	5.03	3.64	1.84	2.34	36.59	42.98	38.64
Mean	5.78	5.11	5.26	3.02	1.59	2.08	50.27	47.65	46.70
SD	**0.26**	**0.24**	**0.19**	**0.44**	**0.18**	**0.19**	**13.12**	**3.37**	**6.29**

Table 6.7: Seasonal Mean Values of the Physico-chemical Characters of Water

Sampling Stations	Ca			Mg			Total Hardness		
	Monsoon	Post Monsoon	Pre Monsoon	Monsoon	Post Monsoon	Pre Monsoon	Monsoon	Post Monsoon	Pre Monsoon
TBL1	55.90	52.37	48.93	25.20	20.28	19.23	96.46	123.12	123.22
TBL2	58.01	51.30	51.08	21.58	20.17	19.06	87.08	95.86	97.54
TBL3	42.30	33.60	39.07	19.78	18.25	18.86	54.23	56.62	60.35
Mean	52.07	45.76	46.36	22.19	19.57	19.05	79.26	91.86	93.70
SD	**6.96**	**8.61**	**5.23**	**2.25**	**0.93**	**0.15**	**18.11**	**27.30**	**25.81**

Contribution of hardness of water mainly by carbonates of calcium and magnesium, the calcium hardness of the water were high during the monsoon season 52.07±6.69 mg/l and during post monsoon it was found to be 45.76±8.61 mg/l and 45.36±5.23 mg/l (Table 6.7). During the pre monsoon season. The value of calcium revealed the fluctuation among the various season of the study site during the study period. Magnesium hardness of water was found to be more in monsoon season 22. 19±2.25 mg/l and there is a slight variation in the magnesium hardness in the post and pre monsoon season 19.57±0.93mg/l and 19.05±0.15 mg/l (Table 6.7). Total hardness of the water was found to be very high during the pre monsoon season 93.70±25.81 mg/l and during post monsoon it was found to be 91.86±27.30 mg/l and during monsoon it was found to be 79.26±18.11 mg/l (Table 6.7). Similar behavior of total hardness was recorded by Khadade and Mule (2003).

Table 6.8: Seasonal Mean Values of the Physico-chemical Characters of Water

Sampling Stations	PO_4			SO_4			Fe		
	Monsoon	Post Monsoon	Pre Monsoon	Monsoon	Post Monsoon	Pre Monsoon	Monsoon	Post Monsoon	Pre Monsoon
TBL1	0.54	0.80	0.64	5.74	7.33	8.33	0.51	0.78	0.76
TBL2	0.58	0.73	0.61	5.99	7.82	8.59	0.68	0.84	0.84
TBL3	0.49	0.68	0.64	5.47	7.10	7.87	0.65	0.81	0.85
Mean	0.54	0.74	0.63	5.73	7.42	8.26	0.61	0.81	0.81
SD	**0.03**	**0.05**	**0.01**	**0.21**	**0.30**	**0.30**	**0.07**	**0.03**	**0.04**

The phosphate content were minimum in monsoon season 0.54±0.03 mg/L and maximum value of phosphate were during the post monsoon season 0.74±0.05 mg/L and during pre monsoon 0.63±0.01 mg/L Table 6.8. The sulphate content was found to be more during pre monsoon season 8.26±0.30 mg/L and less during monsoon season 5.73±0.21 mg/L, 7.42±0.30 mg/L during the post monsoon season Table 6.8. The iron content was found to be more during post and pre monsoon 0.81±0.03 mg/L and 0.81±0.04 mg/L and less during the monsoon season 0.61±0.07 mg/L (Table 6.8). The value of iron reveals the fluctuation among the various seasons of the study site during the study period, similar observation was also done by Panda *et al.* (2004)

All the hydrological and physico-chemical parameter studied showed noticeable seasonal variation. The correlation co-efficient (r) among the various water quality parameter of TBL1 Table 6.9 shown that the pH is significantly correlated with the water temperature; total dissolved solid is also most significantly correlated with the alkalinity. Dissolved oxygen is most significantly correlated with the Chemical oxygen demand, alkalinity, chloride and total hardness. Biological oxygen demand is most significant to phosphate and iron. Acidity is most significantly correlated to the total hardness. Free carbon dioxide is also most significant at phosphate. Chloride is also most significantly correlated to phosphate and sulphate. Total hardness is most significant to sulphate. Correlation co-efficient of Station TBL 2, revealed that air temperature was most significantly correlated with the electrical conductivity and electrical conductivity was most significantly correlated to magnesium hardness. Dissolved oxygen was significantly correlated to iron; biological oxygen demand was most significant to acidity, magnesium hardness, and total hardness. Acidity is most significantly correlated to free carbon di oxide. Free carbon dioxides were most significantly correlated to magnesium hardness, and total hardness, Sulphate I were significantly correlated to iron. Correlation coefficient of station TBL 3 revealed that the p^H was most significantly correlated to water, temperature, chloride, and calcium hardness. Air temperature was most significantly correlated to calcium hardness, electrical conductivity is significantly correlated to total dissolved solids and phosphate. Total dissolved solid is significantly correlated with the biological oxygen demand, chemical oxygen demand, alkalinity, calcium hardness, and magnesium hardness. Dissolved oxygen was significantly correlated with the alkalinity, calcium hardness was most significantly correlated to sulphate, and Alkalinity was also significantly correlated to sulphate.

Table 6.9: Spearman's R S Correlation Co-efficient of the Water Kalhalla Kere

	pH	WT	AT	EC	TDS	DO	BOD	COD	ALK	Acidity	Free CO$_2$	Cl	Ca	Mg	TH	PO$_4$	SO$_4$	Fe
pH	0	**0.91**	0.64	**0.86**	0.20	0.39	0.22	0.33	0.30	0.38	0.51	**0.88**	0.06	0.55	0.37	**0.82**	0.64	0.47
WT		0	0.04	0.43	0.51	0.32	0.29	0.34	0.44	0.11	0.41	0.09	0.63	0.03	0.75	0.66	0.53	0.51
AT			0	0.52	0.16	**0.86**	0.39	0.23	0.68	0.26	0.21	0.36	0.48	0.02	0.20	0.02	0.48	0.06
EC				0	0.28	0.03	0.42	0.07	**0.81**	0.55	0.77	0.67	0.10	0.46	0.34	0.13	0.01	0.43
TDS					0	0.21	0.12	0.81	**0.93**	0.29	0.73	0.39	0.61	0.41	**0.81**	0.26	0.03	0.37
DO						0	0.07	**0.90**	**0.94**	0.31	0.07	**0.96**	0.75	0.55	**0.90**	0.26	0.00	**0.85**
BOD							0	0.56	0.40	0.67	0.01	0.57	0.37	0.43	0.48	**0.94**	0.08	**0.99**
COD								0	0.75	0.54	0.34	0.13	0.02	0.01	0.37	0.55	**0.87**	0.43
ALK									0	0.06	0.37	**0.80**	**0.89**	0.44	0.65	0.48	**0.85**	0.10
Aci										0	0.59	0.21	0.75	0.08	**0.93**	**0.83**	0.62	0.05
Free CO$_2$											0	0.22	0.69	0.15	0.26	**0.95**	0.25	0.41
Cl												0	0.37	0.03	0.64	**0.95**	**0.97**	0.29
Ca													0	0.16	**0.88**	0.33	0.71	0.47
Mg														0	0.59	0.23	**0.89**	0.10
TH															0	0.24	**0.99**	0.21
PO$_4$																0	0.04	0.04
SO$_4$																	0	0.49
Fe																		0

Note: **r** value most significant at 1, significant at 0.80 to 0.90.

Table 6.10: Spearman's R S Correlation Co-efficient of the Water of Kesrhalla

	pH	WT	AT	EC	TDS	DO	BOD	COD	ALK	Acidity	Free CO₂	Cl	Ca	Mg	TH	PO₄	SO₄	Fe
pH	0	0.00	0.25	0.54	0.41	**0.87**	0.69	**0.81**	0.28	0.57	**0.89**	0.14	0.62	0.08	0.25	0.62	0.76	0.50
WT		0	0.01	0.66	0.22	0.69	0.60	0.43	0.43	0.52	0.47	0.55	0.66	0.00	**0.82**	0.27	**0.86**	0.78
AT			0	**0.99**	0.17	0.12	0.69	0.02	0.51	0.15	0.62	0.76	**0.88**	0.00	0.49	0.04	0.65	0.25
EC				0	0.30	0.08	0.69	0.77	0.21	**0.87**	0.26	0.12	0.45	**0.97**	0.45	0.38	**0.87**	0.56
TDS					0	0.60	**0.87**	0.76	**0.84**	**0.80**	0.66	0.43	**0.84**	0.17	0.33	0.02	0.15	0.40
DO						0	0.58	0.19	0.01	0.13	0.08	0.33	**0.84**	0.10	0.21	0.14	0.54	**0.97**
BOD							0	0.46	0.17	**0.90**	0.01	0.19	0.74	**0.99**	**0.95**	**0.86**	0.07	**0.89**
COD								0	0.42	0.46	0.50	0.74	0.76	0.11	0.68	0.15	**0.89**	0.33
ALK									0	0.15	0.05	0.68	0.67	0.63	0.15	0.51	0.45	0.57
Aci										0	**0.93**	0.55	0.45	0.12	0.02	0.44	0.42	0.47
Free CO₂											0	**0.88**	0.63	**0.99**	**0.93**	**0.83**	0.35	0.20
Cl												0	**0.89**	0.65	0.70	0.67	0.14	**0.88**
Ca													0	0.48	0.37	0.69	0.13	0.10
Mg														0	0.46	0.07	0.36	0.70
TH															0	0.71	0.07	0.40
PO₄																0	0.35	0.49
SO₄																	0	0.92
Fe																		0

Note: **r** value most significant at 1, significant at 0.80 to 0.90.

Table 6.11: Spearman's R S Correlation Co-efficient of the Water of Gonimara Hadlae Kere

	pH	WT	AT	EC	TDS	DO	BOD	COD	ALK	Acidity	Free CO_2	Cl	Ca	Mg	TH	PO_4	SO_4	Fe
pH	0	0.97	0.67	0.09	0.58	0.35	0.66	0.68	0.28	0.47	0.60	0.96	0.92	0.49	0.63	0.65	0.40	0.25
WT		0	0.00	0.84	0.07	0.19	0.37	0.21	0.29	0.27	0.06	0.20	0.61	0.15	0.54	0.20	0.60	0.38
AT			0	0.80	0.30	0.33	0.19	0.45	0.36	0.43	0.02	0.09	0.98	0.30	0.42	0.09	0.88	0.59
EC				0	0.90	0.57	0.31	0.77	0.41	0.39	0.32	0.83	0.52	0.33	0.64	0.96	0.58	0.00
TDS					0	0.08	0.97	0.93	0.91	0.08	0.12	0.10	0.96	0.97	0.75	0.80	0.11	0.31
DO						0	0.59	0.31	0.90	0.29	0.18	0.16	0.07	0.24	0.20	0.69	0.63	0.25
BOD							0	0.17	0.15	0.56	0.24	0.56	0.52	0.70	0.01	0.01	0.69	0.67
COD								0	0.33	0.45	0.45	0.21	0.02	0.00	0.11	0.15	0.96	0.21
ALK									0	0.51	0.55	0.47	0.53	0.42	0.60	0.01	0.90	0.79
Aci										0	0.46	0.58	0.83	0.28	0.46	0.82	0.88	0.11
Free CO_2											0	0.00	0.67	0.88	0.51	0.30	0.56	0.56
Cl												0	0.48	0.29	0.46	1.00	0.70	0.76
Ca													0	0.00	0.05	0.29	0.23	0.14
Mg														0	0.24	0.17	0.30	0.08
TH															0	0.19	0.21	0.96
PO_4																0	0.18	0.38
SO_4																	0	0.74
Fe																		0.00

Note: **r** value most significant at 1, significant at 0.80 to 0.90.

Conclusion

As the season changes there were a fluctuation in the physico-chemical characters of the water this will be due to in flow and change in the temperature as season changes. The values of all the parameters have fell in permissible limits standards as the water of Bhadra wildlife sanctuary is potable.

Acknowledgements

We wish to express our gratitude to University Grants Commission for providing financial assistance, Kuvempu University for necessary infrastructure facilities and Karnataka state forest department for their help to carry out field work.

References

APHA, 2005. *Standard Methods for the Examination of Water and Wastewater*, 21ˢᵗ Edn. Washington, DC.

Cowardin, L.M., Carter, V., Golet, F.C. and La Roe, E.T., 1979. Classification of wetlands and deep water habitats of the United States. U.S. Fish and Wildlife Services, Washington, DC, USA, pp. 103.

Dilip, K., Rathore, P., Sharma, G.S., Tyagi, Barupal and Sonle, Krishna Chanda, 2006. Analysis of physico-chemical characteristics to study the water quality indx, algal blooms, and eutrophic conditions of lakes of Udaipur city, Rajasthan. *Indian J. Environ. Ecoplan.*, 12(1): 223–230.

Gupta, S.K. and Gupta, R.C., 2006. *General and Applied Ichthyology (Fish and Fisheries)*. S. Chand and Company Ltd., Ram Nagar, New Delhi, pp. 1130.

Iwama, G.K., Vijayan, M.M. and Morgan, J.D., 2000. The stress response in fish. In: *Ichthyology: Recent Research Advances*. Oxford and IBH Publishing Co. Pvt. Ltd., N. Delhi, pp. 453.

Jadhav, A.R. and Deshmukh, A.M., 2006. Physico-chemical and microbial characteristics of Rankala and Aalamba of Kolhapur district, Maharashtra, India. *Environment and Ecology*, 24(1): 21–27.

Jindal, S. and Gusain, D., 2007. Correlation between water quality parameters and phytoplankton of Bicherli pond, Beawar, Rajasthan. *J. Aqua. Biol.*, 22(2): 13–20.

Kamble, S.M., Kamble, A.H. and Narke, S.Y., 2009. Study of physico-chemical parameter of Ruti dam, tq. Ashti, Dist. Beed, Maharashtra. *J. Aquatic Biology*, 24(2): 86–89.

Khadade, S.A. and Mule, M.B., 2003. Studies on physico-chemical parameters of Pundi water reservoir from Tasgaon Tahsil. I. *J. Environ Prot.*, 23(9): 1003–1007.

Khadade, S.A. and Mule, M.B., 2003. Studies on physico-chemical parameters of Pundi water reservoir from Tasg *J. Aquatic Biology*, 24(2): 86–89.

Koorosh Jalilzadeh, Sadanand, M. Yamakanamardi and Altaf, K., 2009. Physico-chemical parameter of three contrasting lakes of Mysore, Karnataka, India. *J. Aquatic Biol.*, 24(2): 90–98.

Kulakarni, A.S., Tendulkar, Medh, Mavalankar, Sayali and Guhagarkar, A.M., 2009. Study on water quality parameter from Peth-Killa region, Ratnagiri, West Coast of India, Maharashtra. *J. Aquatic Biology*, 24(2): 82–85.

Kulasherstha, H. and Sharma, S., 2006. Impact of mass bathing during Ardhkumbh on water quality status of river Ganga. *J. Environ. Biol.*, 27: 437–440.

Mitsch, W.J. and Gosselink, J.G., 1993. *Wetalnds*, 2ⁿᵈ Edn. Van Nostrand-Reinhold, New York.

Panda, S.P., Bhol, B.N. and Mishra, C.S.K., 2004. Water quality status of 5 major temple ponds of Bhubaneswar city. *Indian J. Environ. Prot.*, 24(3): 199–201.

Prasad, N.R. and Patil, J.M., 2008. A study of physico-chemical parameters of Krishna river water particularly in Western Maharashtra, *J. Rasayan J. Chem.*, 1(4): 943–958.

Rajurkar, N.S., Nongri, B. and Partwardhan, A.M., 2003. Physico-chemical and biological investigation of river Umshyrpi at Shillong, Meghalaya. *Indian J. Environ. Hlth.*, 45(1): 83–92.

Sachidanandamurthy, K.L. and Yajurvedi, H.N., 2004. Monthly variations in water quality parameters (physico-chemical) of a perennial lake in Mysore city. *Indian Hydrobiol.*, 7: 217–228.

Sahu, B.K., Rao, R.J., Behara, S.K. and Pandit, R.K., 2000. Effect of pollutions on the dissolved oxygen concentration of river Ganga at Kanpur. In: *Pollution and Biomonitoring of Indian Rivers*, (Ed.) R.K. Trivedy, ABD Publication, Jaipur, India, p. 168–170.

Sulabha,V. and Prakasam, V.R., 2006.Limnological features of Thirumullavaram temple pond of Kollam municipality, Kerala. *Journal of Environmental Biology*, 27(2): 449–451.

Singabal, S.Y.S., 1973. Diurnal variation of some physico-chemical factors in Zuary estuary of Goa. *Indian Journals of Marine Science*, 2: 90–93.

Chapter 7

Water Quality of GIP Tank in Ambarnath, Thane District, Maharashtra

☆ *Manisha Karpe and Mrinalini Sathe*

ABSTRACT

The present investigation deals with physico-chemical characteristics of GIP Tank and correlation among these characteristics for a period of one year (June 2009–May 2010). The parameters studied were water temperature, pH, turbidity, conductivity, dissolved oxygen, free carbon dioxide, total alkalinity, total hardness, silicates, chlorides, sulphates, phosphates, nitrates, nitrites, NH_3, TDS and BOD. The study on physico-chemical parameters of GIP Tank revealed that the water is suitable for fish culture. Pearson correlation coefficient calculated to study interrelation among water parameters. All the 'r' values are significant at 5 per cent level, $P < 0.05$ (two tailed).

Keywords: Water quality, Correlation coefficient, GIP tank, Thane.

Introduction

Water is one of the abundantly available substances in nature and has great importance for sustenance of life. Water has proved to be most essential commodity on the earth; it has direct bearing on health of all organisms including man. Knowledge of water chemistry is important for understanding biological phenomenon, in the ecosystem and hydro-biological interrelationship (Kiran, 2010). The physical and chemical characteristics of water bodies affect species composition, abundance, productivity and physiological conditions of aquatic organisms. Within recent past decades, there has been considerable interest in the relevance of limnological information to the productivity, development and management of aquatic environments. Regular and periodic investigation of water ecology is useful for monitoring the water body.

In India, in last decades many researchers have studied various water bodies, their ecology, flora and fauna in water bodies. Some of the recent contributions to limnology literature in India are done by Sathe *et al*. (2001), Murlidhar *et al*. (2002), Somani (2002), Lende (2004), Kakvipure (2005), Adarsh Kumar *et al*. (2006), Raut (2006), Balasingh and Shamal (2007), Mishra (2008), Anilkumar *et al*. (2009), Gurav (2009) and Garge *et al*. (2010).The present paper deals with physical and chemical characteristics of surface water of GIP Tank from June 2009 to May 2010. It also discusses the interrelationship of these physico-chemical parameters.

Description of Study Area

Ambernath in Thane district of Maharashtra is surrounded by Malang-gadh hilly ranges and lies between $19° 12'$ North latitude and $73° 10'$ East Longitude. It is 61 feet high from sea level. To the South West of Ambernath, nearly 2kms away Kakuli reservoir [GIP Tank] is situated. After 1856 (Mumbai - Pune Railway) GIP Tank got great significances as it was used to provide water for steam engines of railway at Kalyan. The Great Indian Peninsular Railway Company constructed dam on Kakuli Reservoir and hence it is called as GIP Tank. The GIP Tank receives the runoff water from a cattle shade present at the bank of the reservoir on one side and the nutrients thus received make it suitable for the evaluation for fisheries potential.

The average annual rain fall at Ambernath is 1820.38 mm. Rain fall during south-west monsoon season, June to September constitutes maximum percentage of the annual rain fall. The average mean air temperature ranged between 24°C to 34°C. Climate is less humid (average humidity 79 per cent), the average mean daily minimum temperature is slightly lower in winter season and the average mean daily maximum temperature is higher in summer season.

Materials and Methods

For the study of physico-chemical parameters water samples were collected from GIP Tank every month from June 2009 to May 2010, in morning hours between 7a.m. and 9 a.m. Clean glass bottles and plastic bottles were used for sample collection. Water temperature was recorded on the spot. Water analysis was performed as per the methods prescribed in Standard methods APHA, (2002), 18th Edition, Trivedy and Goyal (1984), Gupta (2007). To study interrelation among the physico-chemical parameters Excel sheet (Windows 2007) was used and Pearson's 'r' was calculated.

Results and Discussion

Water Temperature

Temperature variations of a water body have a great bearing upon its productivity. All the metabolic and physiological activities and life processes are greatly influenced by water temperature. During present study, surface water temperature for GIP Tank ranged between 23 to 30.5°C. The minimum temperature was noted in January 2010 and maximum in May 2010 (Table 7.1). Balasingh and Shamal (2007) reported similar range of water temperature for Tungbhadra Reservoir in Karnataka and for a Perennial Pond in Tamil Nadu, respectively. The surface water temperature showed positive correlation with free CO_2 (r = 0.566), Total Alkalinity (r = 0.492) and Total hardness (r = 0.484). The water temperature showed significant negative correlation with DO (r = -0.760), negative correlation with Silicates (r = -0.489), (Table 7.2). Tiwari and Sharma (2011) reported that water temperature exhibited positive correlation with pH and negative correlation with total hardness in Lony dam.

Table 7.1: Physico-chemical Characteristics of Water of GIP Tank during 2009-10

Month/Year	Temperature	pH	Turbidity	Conductivity	Dissolved Oxygen	Free CO₂	Total Alkalinity	Total Hardness	Silicates	Chlorides	Sulphates	Phosphates	Nitrates	Nitrites	NH₃	TDS	BOD
Jun-09	29.25	6.87	8.7	275.6	3.45	51.6	127	131	1.485	18.995	9.705	0.115	3.76	5.48	1.165	322.5	32.5
Jul-09	26.6	7.285	7.5	117.5	6.1	8.15	163	146.5	2.06	114.2	11.35	0.288	0.1475	3.05	0.785	215	118.8
Aug-09	25.75	7.555	11.5	0.07	5.95	13.15	72	88	4.15	61.99	7.9375	0.57	1.195	0.165	0.073	141	20.9
Sep-09	26.25	8.745	5.5	0.08	6.45	20	73.5	82.5	2.185	18.155	9.415	0	0.3035	0.0595	0	173.5	3.9
Oct-09	26.4	7.99	8	0.93	6.6	5	87.5	88.5	2.33	19.195	10.83	0	0.3225	0.0735	0	147	7.5
Nov-09	26.75	8.13	4.5	104.5	4.55	1.9	95	77.5	2.33	16.615	9.485	0	0.355	0.063	0.065	146	11.5
Dec-09	24.3	7.455	25.5	113.4	7.25	5.25	66.5	111	3.225	15.1	10.63	0	2.17	0.073	0.075	125.5	7.1
Jan-10	23	7.63	10.5	157.35	6.35	0	86.5	92	5.29	15.7	22.14	0	2.685	0.0445	0	207.5	2.9
Feb-10	26.65	7.335	3.5	162.95	6.3	4.3	74.5	94.5	3.19	14.055	13.985	0	0.25	0.02	0	190	15.5
Mar-10	27.6	6.71	11	129	7	61.6	92	42	3.84	51	9.72	0.0119	2.97	0.023	0	161	4
Apr-10	28.1	6.57	3	138.2	5.2	8.8	108	104	3.32	16	27.17	0	2.03	0.1062	0	87	2.8
May-10	30.5	7.15	42	248.5	0	26.4	116	198	3.21	16	0.799	0	0	0.1189	0	342	0
Min	23	6.57	3	0.07	0	0	66.5	42	1.485	14.055	0.799	0	0	0.02	0	87	0
Max	30.5	8.745	42	275.6	7.25	61.6	163	198	5.29	114.2	27.17	0.57	3.76	5.48	1.165	342	118.8
Avg	26.763	7.452	11.767	120.673	5.433	17.179	96.792	104.625	3.051	31.417	11.931	0.082	1.349	0.773	0.18	188.167	18.95

Table 7.2: Intercorrelation Matrix for Various Physico-chemical Paramers of GIP Tank during 2009-10

	Temperature	pH	Turbidity	Conductivity	Dissolved Oxygen	Free CO$_2$	Total Alkalinity	Total Hardness	Silicates	Chlorides	Sulphates	Phosphates	Nitrates	Nitrites	NH$_3$	TDS	BOD
Temperature	1.000																
pH	-0.435	1.000															
Turbidity	0.285	-0.164	1.000														
Conductivity	0.506	0.642	0.373	1.000													
Dissolved Oxygen	*-0.760	0.231	*-0.595	*-0.630	1.000												
Free CO$_2$	0.566	-0.454	0.144	0.383	-0.228	1.000											
Total Alkalinity	0.492	-0.413	0.027	0.471	-0.419	0.213	1.000										
Total Hardness	0.484	-0.229	*0.658	0.554	*-0.758	-0.057	0.572	1.000									
Silicates	-0.489	-0.179	0.172	-0.050	0.202	-0.173	-0.436	-0.249	1.000								
Chlorides	-0.036	-0.165	-0.132	-0.196	0.208	0.087	0.565	0.086	-0.084	1.000							
Sulphates	-0.389	-0.245	-0.559	-0.014	0.416	-0.353	-0.037	-0.321	0.355	-0.150	1.000						
Phosphates	-0.084	-0.054	-0.080	-0.300	0.071	-0.011	0.155	0.064	0.083	*0.681	-0.208	1.000					
Nitrates	-0.084	-0.536	-0.086	0.396	0.152	0.534	-0.018	-0.256	0.252	-0.151	0.350	-0.054	1.000				
Nitrites	0.348	-0.313	-0.130	0.489	-0.240	0.420	*0.687	0.369	-0.574	0.317	-0.112	0.254	0.379	1.000			
NH$_3$	0.298	-0.290	-0.136	0.48	-0.198	0.367	*0.708	0.369	*-0.579	0.397	-0.126	0.304	0.336	*0.993	1.000		
TDS	0.535	-0.175	0.501	*0.706	*-0.740	0.426	0.488	*0.698	-0.235	-0.009	-0.503	-0.027	0.046	0.563	0.523	1.000	
BOD	0.017	-0.118	-0.200	0.043	0.086	-0.066	*0.752	0.321	-0.392	*0.84	-0.076	0.493	-0.189	*0.601	*0.676	0.181	1.000

*: Values significant at 5 per cent level, $P < 0.05$ (two tailed).

pH

pH values ranged between 6.57 - 8.75 during present study of GIP Tank. The minima was recorded in April 2010 and maxima in September 2009 (Table 7.1). The acidic pH recorded in April 2010 can be correlated with high concentration of carbon-dioxide in GIP Tank. In April 2010 maximum fish production was recorded in GIP Tank. According to Krishnamoorthi *et al.* (2011), high temperature enhances microbial activity, causing excessive productivity leading to increased production of CO_2 and reduced pH. They studied Veeranam lake and investigated water quality parameters of same lake for aquaculture. Lokhande *et al.* (2010) indicated pH range (7.8 – 9.5) of Gharni reservoir suitable for the survival of fish. pH showed significant negative correlation with Conductivity (r = -0.642) and, negative correlation with Nitrate (r = -0.536), temperature (r = -0.435), free CO_2 (r = -0.454) and total alkalinity (r = -0.413) (Table 7.2). According to Kamble *et al.* (2009), the reduced rate of photosynthetic activities reduces the assimilation of carbon dioxide and bicarbonates which is ultimately responsible for increase in pH, the low oxygen values coincided with high temperature during the summer months. Laskar and Gupta (2009) indicated positive correlation of pH with total hardness from Chatla floodplain lake of Assam. Ayoade *et al.* (2009), reported negative correlation between water temperature and dissolved oxygen from Tehri dam. Acidic water reduces the appetite of the fish, their growth and tolerance to toxic substances, and fish become prone to attacks of parasites and diseases (Jhingran, 1982).

Turbidity

Turbidity is an important limiting factor in the productivity of a pond, it may be either due to suspended inorganic substances or due to planktonic organisms. Turbidity due to planktonic organisms is an indication of pond fertility while silt or mud causes harmful turbidity which may affect life of fish and fish food organisms. During present study turbidity ranged between 3 - 42 NTU. Minimum turbidity was noted in April 2010 and maximum was in May 2010 (Table 7.1). Jalizadeh *et al.* (2009) reported 2.4 – 52 NTU from Hebbal lake, Karnataka. They studied three contrasting lakes from Karnataka to reveal their status with respect to physico-chemical parameters. During present study, turbidity showed positive correlation with total dissolved solids (r = 0.501) and Conductivity (r = 0.373) (Table 7.1).

Conductivity

Conductivity is the ability of a substance to conduct the electric current and depends on presence of various ions in the water. As most of the salts in water are present in the ionic forms, able to conduct current, therefore conductivity is rapid measure of total dissolved solids. During present study conductivity was found to be varying between 0.07 to 275.6 µS. Maxima noted in the month of June 2009 while minimum conductivity was noted in August 2009 (Table 7.1). Conductivity showed significant positive correlation with total dissolved soilds (r = 0.706) and positive correlation with Nitrites (r = 0.489), NH_3 (r = 0.448) and Nitrates (r = 0.396) (Table 7.2). Laskar and Gupta (2009) indicated positive correlation of conductivity with water temperature and BOD from Chatla floodplain lake of Assam. Ayoade *et al.* (2009), reported positive correlation between conductivity and TDS from Tehri dam. Singh *et al.* (2010) reported negative correlation of conductivity with BOD, water temperature, total hardness and positive correlation with dissolved oxygen from Mansagar Lake, Rajasthan.

Dissolved Oxygen

Dissolved oxygen in GIP Tank ranged between nil to 7.25mg/L. The minimum concentration was noted in May 2010 (Table 7.1). Increased temperature increases metabolic activities of aquatic lives which lead to use of dissolved oxygen. In the present study, the maximum dissolved oxygen in the water of GIP Tank was recorded during winter, there after it started declining gradually and in summer it reached to lowest concentration. This is in accordance with observations by Sisodia and Moundiotiya (2006). According Sharma (2001) low values of dissolved oxygen during rainy season may be due to heavy influx of the water and poor photosynthesis. DO showed significant negative correlation with temperature (r = -0.760), total hardness (r = 758), total dissolved solids (r = -0.740), conductivity (r = -0.630) and turbidity (r = -0.595) (Table 7.2). Kamble *et al.* (2009) indicated inverse correlation between O_2 and CO_2 and supported that concentration of O_2 and CO_2 depends upon plants, aquatic animals as well as alkalinity and hardness of water. They studied Ruti dam water parameters and their interrelation. Laskar and Gupta (2009) indicated positive correlation of DO with total alkalinity, negative correlation with conductivity, BOD from Chatla floodplain lake of Assam. Singh *et al.* (2010) reported negative correlation of DO with BOD from Mansagar Lake, Rajasthan. Dissolved oxygen of GIP Tank provides suitable environment for fish culture as submerged weeds are absent

Free CO_2

The atmosphere, respiration of animals and plants, bacterial decomposition, ground water are the main sources of Free CO_2 in water body. In photosynthesis CO_2 is utilized by chlorophyll bearing organisms inhabiting the water body. During present study Free CO_2 ranged between nil to 61.6mg/l in the water samples from GIP Tank (Table 7.1). Subba Rao and Govind (1967) also recorded absence of CO_2 from January to June in Tungbhadra reservoir. Free CO_2 exhibited positive correlation with Temperature (r = 0.566), Nitrates (r = 0.534) (Table 7.2).

Total Alkalinity

During present study, total alkalinity ranged between 66.5 to 163 mg/l for GIP Tank. The lowest total alkalinity was observed in December 2009, the highest was noted in July 2009 (Table 7.1). According to Wurts and Durborow (1992) alkalinity in water is increased due to rain water which gets saturated with CO_2 while in pond water lime stone reacts with CO_2, the resultant water has increased alkalinity. Garg *et al.* (2009) reported 64.25–146.25 mg/L total alkalinity from Ramsagar reservoir, MP.Total Alkalinity showed significant positive correlation with NH_3 (r = 0.708), positive correlation with total hardness (r = 0.572), temperature (r = 0.492), total dissolved solids (r = 0.488), conductivity (r = 0.471) (Table 7.2). Sharma *et al.* (2008) depicted positive correlation of total alkalinity with conductivity, hardness, chlorides, nitrates, phosphates from Jaisamand lake, Rajasthan. Laskar and Gupta (2009) indicated negative correlation of total alkalinity with water temperature and BOD from Chatla floodplain lake of Assam. Ayoade *et al.* (2009), reported positive correlation between total hardness and alkalinity from Tehri dam.Deviprasad *et al.* (2009) described 120 to 360 mg/l range of total alkalinity, good enough for fish productivity. Wurts and Durborow (1992) indicated a desirable range of total alkalinity for fish culture is between 75 and 200 mg/l $CaCO_3$. GIP Tank is suitable for aquaculture with respect to total alkalinity.

Total Hardness

Total hardness of GIP Tank water ranged between 42–198 mg/L. Its minima was noted in March 2010, and maxima in May 2010 (Table 7.1). Maximum total hardness in GIP Tank reported during May 2010 may be attributed to lower water level and high rate of evaporation due to high temperature. Similar observation is made by Sisodia and Moundiotiya (2006) and Jayabhaye *et al.* (2008).Total hardness was significantly positively correlated with TDS (r = 0.698) and Turbidity (r = 0.658).Total

Hardness showed positive correlation with conductivity (r = 0.554), temperature (r = 0.484), total hardness showed significant negative correlation with dissolved oxygen (r = -0.758) (Table 7.2). Sharma *et al.* (2008) indicated positive correlation of total hardness with nitrates and phosphates, they studied Jaisamand lake with respect to its trophic status and zooplankton diversity. The total hardness of GIP Tank showed optimum range which is suitable for fish production. Sakhare (2007) reported 82 to 148 mg/l range of total hardness as suitable for fish production. He further indicated that Total hardness below 20mg/l produces disease in fishes and very hard water (>300mg/l) is uncongenial for fish production. Bhakta and Bandopadhyaya (2008) reported 92 to 206 mg/l total hardness congenial for fish production.

Silicates

Silicates are most abundant in sedimentary rocks and come in water in undissociated condition. Silica is utilized by the diatoms. Silicate forms an important parameter of phytoplankton distribution. Silica mostly used by diatoms for formation of their wall and hence related to fertility of pond. During present study silicates ranged between 1.49 to 5.29mg/l, the lowest value for silicate was recorded in June 2009 and the highest value in January 2010 (Table 7.1). Subbarao and Govind (1963) reported 6.1 – 9 mg/l silicates from Tungbhadra reservoir. Silicates showed significant negative correlation with NH_3 (r = -0.579) and negative correlation with nitrites (r = -0.574), Temperature (r = -0.489), BOD (r = -0.392) and sulphates (r = 0.355) (Table 7.2). Nadoni *et al.* (2001) indicated positive correlation between silicates and BOD, silicates and Sulphates from Amani tank Tumkur, Karnataka which is not coinciding with the present study.

Chlorides

Chlorides are essential for maintenance of normal physiological functions in aquatic organisms. But high chloride content in water has deleterious effect on aquatic life. Exposure to higher levels of chloride interferes with osmoregulation in aquatic organisms. Chlorides are also important for their significant role in nutrition, neurophysiology and renal functions. Chloride ions are highly permeable across plasma membranes and responsible for maintenance of osmotic pressure, water balance and acid base balance in animal tissues. During present study, chloride concentration ranged between 14.06 -114.2 mg/L in GIP Tank. Maxima was observed in July 2009 and minima in February 2010 (Table 7.1). Zutshi *et al.* (1984) reported 47 – 114 mg/L chlorides from Perennial pond, Srinagar. Chlorides showed significant positive correlation with BOD (r = 0.854), Phosphates (r = 0.681) and positive correlation with NH_3 (r = 0.397), nitrites (r = 0.317) (Table 7.2). Sharma *et al.* (2008) recorded positive correlation of chloride with dissolved oxygen, from Jaisamand lake, Rajsthan.

Sulphates

Sulphate ions usually occur in natural waters. In freshwaters it occurs in combination with cations like calcium, potassium, iron, etc. Many sulphate compounds are readily soluble in water. sulphates along with calcium and magnesium are an important constituent of hardness. Rain water, biological oxidation of reduced Sulphur, dissolution of sulphate compounds present in the sedimentary rocks of the catchment area are the sources of sulphates in water. Sulphate is important for growth of plants and short supply of Sulphate may inhibit the development of plankton. Sulphates of GIP Tank ranged between 0.80 to 27.17 mg/l. The maxima was noted in April 2010 and minima was recorded in May 2010 (Table 7.1). Sharma and Hussain (2001) reported 4.2 – 33.4 mg/l sulphates from tropical flood plain lake Assam. During present study sulphate showed positive correlation with dissolved oxygen (r = 0.416) and negative correlation with Turbidity (r = -0.559),total dissolved soilds (r = -0.503) (Table 7.2).

Phosphates

Phosphates recorded in the water samples of GIP Tank varied between nil and 0.57mg/L. The highest value for phosphates was noted in August 2009 and the lowest (nil) during September to February 2010 and in April, May 2010. In March 2010 phosphate content was 0.01 mg/L (Table 7.1). Johri *et al.* (1989) reported nil – 0.45 mg/L phosphates from Nainital lake of Kumaon Hills. Phosphate showed positive correlation with BOD (r = 0.493) (Table 7.2). Sharma *et al.* (2008) noted negative correlation of phosphates with chlorides, alkalinity, dissolved oxygen from Jaisamand lake, Rajasthan. Jhingran (1982) stated that, natural waters having a phosphorus content of more than 0.2 ppm are likely to be more productive. In this sense GIP Tank water is more productive for fish production.

Nitrates

Nitrates varied from 0.00 to 3.76 mg/l in GIP Tank. The lowest value (0.00mg/L) of nitrates was recorded in May 2010 and the highest was recorded in June 2009. Decreased concentration of nitrate was noted from September 2009 to December 2009 and increased concentration was noted from December 2009 to April 2010 (Table 7.1). Song *et al.* (2010) reported 0.44 – 3.39 mg/L Nitrates from Taihu lake, China. Nitrates in GIP tank showed positive correlation with pH, conductivity, free CO_2 and dissolved oxygen (Table 7.2). According to Bhuiyan and Gupta (2007), phosphates showed positive correlation with nitrates which enhanced growth of phytoplankton which in turn produced more dissolved oxygen. Sharma *et al.* (2008) reported negative correlation of nitrates with total dissolved soilds, conductivity, dissolved oxygen, chlorides, from Jaisamand lake, Rajasthan.

Nitrites

Nitrites from surface waters of GIP Tank varied between 0.02 to 5.48 mg/L and maxima was noted in June 2009 while minima in February 2010 (Table 7.1). Brahma *et al.* (2008) reported 0.4 – 4.869 mg/L nitrites from Chilka lake, Orissa. Nitrites showed significant positive correlation with NH_3 (r = 0.993) and positive correlation with TDS (r = 0.563) (Table 7.2).

NH_3

Fish are very sensitive to unionized ammonia (NH_3) and the optimum range is 0.02-0.05 mg/L in the pond water. Normally in the case of high dissolved oxygen and high carbon dioxide concentrations, the toxicity of ammonia to fish is reduced. Aeration can also reduce ammonia toxicity. Healthy phytoplankton populations remove ammonia from water. Pond water shows decreased pH just before the period of darkness ends and increased pH in afternoon. In decreased pH unionized NH_3 become ionized and has low toxic effect on aquatic life before dawn. NH_3 in the surface water samples of GIP Tank ranged between 0.00- 1.17mg/L Maxima was noted in June 2009. NH_3 remained absent in GIP Tank except for June–August and November, December 2009 and February 2010 (Table 7.1). Garg *et al.* (2010) reported nil – 0.84 mg/L NH_3 from Ramsagar reservoir.

Total Dissolved Solids (TDS)

It varied between 87- 342 mg/L in GIP Tank. The highest value for total dissolved solids was noted in June 2009 and the lowest was noted in April 2010 (Table 7.1). Sakhare (2007) reported 200-385 mg/l TDS value suitable for fish culture in Wan reservoir. Johri *et al.* (1989) reported 80– 200 mg/l TDS from Bhimtal lake of Kumaon Hills. TDS showed positive correlation with NH_3 (r = 0.523) (Table 7.2). Singh *et al.* (2010) reported negative correlation of TDS with dissolved oxygen and positive correlation with conductivity, BOD, water temperature and total hardness from Mansagar Lake, Rajasthan.

BOD

During present study BOD of GIP Tank water ranged between 0.00 - 118.8 mg/L (Table 7.1). Higher values of BOD were noted in July 2009 which could be due to higher microbial activity when the water level was very low as heavy rainfall was recorded mainly in August 2009. BOD showed significant positive correlation with NH_3 (r = 0.676) and Nitrites (r = 0.601) (Table 7.2).

Summary of Interrelation

☆ Temperature positively correlated with conductivity, free CO_2, total hardness, nitrites, NH_3, and total dissolved soilds and correlation between Temperature and Free CO_2 was significant. Temperature significantly correlated to dissolved oxygen.

☆ pH showed negative correlation with conductivity, free CO_2, total alkalinity, silicates, sulphates, nitrates, nitrites, and NH_3.

☆ Correlation between conductivity and total hardness was positive, conductivity and TDS were significant and positive. Conductivity significantly and negatively correlated to pH and dissolved oxygen.

☆ Correlation between turbidity and total hardness was positive and significant.

☆ Correlation of dissolved oxygen with total hardness and TDS were significant and negative.

☆ Free CO_2 positively correlated with total alkalinity, nitrates, nitrites, NH_3, and TDS.

☆ Total alkalinity positively correlated with total hardness, chlorides, nitrites, NH_3, and BOD.

☆ Total hardness positively correlated with TDS and correlation was significant.

☆ Silicates positively correlated with sulphates, nitrates and negatively correlated with nitrites, NH_3.

☆ The correlation of chlorides with phosphates and BOD were significant and positive.

☆ Phosphates positively correlated with NH_3 and BOD.

☆ Nitrates positively correlated with nitrites and NH_3.

☆ Nitrites positively and significantly correlated with NH_3, TDS, and BOD.

☆ NH_3 positively correlated with TDS and BOD.

Conclusion

Seasonal variation in parameters like water temperature, pH, conductivity, dissolved oxygen, and carbon dioxide were observed. During present study, total alkalinity, total hardness formed limiting factors for fish culture along with phosphates and nitrites. GIP Tank showed water characterstics suitable for the fish survival and growth (fish culture).

References

APHA, AWWA, WPCF., 2002.International Standard Methods for the examination of Water and Wastewater 18th Edition, Washington D.C.

Ayoade, A.A., Agrarwal, N.K. and Chandala–Sakalani, A., 2009.Changes in physico–chemical features and plankton of two high altitude rivers Garhwal Himalaya, India. *Europian Journal of Scientific Research*.27 (1): 77–92.

Balasingh, Regini, G.S. and Selva Shamal, V.P., 2007. Phytoplankton diversity of a perennial pond in Kanyakumari district. *J. Basic and Applied Biology*, 1: 23–26.

Bhakta, J.N. and Bandyopadhyay, P.K., 2008. Fish diversity in freshwater perennial water bodies in east Midnapore district of West Bengal, India. *International Journal of Environmental Research,* 2(3): 255–260.

Bhuiyan, J.R. and Gupta, S., 2007. A comparative hydrobiological study of a few ponds of Barak Valley, Assam and their role as sustainable water resources. *Journal of Environmental Biology,* 28(4): 799–802.

Bramha, Satyanarayan, Panda Umesh Chadra, Bhatta Kripasindhu and Sahu Bijay Kumar, 2008. Spatial variation in hydrological characteristics of Chilka: A coastal lagoon of India. *Indian Journal of Science and Technology,* 1(4): 1–7.

Garg, R.K., Rao, R.J., Uchchariyya, D., Shukla, G. and Saksena, D.N., 2010. Seasonal variations in water quality and major threats to Ramsagar reservoir, India. *African Journal of Environmental Science and Technology,* 4(2): 61–76.

Gupta, Modadugu V., 2007. Tilapia farming: Prospects and constraints. *Fishing Chimes,* 27(1): 35–38.

Gurav, Meenakshi N., 2009. Limnology and biodiversity of river Gadhi and Dehrang reservoir. *Ph.D. Thesis,* Mumbai University.

Jalilzadeh, Koorosh, Sadanand, M., Yamakanamardi and Altaff, K., 2009. Physico-chemical parameters of three contrasting lakes of Mysore, Karnataka, India. *J. Aqua. Biol.,* 24(2): 90–98.

Jawale, A.K. and Patil, S.A., 2009. Physico-chemical characteristics and phytoplankton abundance of Mangrul dam, Dist. Jalgaon, Maharashtra. *J. Aqua. Biol.,* 24(1): 7–12.

Jayabhaye, U.M., Pentewar, M.S. and Hiware, C.J., 2008. A study on physico-chemical parameters of Minor reservoir, Sawana, Hingoli district, Maharashtra. *J. Aqua. Biol.,* 23(2): 56–60.

Jhingran, V.G., 1982. *Fish and Fisheries of India.* Hindustan Publishing Corporation (India), Delhi, p. 645.

Johri, V.K., Awasthi, S.K., Sharma, S.R. and Tandon, N.K., 1989. Observations on some limnological aspects of four important lakes of Kumaon hills of U.P. and suggestions for their proper exploitation. *Indian J. Fish.,* 36(1): 19–27.

Kakavipure, Dilip, 2005. Environmental studies in relation to aquaculture of Khativali–Vehloli lake near Shahapur. *Ph.D. Thesis,* Mumbai University.

Kamble, S.M., Kamble, A.H. and Narke, S.Y., 2009. Study of physio-chemical parameters of Ruti dam, TQ, Ashti, Dist. Beed, Maharashtra. *J. Aqua. Biol.,* 24 (2): 86–89.

Krishnamoorthi, A., Senthil Elango, P. and Selvakumar, S., 2011. Investigation of water quality parameters for aquaculture: A case study of Veeranam lake in Cuddalore district, Tamil Nadu. *International Journal of Current Research,* 3(3): 13–17.

Kiran, B.R., 2010. Physico-chemical characteristics of fish ponds of Bhadra project at Karnataka. *Rasayan J. Chemistry,* 3(4): 671–676.

Kumar, Adarsh, Qureshi, T.A., Parashar, Alka and Patiyal, R.S., 2006. Seasonal variation in physico-chemical characteristics of Ranjit Sagar reservoir, Jammu and Kashmir. *The Academy of Environmental Biology,* India.

Kumar, Anil, Sharma, L.L. and Aery, N.C., 2009. Physico-chemical characteristics and diatom diversity of Jawahar Sagar lake: A wetland of Rajasthan. *Sarovar Saurabh,* 5(1): 8–14.

Lendhe, Rajendra S., 2004. Ecophysiology of Phirange Kharbav Lake. *Ph.D. Thesis*, Mumbai University.

Lokhande, M.V.,Waghmare, V. N. and Bais, U.E., 2010. Studies of physical parameters of Gharni Reservoir at Shivpur, Tq. Nalegoan Dist. Latur, Maharashtra. *Shodh, Samiksha aur Mulyankan*, 2: 11–12.

Mishra, Ashutosh, 2008. Study of biodiversity of selected reservoirs of Uttaranchal. *Ph.D. Thesis* submitted to CIFE.

Muralidhar, V.N., Narayana, J., Puttaiah, E.T. and Narayana, Laxmi, 2002. Water quality of Gubbi tank, Tumkur in relation to physico-chemical characteristics, diversity and periodicity of phytoplankton limnology of lakes, reservoirs, wetlands.

Raut, Nayana, 2006. Comparative study of water chemistry and biodiversity from some macrophyte infested and non-infested lakes from Thane city, Maharashtra. *Ph. D. Thesis*, Mumbai University.

Sakhare, V.B., 2007. Wan reservoir (Beed Dt., Maharashtra): Ecology and fisheries. *Fishing Chimes*, 27(6): 23–26.

Sathe, S.S., Khabade, S.A. and Hujare, M.S., 2001. Hydrobiological studies on two man-made reservoirs from Tasgaon Tahsil (Maharashtra), India. *Ecol. Env. and Cons.*, 7(2): 211–217.

Sharma, Dushyant, 2001. Seasonal variations in different physico-chemical characteristics in Makroda reservoir of Guna district (M.P.). *Ecol. Env. and Cons.*, 7(2): 201–204.

Sharma, Vipul, Sharma, Madhusudan, Malara, Heena, Sharma, Riddhi and Baghela, Brijraj Singh, 2008. Trophic status and zooplankton diversity of Lake Jaismand in relation to its physico-chemical characteristics. *Proceedings of Taal 2007: The 12th World Lake Conference*, p. 490–495.

Singh, Meenakshi, Lodha, Payal and Singh, Gajendra Pal, 2010. Seasonal diatom variations with reference to physico-chemical properties of water of Mansagar lake of Jaipur, Rajasthan. *Research J. of Agricultural Sciences*, 1(4): 451–457.

Sisodia, Rashmi and Moundiotiya, Chaturbhuj, 2006.Assessment of the water quality index of wetland Kalakho lake, Rajasthan India. *Journal of Environmental Hydrology*, 2(14): Paper.

Somani, Vaishali, 2002. Ecological studies on Kacharali and Masunda lakes of Thane city with reference to bacterial treatment of Kacharali for lake beautification. *Ph. D. Thesis*, Mumbai University.

Subba Rao, D. and Govind, B.V., 1967. Hydrology of Tugabhadra reservoir. *Indian Journal of Fisheries*, p. 321–344.

Sultana, Laskar Hafsa and Gupta, Susmita, 2009. Phytoplankton diversity and dynamics of Chatla floodplain lake, Barak Valley, Assam, North East India: A seasonal study. *Journal of Environmental Biology*, 30(6): 1007–1012.

Tiwari, Rohini Prasad and Sharma, Chandan, 2011. Studies on monthly population of total rotifer zooplanktons and their correlation coefficient with physico-chemical factors of Lony dam, Theonthar, Rewa (M.P.). *International Journal of Pharmacy and Life Sciences*, 2(3): 617–619.

Trivedy, R.K. and Goyal, P.K. 1984. *Chemical and Biological Methods for Water Pollution Studies*. Environmental Publications, Karad, India, pp. 244.

Wurts, William A. and Durborow, Robert M., 1992. Interactions of pH, carbon dioxide, alkalinity and hardness in fish ponds. SRAC publication No. 464.

Zutshi, D.P., Vishin, N. and Subla, B.A., 1984. Nutrient status and plankton dynamics of a perennial pond. *Proc. Indian National Sci. Acad.*, b 50 No. 6: 577–581.

Chapter 8

Physico-chemical Characteristics of Coastal Environment of Thootukudi

☆ *V. Santhi, V. Sivakumar, R.D. Thilaga and S. Kathiresan*

ABSTRACT

The present study was attempted on the physico-chemical study of Thootukudi coastal water, South east coast of India to evaluate its suitability for the growth of molluscan animals. Seasonal variations study was carried out to examine the level of varying physico- chemical parameters such as rainfall, temperature, dissolved oxygen, pH and salinity for a period of one year from April 2010 to March 2011. The Physico-chemical parameters have exhibited considerable seasonal and spatial variations which were found to be conducive for the survival of animals.

Keywords: Physico-chemical characteristics, Coastal environment, Thootukudi.

Introduction

The total life of the world depends on water and hence the hydrological study is very much essential to understand the relationship between its different trophic levels and food webs (Soundara Pandiyan *et al.,* 2009). Oceans are very complex environment with great variations in temperature, salinity, pH, dissolved oxygen and nutrients. Environmental conditions such as topography, water movement and stratification, salinity, oxygen, temperature and nutrients of particular water mass also determine the composition of its biota (Karande, 1991) and it also play a major role in promoting the occurrence and abundance of commercially exploitable marine resources (Iveiev,1966). The coastal zones are, by and large, of prime importance for human beings. In many countries coastal and marine environs are under special jurisdiction and hence, it is imperative to know the interrelationships between the organisms and the environmental parameters in order to evaluate the suitability and function of an ecosystem (Baskara Sanjeevi, 2001). The physico-chemical parameters play a major role in the ecosystems to maintain the biodiversity profile of the marine environment. Hence the present study was designed to investigate the seasonal changes to physico-chemical parameters at the collection area of the molluscan animals.

Materials and Methods

The Thootukudi coast lies on the southeast coast of India (Latitude 850′-910′N and Longitude 7810′-7910′E). The present investigation was carried out at monthly intervals for one year from April 2010 to March 2011. The water samples were collected in clean glass containers and brought to the laboratory. For the estimation of dissolved oxygen the samples were fixed on the spot and then estimated following Winkler's method as described by Strickland and Parsons (1968). Temperature was recorded with the help of Celsius thermometer. The salinity was estimated using a Refractometer (ATAGU, Japan). The hydrogen-ion-concentration (pH) of the water was measured by pH Pen (Hanna, Italy). Data on monthly rainfall at the study area was obtained from the Indian Meteorological Department at Thootukudi.

Results and Discussion

The hydrographical parameters recorded from the study area for one year are presented in Figures 8.1–8.5.

Rainfall

Monthly variations in the rainfall during the study period (April 2010 -2011) are presented in Figure 8.1.The maximum rainfall of 300.9mm was observed during November 2010 and the minimum of 4.4mm in June 2010.No rainfall was recorded during February and March 2011. The rainfall in India is largely influenced by two monsoons *viz.*, south east and north east monsoon. North east monsoon was responsible for rainfall at Tuticorin coastal region in Gulf of Mannar (south east coast of India). Seasonal changes in rainfall influence the density of lower invertebrate population of the intertidal areas (Odum 3rd Ed.).

Temperature

Monthly variations in atmospheric temperature are depicted in Figure 8.2. Atmosphere temperature fluctuated between 30°C and 37°C. The highest temperature (37°C) was observed during summer season (April, May 2010) and low temperature (30°C) in monsoon season (November and December 2010).

Water Temperature

The water temperature varied from 27°C to 32°C (Figure 8.3) Maximum surface water temperature (32°C) was noted during summer season (April, May and June 2010) and minimum was recorded during monsoon season (November 2010). Atmospheric and water temperature fluctuated seasonally. The low temperature could be due to strong land breeze, rainfall, river run-off, and cloudy sky as reported by Ananthan (1994). The temperature variation is one of the factors in the coastal estuarine system, which may influence the physico-chemical characteristics and also influence the distribution and abundance of flora and fauna (Soundarapandian *et al.*, 2009). In the present study high temperature observed during April to June might be brought down by cold breeze, cloudy sky and rain fall during November.

Dissolved Oxygen

Monthly variations in oxygen values are shown in Figure 8.4. The higher dissolved oxygen content of 6.398ml/l was recorded during monsoon season (November 2010) and the lower value

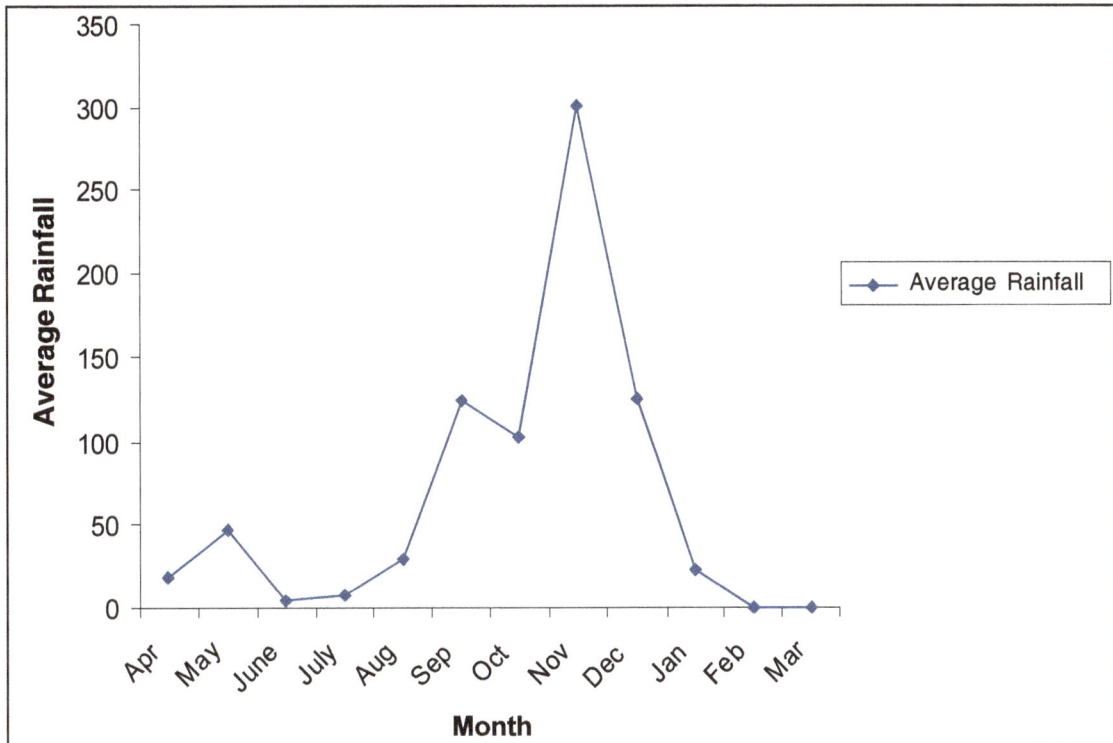

Figure 8.1: Monthly Variation of Average Rainfall during the Study Period
(April 2010 to March 2011)

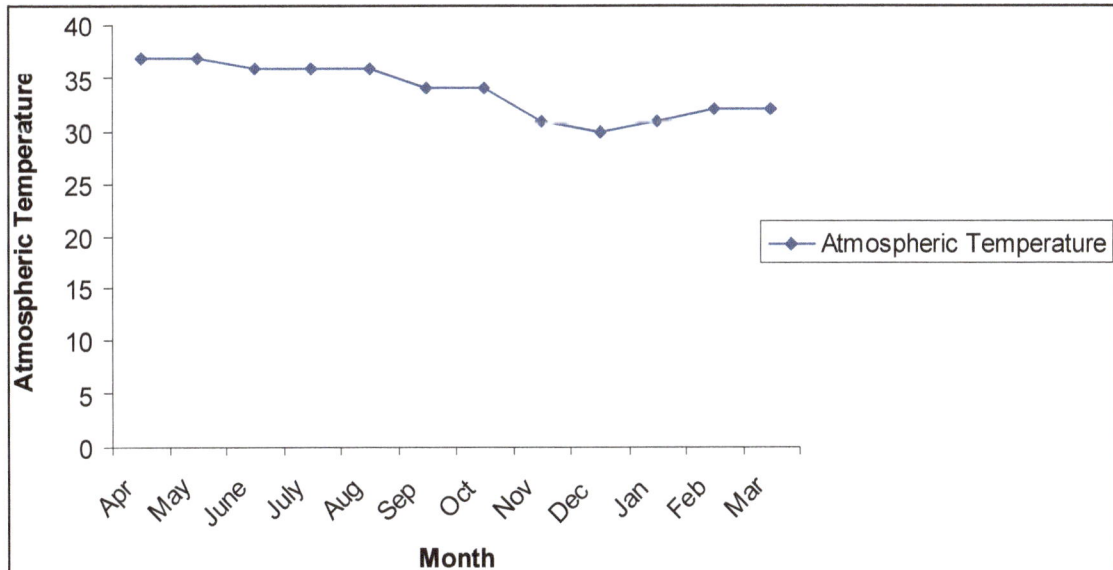

Figure 8.2: Monthly Variation of Atmospheric Temperature during the Study Period
(April 2010 to March 2011)

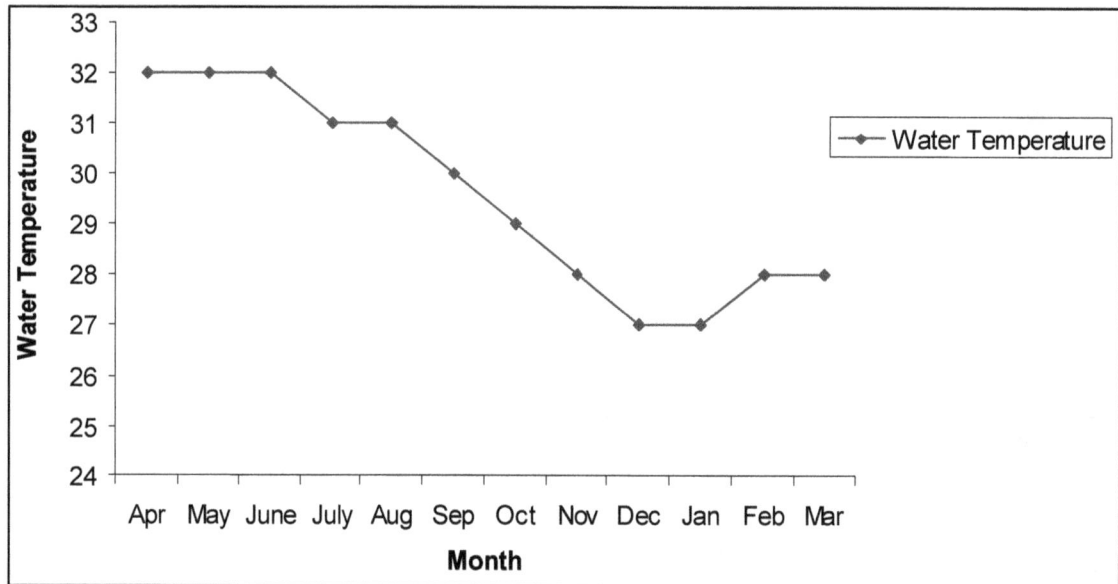

Figure 8.3: Monthly Variation of Water Temperature during the Study Period
(April 2010 to March 2011)

(3.7477ml/l) was recorded during summer season (May 2010). Dissolved oxygen content showed seasonal fluctuation, and it varies due to photosynthesis and respiration by organisms. Percent saturation of oxygen depends strongly on temperature as cold water holds more oxygen. The highest

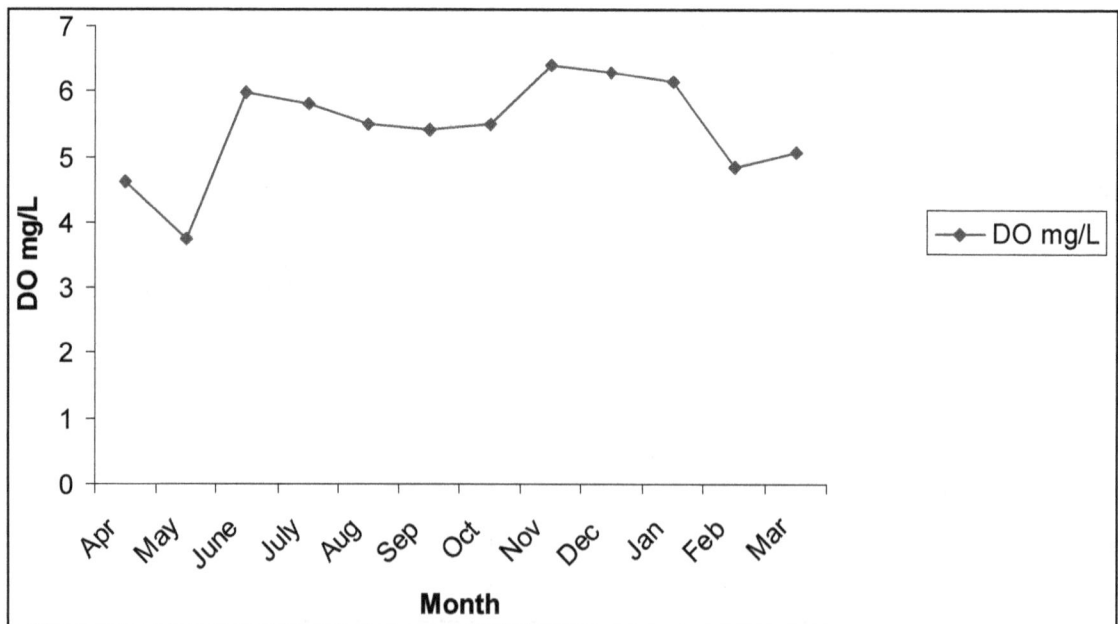

Figure 8.4: Monthly Variation of DO mg/L during the Study Period
(April 2010 to March 2011)

level of dissolved oxygen observed during monsoon was due to phytoplankton production and photosynthesis as reported by Kannan (1996) and Balasubramanian and Kannan (2005). An earlier report from these areas coincides with the present study, revealing the high contents of dissolved oxygen during monsoon (Thilaga, 2005). Low dissolved oxygen observed during the summer season could be attributed to the lesser input of freshwater into the study area as reported by Thilaga (2005).

pH

Water pH values were ranging from 6.7 to 8.3 (Figure 8.5) and the higher (8.3) pH value was observed during April, May and June 2010 (summer) and the lower value was recorded during November and December 2010 (6.8) (monsoon). pH showed only slight variation during the present observation. The highest pH values during summer could be due to uptake of CO_2 by photosynthesizing organisms (Santhoshkumar and Perumal, 2011). Generally pH is quite lower in rainy season as compared to winter and summer may be due to the influence of freshwater influx, dilution of sea water, low temperature and organic matter decomposition as suggested by Santhoshkumar and Perumal (2011). Similar trend in pH was reported by Seenivasan (1998) from the Vellar estuarine system, and Ananthan (1994) from Pondicherry coastal water. Mohamed Meeran *et al.* (2011) also stated that pH profile altered with seasons and the pH of waters also gets drastic change with time due to exposure to biological activity and temperature.

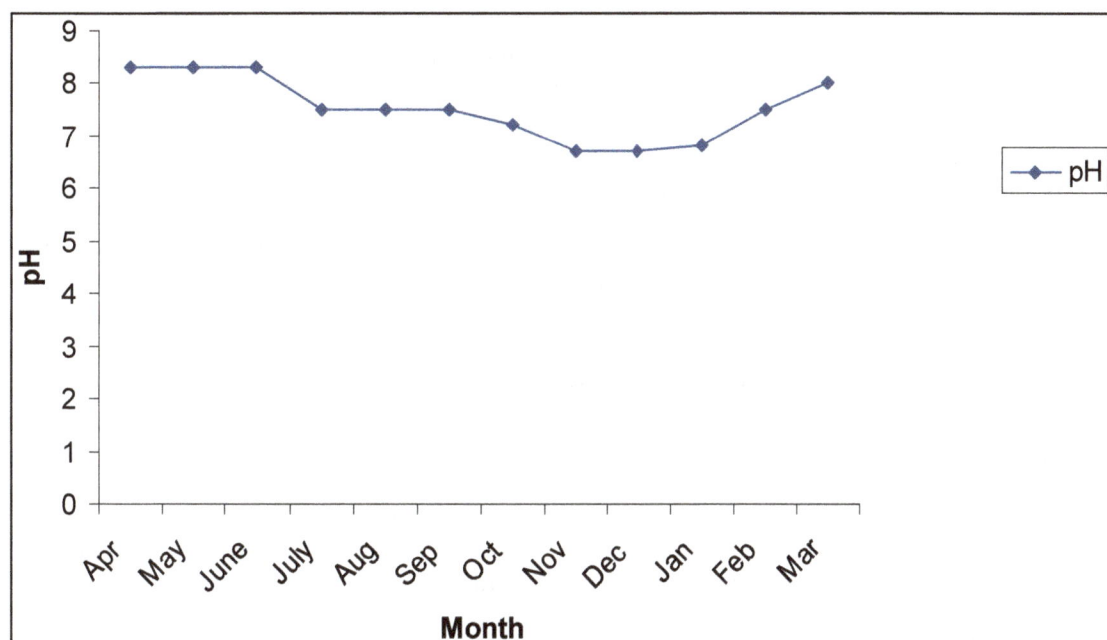

Figure 8.5: Monthly Variation of pH during the Study Period
(April 2010 to March 2011)

Salinity

In the present study the salinity values of the water ranged between 31 and 36.5 per cent (Figure 8.6). Maximum salinity was recorded during summer (April and May 2010) and minimum salinity was recorded during monsoon (October and November 2010). Salinity in the study area was higher

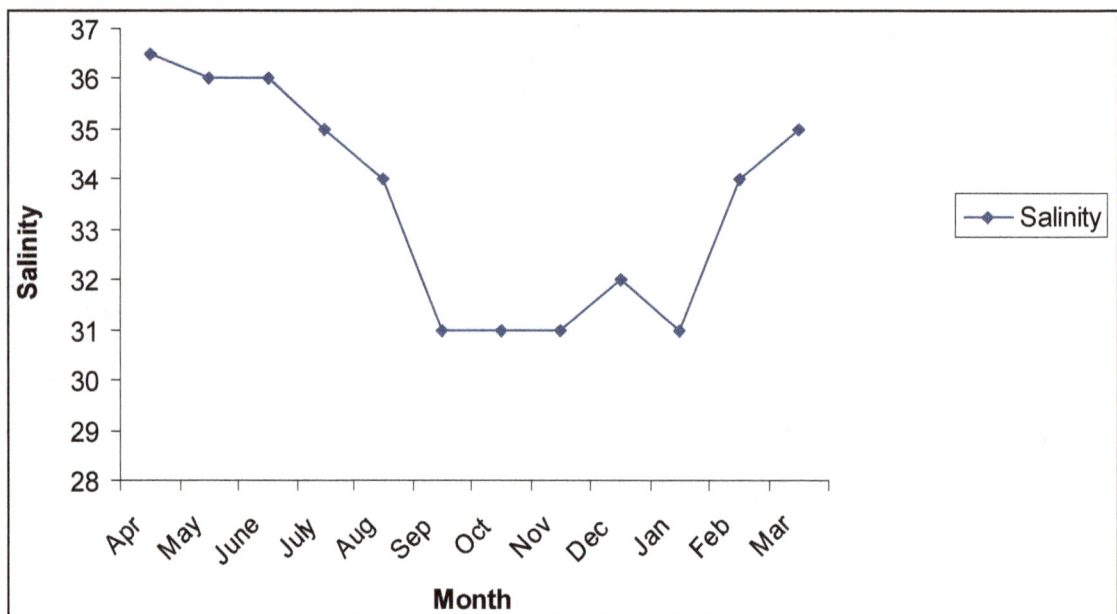

**Figure 8.6: Monthly Variation of Salinity during the Study Period
(April 2010 to March 2011)**

during summer which could be described to the higher degree of evaporation and less tidal action. The present findings are in affirmative with that of Mohamed Meeran *et al.* (2011). Similar trend in the salinity values were also observed from various parts of south east coast of India (Palanichamy and Rajendren, 2000; Sulochana and Muniyandi, 2005; Sundaramanickam *et al.*, 2008; Soundarapandian *et al.*, 2009). The monsoon driven current systems govern the seasonal distribution of salinity in these waters. The lower salinity during monsoon could be due to inflow of freshwater through rainfall as reported by Mitra *et al.* (1990). Salinity is known to play a key role in the distribution of marine organism in near shore and estuarine region, where fluctuations in salinity are well pronounced (Ajmal Khan and Natarajan 1981).

The environment condition such as temperature, oxygen, pH, salinity and also nutrients influence not only the composition of its biota but also the growth and distribution of species of flora and fauna (Karanade, 1991 and Swami *et al.*, 2000). Extrinsic and intrinsic factors are known to influence the distribution and abundance of the animals in the Gulf of Mannar. The existing physico-chemical characteristics from the study area were found to be conducive for the test animals to survive.

References

Ajmal Khan, S. and Natarajan, R., 1981. Salinity tolerance of hermit crabs. *Indian J. Mar. Sci.*, 10: 393–395.

Ananthan, G., 1994. Plankton ecology and heavy metal studies in the marine environments of Pondicherry, India. *Ph.D. Thesis*, Annamalai University, India, 125 pp.

Balasubramanian, R. and Kannan, L., 2005. Physico-chemical characteristics of the coral reef environs of the Gulf of Mannar biosphere reserve.

Baskarsanjeevi, S., 2001. Studies on Eco-biology of the spider conch *Lambis lambis* (Linn. 1758) (Gastropoda : Prosobranchs Strombidae) from the Madapam waters, South east coast of India. *Ph.D. Thesis,* Annamalai University.

Kannan, R. and Kannan, L., 1996. Physico-chemical characteristics of sea weed beds of the Palk Bay, South-east coast of India. *J. Mar. Sci.,* 25: 358–362.

Karande, A.A., 1991. Use of epifaunal communities in pollution monitoring. *J. Environ. Biol.,* p. 191–200.

Mitra, A., Patra, K.C. and Panigraty, R.C., 1990. Seasonal variations of some hydrographical parameters in tidal creek opening into the Bay of Bengal. *Mahasagar Bull. Natl. Inst. Oceanogr.,* 23: 55–62.

Mohammed Meeran, M., Joseph Antony, Jerland, I., Ajmal Khan, S., Lyla, P.S. and M. Ashiqur Rahman, 2011. Seasonal variations of physico-chemical properties of Thondi coast (Palk Bay), South-east coast of India. *International Journal of Currrent Research,* 2(1): 170–177.

Odum, E.P.. *Fundamentals of Ecology,* 3rd edn. W.B. Saunders company, Philadelphia London, Toronto, p. 574.

Palanichamy, S. and Rajendran, A., 2000. Heavy metal concentration in seawater and sediments of Gulf of Mannar and Palk Bay, south-east coast of India. *Ind. J. Mar. Sci.,* 29: 116–119.

Santhoshkumar, C. and Perumal, P., 2011. Hydrobiological investications in Ayyampattinam coast (South-east coast of India) with special reference to zooplankton. *Asian Journal of Biological Sciences,* 4: 25–34.

Seenivasan, R., 1998. Spectral reflectance properties of the Vellar estuarine environment, South-east coast of India. *Ind. J. Mar. Sci.,* 2: 221–224.

Sudaramanickam, A., Sivakumar, T., Kumaran, R., Ammaiappan, V. and Velappan, R., 2008. A comparative study of physico-chemical investigation along Parangipettai and Cuddalore coast. *J. Environ. Sci. Tech.,* 1(1): 1–10.

Sulochana, B. and Muniyandi, K. 2005. Hydrographic parameters of Gulf of Mannar and Palk Bay during an year of abnormal rainfall. *J. Mar. Biol. Ass., India,* 47(2): 198–200.

Soundarapandian, P., Premkumar, T., and Dinakaran, G.K., 2009. Studies on the physico-chemical characteristics and nutrients in the Uppanar estuary of Cuddalore, South-east coast of India. *Curr. Res. J. Biol. Sci.,* 1(3): 102–105.

Strickland, J.D.H. and Parsons, T.R., 1972. *A Practical Handbook of Sea Water Analysis,* 2nd Edn. Bull.: 167. Fisheries Research Board, Canada, Ottawa, Onto, pp. 331.

Swami, B.S., Suryawanshi, U.G. and Karande, A.A., 2000. Water quality of Mumbai Harbour: An update. *Indian J. Mar. Sci.,* 29: 111–115.

Thilaga, R.D., 2005. Studies on some ecological aspects of *Babylonia spirata* (Linn.) among the Tuticorin Coast, *Ph.D. Thesis.* Manonmaniam Sundaranar University, Thirunelveli.

Chapter 9

Ecological Study of Mankeshwar Beach, Uran, Maharashtra

☆ *Chhaya Panse, Vaishali Somani, Nitin Walmiki*
and Amol Kumbhar

ABSTRACT

Hydrological parameters of Mankeshwar beach, Uran were studied. Alkaline pH was recorded. High Dissolved Oxygen was observed in post monsoon season. Silicates, phosphates and nitrates exhibited higher concentration during monsoon. Four species of macroalgae and twelve species of phytoplankton were recorded.

Keywords: Mankeshwar beach, Hydrology.

Introduction

Uran is primarily a fishing village in Navi Mumbai, Maharashtra. The coast is characterized with port activities of Jawaharlal Nehru Port Trust. The area is also under development as special economic zone. Due to these factors, this ecosystem is under influence of anthropogenic activities. Present study was undertaken to collect base line information of hydrological parameters, phytoplankton and macroalgal diversity of Mankeshwar beach at Uran (Maharashtra).

Materials and Methods

Water samples from the coastal areas of Mankeshwar beach in Uran for two seasons–Pre-monsoon and Monsoon in year 2010. Analysis of hydrological parameters was carried out as per standard methods (APHA, 1992). Samples were preserved separately for study of phytoplankton and identified using standard keys. Macroalgae were collected from prefixed sampling points in quadrants of size of 0.25 sq.mt. and herbaria were prepared for identification.

Results and Discussion

Hydrological parameters reflect the chemical and biological processes in aquatic ecosystems. These parameters also influence the occurrence of various components of food webs including the producers.

In the present study, water temp recorded during pre monsoon period was 29 °C and in monsoon, it was slightly lower, 27 °C. pH was recorded in the range of 7. 4 to 8.04. This is favourable for growth of algae. (Middelboe and Hansen, 2007). Salinity is typically a local rather than global parameters and is highly variable in coastal regions. During the present study salinity fluctuated between 18.08ppt to 36.13 ppt.Higher O_2 level was recorded during pre monsoon. This may be attributed to increased photosynthetic activity of macroalgal community on this coast. Dissolved oxygen showed lower range during monsoon.

Low CO_2 values in pre-monsoon season may be due to utilization of CO_2 by the macroalgae which are primary producers and exhibited significant abundance during this season.

Low values of PO_4, SiO_3 and NO_3 in pre-monsoon season may be due to utilization for growth of phytoplankton and macro algae.

The macroalgal community was represented by 4 genera, 3 belonging to Chlorophyta – *Ulva sps, Bryopsis* sps and *Chetomorpha* sps. Division Rhodophyta was represented by *Gracilaria sps*. However the macroalgal growth was observed to be at lower ebb.

During present investigation twelve species of diatoms were recorded.Those species are *Biddulphia sinensis, Biddulphia alternuns, Melosira* sps, *Pleurosigma* sps, *Skeletonema* sps, *Bacillaria* sps, *Cyclotella* sps, *Navicula* sps, *Rhizosollenia* sps, *Cossinodiscus raditus, T. halassosira baltica* and *Nitzschia* sps.

Observations

Particulars	Pre-mosoon	Monsoon
Temperature	29 °C	27 °C
pH	8.04	7.43
CO_2	0.0 ml	9 mg/l
PO_4	22 µg atoms/l	32 µg atoms/l
SiO_3	0.5 µg atoms/l	2.2 µg atoms/l
NO_3	6.0 µg atoms/l	9.04 µg atoms/l
Dissolved oxygen	7.2 mg/l	3.6 mg/l
Salinity	36.13 °/100	18.08 °/00

Acknowledgements

The Authors wish to thank the University of Mumbai for funding this Study. Sincere thanks to Principal, M.D College and teaching and nonteaching members of Zoology Department.

References

APHA, AWWA, WPCF, 1992. *Standard Methods of Water and Wastewater*, pp. 824.

Middleboe, Anne Lise and Hansen Per Juel, 2007. Direct effects of pH and inorganic carbon on macroalgal photosynthesis and growth. *Marine Biological Research*, 3(3): 134–144.

Chapter 10

Seasonal Variation in Water Quality of Rankala Lake Kolhapur

☆ *S.A. Deshmukh and D.S. Patil*

Introduction

Rankala Lake is a popular evening spot and recreation centre in Kolhapur city. In recent years it has been observed that the lake water was covered by *Eichhornia crassipes* (Mart.) Solms and by an aquatic pteridophyte *Salvinia, Seguir* sp. Alongwith these plants, the mass of certain pollution indicating algae has been observed in water body indicating increased water pollution.

According to Agbaire *et al.* (2009) water bodies are constantly used as receptacles for untreated wastewater or poorly treated effluents accrued from industrial activities. Rankala Lake is facing the same problem of pollution because untreated domestic waste sewage from various areas of Kolhapur city, directly mix within it. As per Fakayode (2005), water is most poorly managed resources in the world. Developments in science and technology have brought improved standard of living, but have also unwittingly introduced some pollution into our environment (Dinrifo *et al.*, 2010). Balanced physico-chemical properties play vital role in the functioning of an aquatic ecosystem and its stability to support life forms.

Materials and Methods

Water sample was collected in pre washed plastic bottles during monsoon and summer season respectively. After that immediately the samples were transported to laboratory and were analyzed for physico-chemical properties. Winkler's method was used to measure Dissolved Oxygen (DO); various metals were analyzed with the help of Atomic Absorption Spectrophotometer (AAS).

Results and Discussion

The water samples were analyzed with respect to 10 different parameters. It was found that the amount of two parameters *i.e.*, total solids and total dissolved solids was increased in monsoon season, while amount of other parameters was found to be less in monsoon season than the summer one.The variability in physico-chemical properties of water collected in two different seasons is tabulated in Table 10.1.

Table 10.1: Physico-chemical Properties of Rankala Lake Water

Sl.No.	Parameter	Summer Season	Monsoon Season	WHO Limits
1.	pH	7.34	7.1	6.5-8.5
2.	Temperature	27.7	26.2	20-32
3.	Dissolved Oxygen (DO) mg/l	6.7	6.1	5.0
4.	Chlorides mg/l	44	39	200-300
5.	Magnesium mg/l	20	13	30
6.	Sulphate mg/l	40	36	200
7.	Nitrate mg/l	0.004	0.003	45
8.	Calcium mg/l	40	35	100
9.	Total solids mg/l	878	989	500
10.	Total dissolved solids mg/l	654	732	500

Discussion

Urbanization and development in most of the fields are unending and necessary processes, but it causes major problems of water pollution, air pollution, sound pollution etc. which creates hazardous effects on all living organisms. In above case the domestic waste and sewage water should have to be treated before it mixes with water body for reducing the pollution level.

References

Agbaire, P.O. and Obi, C.G., 2009. Seasonal variations of some physico-chemical properties of River Ethiope Water in Abraka, Nigeria. *Journal of Appl. Sci. Environ. Manage.*, March, 13(1): 55–57.

Dinrifo, R.R., Babatunde, S.O.E., Bankole, Y.O. and Demu, Q.A., 2010. Physico-chemical properties of rain water collected from some industrial areas of Lagos state Nigeria. *European Journal of Scientific Research*, 41(3): 383–390.

Fakayode, S.O., 2005. Impact assessment of the industrial effluent on water quality of the receiving Alaro river in Ibadan, Nigeria. *Ajeam–Ragee*, 10: 1–13.

Chapter 11

Physico-chemical and Microbiological Examination of Water from Raviwarpeth Lake Ambajogai

☆ *V.S. Hamde*

ABSTRACT

Physico-chemical quality of water is studied by systematic collection and analysis of water samples which enable us to properly manage the resources. Present study was undertaken to investigate the physico-chemical characteristics and microbiological examination of Raviwarpeth lake, Ambajogai. Physico-chemical analysis of lake water exhibited richness in inorganic nitrogen and phosphate which favour growth of phytoplankton.

Physico-chemical and bacteriological profile indicate a very high population of coliforms *i.e.*, $2.8 \times 10^3/100$ ml, high bacterial density and presence of pathogenic bacteria like *E. coli, Salmonella, Shigella* are also reported. The samples were collected during month of January 2009 to December 2009 and were analysed for temperature, pH, total solids, alkalinity, dissolved oxygen, nitrite ammonia and phosphate. It is observed that values of several parameters exceeded the permissible limits pointing out to the necessity of proper measures to control pollution of water.

Keywords: Microbial examination, Ravivar Peth Lake, Ambajogai.

Introduction

The microbial population of Lake derived largely from the soil. Lakes have characteristic zonation and stratification. Photosynthetic activity decreases progressively in the deeper region of open water. The greatest variety of physiological types is found in the limnetic and littoral zones and in addition they constitute the most productive regions. Productivity is affected by chemical nature of the basin and nature of imported material (community waste etc.) from streams and rivers. The temperature of lake shows seasonal variations and it affects on the seasonal change in the microbial population. Such satification acts as a barrier to nutrient and oxygen exchange especially in still water. In the summer, the top layers tend to be warmer than the lower regions, but in the winter, it becomes cool.

A variety of bacteria are regarded as nuisance bacteria in water system because they create problems of odor, color and taste as well as precipitates insoluble compounds etc. Algae may also be responsible for the development of odors, discolouration and other objectionable characteristics.

Lakes are rivers provide the major freshwater bodies which are used for potable water. However due to growth of phytoplanktons and decompositions of organic matter nutrient status varies throughout the year. In addition animal activities (including man) also influence the nutrient level of lakes and rivers. The nutrient poor lakes are known as oligotrophic lakes, where as the nutrient rich lakes are called eutrophic lakes.The nutrient rich lakes contain high amount of bottom sediments containing organic matter. Eutrophication (enrichment of lake due to high concentration of nutrients) of lakes occur by multifarious ways where anthropogenic activities are of much importance size. The eutrophic lakes support luxuriant growth of bacteria and algae. Some of the fast growing algae at optimum condition bloom well showing their maximum population. This phenomenon is known as water blooming and microorganisms associated with it are called water blooms. The two most important lake of India which are fully eutrophicated are Dal Lake and Naina Lake.

Microorganisms showing in lakes are the genera *Anabaena, Microcystis, Nostoc, Oscillatoria, Spirulina, Oedogonium,* diatoms, protozoa etc.

Mycoflora of freshwater bodies also show great diversity of water borne condial fungi which are characterized by magnificent spore types. Water borne conidial fungi, which are characterized by magnificent spore types. Water borne conidial fungi were previously described by Ingold (1942) in an "Aquatic hyphomycates ". Belwal and Sati (2009) reported that quantitative estimation of water borne conidial fungal spores in the running freshwater streams of western Himalaya. Hashmi *et.al.* (2000) reported the physico-chemical quality of water sample of Khazana well at Beed District at Maharashtra.

Materials and Methods

Physico-chemical and biological characteristics of lake were studied seasonally throughout the year 2009. For these studies four different sites were selected on the basis of substratum structure, algal occurrence and human activities.

Physico-chemical analysis of water was done as per standard methods recommended by APHA (1995) and Trivedy and Goel (1984). Analysis of water also carried according to Manual of Water and Wastewater Analysis NEERI (1988).

Results

Quality of water is dependent on the physical and chemical qualities of water and also biological diversity of the system. Cairns and Dickson (1971) stated that the analysis of biological materials along with chemical characteristics of water forms a valid method for water quality assessment. Hence the physico-chemical characteristics and bacteriological studies during different seasons of year observed in the present study have been discussed below.

Mc Combie (1953) stated that temperature may affect the seasonal cycle of the phytoplankton in temperate zones. Similarly, Hutchinson (1957) mentioned that temperature in controlling both the quality and quantity of planktonic flora. Jana (1973) and Chari (1980) observed that temperature is critical factor for seasonal periodicity of phytoplankton. Present paper states that the change in temperature affects on microbial population.

It is well known that transparency of water is negatively proportional to primary productivity. Present study states that there is little fluctuation in microbial population with respect to transparency.

It was observed that higher concentration of bicarbonate during summer may be due to the decrease in water level by evaporation. It was also noted that alkalinity decreases during rainy season. The correlation of alkalinity was positive with microbial population.

Total dissolved solids varied significantly along with microbial count. Value of total dissolved solids was positively significant with microflora of water.

It was observed that there was no remarkable change in pH of water throughout the year. Oxygen content is important for direct need of many organisms and affects the solubility and availability of many nutrients and therefore the productivity of aquatic ecosystem (Wetzel, 1983). The factors affecting the oxygen balance in water bodies are input due to atmosphere and photosynthesis and output from respiration, decomposition and mineralization of organic matter as well as losses to atmosphere. Hence, the oxygen balance in water becomes poor as the input of oxygen at the surface and photosynthetic activity decreases and as the metabolic activities of heterotrophs are enhanced.

Table 11.1: Physico-chemical Characteristics of Raviwarpeth Lake Water

Factors	Duration		
	June-Sept	Oct-Jan	Feb-May
Temperature (°C)	25	16.0	26
Transparency (cm)	98	99	95
Total solids	210	190	230
pH	7.2	7.3	7.4
Alkalinity (mg/l)	170	181	193
Dissolved oxygen (mg/l)	7.2	7.2	6.0
Nitrite (mg/l)	0.0012	0.0013	0.0014
Ammonia (mg/l)	0.0061	0.0071	0.001
Orthophosphate (mg/l)	0.04	0.05	0.03

The bacteria play a significant role as decomposers and are decisive in determining the biological quality of water bodies. They varied from season to season and pond to pond and are given in Table 11.2.

Coliforms

The present study indicates that there was a great variations in number of coliform bacteria. It was also noted that the population was highest in summer season. Their maximum population was 28.1×10^3/100 ml in May. Its population was low during winter and moderate in rainy season. It indicates that the population in this lake is due to civic population near the lake.

Faecal Coliforms

It was observed that the load of faecal coliform was maximum during summer season. Its maximum count was $11.3 \times 10^3/100$ ml in June.

Faecal Streptococci

Streptococci are used as indicator of faecal habitat. Their maximum density was recorded in June.

Escherichia coli (MPN)

It was observed that high MPN of Raviwarpeth Lake during the period of investigation.

This high population of coliform is due to intense civic population activity of man and cattle throughout the year.

Bacterial Density

Total bacterial density showed identical pattern of fluctuations in lake. Their maximum density was recorded in June–July (3.2×10^7/lit) where as maximum count was recorded in January.

Table 11.2: Bacterial Count of Raviwarpeth Lake from January to December 2009

Month	Coliforms	Faecal Coliforms ($x\ 10^3/100$ ml)	Faecal Streptococci ($x\ 10^3/100$ ml)	E. coli ($x\ 10^3/100$ ml)	Total Bacterial Density ($x10^7$/L)	FC/FS Ratio
Jan	5.1	2.2	0.31	1.80	0.51	7.09
Feb	5.3	2.2	0.37	1.81	0.71	7.09
Mar	16.2	6.1	0.39	2.1	1.0	15.64
Apr	26.4	10.3	0.28	8.3	1.6	36.78
May	28.10	10.3	0.31	18.9	1.8	33.22
June	26.0	11.2	0.33	19.1	3.2	33.93
July	27.1	8.9	0.29	19.3	3.2	30.68
Aug	16.1	7.3	0.28	20.1	3.1	26.07
Sep	18.2	7.3	0.28	18.1	2.3	26.07
Oct	19.8	8.1	0.11	18.2	2.1	73.63
Nov	6.9	3.1	0.11	5.8	0.78	46.36
Dec	6.2	2.1	0.11	2.8	0.8	19.09

Discussion

The bacterial analysis also shows some interesting results regarding the quality assessment of water of lake. Bacterial population was lower in the monsoon, lowest in winter and highest during summer. It is obvious that the lake water is highly polluted in summer. Highest count of coliforms in summer has greatly lowered in monsoon months. This may be due to excessive dilution by rain water that drained into the ponds, although rain drain was the important source of bacterial population in the water body (William and Evns, 1972). However, this does not agree with observations made by Shasty et.al. (1970) and Aboo (1968), who noted that the maximum bacterial densities observed during the rainy season, and smaller counts in the winter.

The higher values in both the ponds are explained on the basis of their being used as bathing and washing ghats making them more polluted. Saxena *et al.* (1966) attributed summer high coliform organisms to the less available dilution in the summer. The low winter counts can be explained on the basis of lower multiplication and poor growth following low temperature, where as variations in the number of coliforms in water can be attributed to the intensity and age of pollution in addition to the air temperature and runoff. Waters (Panicker, 1976).

The FC/FS (faecal coliform/faecal streptococci) ratio is also used to distinguish whether the suspected contamination derives from human or from animal excreta. Metcalf and Eddy (1979), Gupta (1994) and Singh and Singh (1995) proposed that if the values of FC/FS ratio is lesser than contamination derives from domestic animals and if the value is greater than 4, pollution derives from human excreta. The FC/FS ratio was in the range of 36.78 in water, The above ratio reflects that the pollution ocured in lake is mainly due to cattle and human based sources.

The values for both total and faecal coliform in summer in lake were excessive and greater than all recommended standards. This summons towards the danger of health hazard when using the pond water for whatsoever purpose. Thus, the values of species diversity and bacterial density computed within certain time period can be used as bioindicator for the assessment of quality of freshwater ecosystem.

References

Aboo, K.M., 1968. A study of well waters in Bhopal city. *Environ, Health*, 10: 189–203.

Belwal, M. and Sati, S.C., 2009. Qualitative estimation of water borne conidial fungal spore in the running water streams of western Himalaya. *J. Microb. World*, 11(1): 42–48

Cairns, J. Jr. and Dickson, K.L., 1971. A simple method for biological assessment of the effect of waste discharges on the aquatic bottom dwelling organisms. *J. Wat. Poll. Control Fed.*, p. 755.

Chari, M.S., 1980. Environmental variation in the physico-chemical characteristics of freshwater ponds. *M.Phil.* Thesis, Aligarh Muslim University, Aligarh.

Gupta, A.K., 1994. Impact of sewage and human activities at Veruna river corridor. In: *International Conference on Recent Trends in Water Pollution and Research*, Dobbi, Janpur, 19–21 Nov.

Hashmi, Seema, Fazi, Ilyas, Shakir, N.and Musaddiq, M., 2000. Water quality of 'Khazana Well' at Beed district in Maharashtra. *J. Microb. World*, 2(2): 29–33.

Hutchinson, G.E., 1957. *A Treatise on Limnology*, Vols. I and II. John Wiley and Sons Inc., New York.

Jana, B.B., 1973, Seasonal periodicity of plankton in freshwater pond in West Bengal, India. *Hydrobiology*, 58: 127–143.

Metcalf and Eddy, 1979. *Wastewater Engineering: Treatment, Disposal and Reuse*. Tata McGraw-Hill Co. Ltd., New Delhi.

Mc Combie, A.M., 1953. Factors influencing growth of phytoplankton. *J. Fish. Res. Bd. Canada*, 10: 253–282.

Panicker, P.V. 1976. Coliform spectra of raw water sources of Nagpur Water supply. *Environment Hlth.*, 8: 266–296.

Shastry, C.A., Aboo, K.H. and Khare, G.H., 1970. Reduction of microorganisms at different stages of water treatment. *Environ. Health*. 12: 66–79.

Singh, T.N. and Singh, S.N., 1995. Impact of river Varuna on the river Gangas water quality at Varanasi. *Indian J. Env. Hlth.,* 37: 272–277.

Saxena, K.L., Chakrabarty, R.N., Khan, R.Q., Chatopadhya, S.N.N. and Chandra, H., 1966. Pollution studies of the river Ganga near Kanpur. *Environ. Hlth.,* 8: 270–285.

William, H.C. and Evans, M.J.C., 1972. A hydrological study of the polluted river lieve (Ghent, Belgium). *Hydrobiologia,* 39: 99–154.

Wetzel, R.G., 1983. *Limnology,* 2nd Edn. Saunders College Publishing, USA, p. 767.

Chapter 12

Effect of Varying Concentrations of Dietary Protein on Growth, Survival and Biochemical Aspects of Banana Prawn, *Fenneropenaeus merguiensis* (De Man, 1888)

☆ *Medha Tendulkar and A.S. Kulkarni*

ABSTRACT

The effect of dietary protein (40 per cent, 45 per cent, 50 per cent, 55 per cent and 60 per cent) on growth, survival, feed conversion ratio (FCR), feed conversion efficiency(FCE), specific growth rate(SGR) and body weight gain (BWG) were investigated for juveniles of *Fenneropenaeus merguiensis*. The best growth was obtained with the diet containing 60 per cent dietary protein. Feed conversion ratio decreased with increasing weight of prawn and is negatively correlated with dietary protein level. The FCR was found to be minimum (0.31) in 60 per cent dietary protein and maximum (1.274) in 40 per cent dietary protein. FCR and FCE were found to differ significantly between prawn fed 40-45 per cent and 50-60 per cent dietary protein. BWG differs significantly among each treatment but SGR differs between prawns fed 40-45 per cent and 50-60 per cent dietary protein. Percentage body protein of prawn fed 50-60 per cent was higher than that of prawn fed 40-45 per cent dietary protein, whereas lipid content decreased with increasing dietary protein level. Overall growth performance of *Fenneropenaeus merguiensis* was significantly higher (P>0.5) in the diet containing 60 per cent dietary protein when compared with 40 per cent, 45 per cent, 50 per cent and 55 per cent dietary protein.

Keywords: Fenneropenaeus merguiensis, Dietary protein, Food Conversion Ratio (FCR), Specific Growth Rate (SGR).

Introduction

One of the major factors limiting the economic success in any commercial culture of a species is the food requirement. Protein being an important dietary constituent among animals directly influences the formulation of diets and consequently affects the cost of production. Accumulated knowledge of the nutrient requirements of the prawn is limited and the lack of standard techniques among researchers resulted to wide variations of findings thereby making direct comparisons difficult. Several workers have tried to develop artificial diets capable of sustaining good growth using a variety of foodstuffs. (Cowey and Forster, 1971; Sick *et al.,* 1972; Andrews *et al.,* 1972; Forster, 1976; Balazs *et al.,* 1973). Among the foodstuffs used, flesh of molluscs and crustaceans were found the most acceptable producing the best growth especially among the marine prawns.

A range of feeding habits from carnivorous to herbivorous, has been suggested as one possible reason for the wide range in protein requirements among penaeid shrimp species. (Deshimaru and Yone, 1978) Dietary protein levels ranging from 30 to 60 per cent are recommended for various species and sizes of marine shrimp. (Akiyama *et al.,* 1992) As the principal and most expensive component of the diet, protein has received much attention in nutrition requirements studies. Animals including shrimps, must consume protein to furnish a continuous supply necessary for replacing worn-out tissues and for synthesizing new tissues. (Lim and Akiyama, 1997) Such studies however are completely lacking in another closely related and culturable species of marine prawn *i.e. F. merguiensis.* It is a culturable and palatable species and is readily available in Konkan area. For raising the successful culture of any species; knowledge of its nutritional requirement is of prime importance. Thus, in view of this background, present work has been designed to evaluate effect of varying levels of protein on the growth and biochemical aspects of *F. merguiensis.*

Materials and Methods

Five isocaloric diets with varying levels of proteins were prepared. Various ingredients like clam meat, rice bran and Groundnut oil cake were used for feed formulation. The proximate composition of different ingredients was determined. After knowing the proximate composition, five feeds with varying protein levels (40 per cent, 45 per cent, 50 per cent, 55 per cent and 60 per cent) were formulated using Pearson's square method. Proximate analysis of the ingredients used and percentage proportion of different ingredients used in feed formulation are depicted in Tables 12.1 and 12.2 respectively.

Healthy juveniles of *F. merguiensis* were collected from a local estuary at Zadgaon, Ratnagiri and brought to the Department of Zoology where laboratory experiments were conducted. 20 prawns each were stocked in plastic troughs of 20 litres capacity. Diet was given twice a day at the rate of 10 per cent body weight. Growth in terms of per cent Body Weight Gain (BWG), Food Conversion Ratio (FCR), Food Conversion Efficiency (FCE) and per cent Specific Growth Rate (SGR) was calculated. Prawns from each trough were weighted every week and biochemical analysis was done monthly.

Biochemical analysis of flesh of prawns was done by using standard methods protein by Lowry's *et al.* (1951) and total lipids by Barne's *et al.* (1973). Standard statistical methods were adopted for testing significance. One way analysis of variance was carried out to find out the difference among mean values of growth and survival of prawns. F value ($P>0.05$) was employed using LSD test to determine significant differences between the feeds. (Snedecor and Cochran; 1967)

Results and Discussion

The effects of dietary protein (40 per cent, 45 per cent, 50 per cent, 55 per cent and 60 per cent) on growth survival and biochemical aspects of *F. merguiensis* were investigated and best growth was obtained with the diet containing 60 per cent dietary protein. The growth rate of prawn fed 50 per cent, 55 per cent and 60 per cent dietary protein did not differ significantly from each other and was higher (P>0.5) than prawn fed with 40 per cent dietary protein (Table 12.3). It is evident from Table 3 that FCR decreases progressively with increasing dietary protein and FCE increases with an increase in dietary protein level. FCR and FCE were found to differ significantly between prawn fed 40 - 45 per cent and 50- 60 per cent dietary protein. BWG and SGR increases with an increase in dietary protein level. BWG differs significantly among each treatment but SGR differs between prawns fed 40-45 per cent and 50-60 per cent dietary protein. Whole body composition of *F. merguiensis* was also significantly influenced by dietary protein content. Percentage body protein of prawn fed 50-60 per cent protein was higher than that of prawn fed 40-45 per cent dietary protein, whereas lipid content decreased with increasing dietary protein level.

Table 12.1: Proximate Composition of Ingredients Used in Feed Formulation

Ingredient	Proteins	Lipids	Carbohydrates
Rice bran (RB)	14 per cent	10 per cent	4.5 per cent
Groundnut Oil Cake (GOC)	49.96 per cent	8.66 per cent	28.31 per cent
Clam meat	46.93 per cent	5.00 per cent	18.03 per cent

Table 12.2: Quantity of Individual Ingredient to be Used in Feed Formulation

Ingredient	40 per cent	45 per cent	50 per cent	55 per cent	60 per cent
Rice bran (RB)	24.50	10	4.16	13.80	20.08
Groundnut Oil Cake (GOC)	37.75	45.00	47.92	43.10	39.96
Clam meat	37.75	45.00	47.92	43.10	39.96

The FCR was found to be minimum (0.31) in 60 per cent dietary protein and maximum (1.274) in 40 per cent dietary protein FCE varied in the range of (0.785 – 3.225) with maxima and minima obtained with 60 per cent and 40 per cent dietary levels respectively. SGR also increases with the dietary protein level, being maximum (2.934 per cent) and minimum (1.588 per cent) in prawn fed 60 per cent and 40 per cent dietary protein respectively. BWG also increases with increasing dietary protein level. A minimum of 317.77 and maximum of 1302.22 was observed in case of 40 per cent and 60 per cent dietary protein respectively (Table 12.3).

Table 12.3: Growth, Survival and Feed Conversion Ratio of
***F. merguiensis* Fed Varying Levels of Dietary Protein**

Parameters	40 per cent	45 per cent	50 per cent	55 per cent	60 per cent
Initial av. Weight	0.45 g	0.45 g	0.45 g	0.45 g	0.45 g
Final av. Weight	1.88 g	2.10 g	3.67 g	5.17 g	6.31 g
Mean Weight Gain	1.43 g	1.65 g	3.22 g	4.72 g	5.86 g
BWG	317.77[a]	366.66[b]	715.55[c]	1048.88[d]	1302.22[e]
FCR	1.274[a]	1.104[a]	0.46[b]	0.38[b]	0.31[b]
FCE	0.785[a]	0.905[a]	2.173[b]	2.631[b]	3.225[b]
Per cent SGR	1.588[a]	1.711[a]	2.331[b]	2.712[b]	2.934[b]

Values sharing similar superscripts do not differ significantly (P > 0.05) from each other.

Table 12.4: Quantitative Analysis of Muscle of Prawn Showing CP by Lowry's Method
(Per cent composition) CP: Initial reading = 5. 25 per cent

Diet	30 Days	60 Days	90 Days
A	8.75	9.75	13.75
B	8.20	11.70	14.20
C	8.3	12.00	14.55
D	7.65	12.30	15.00
E	8.15	13.05	16.22

Table 12.5: Quantitative Analysis of Muscles of Prawn Showing Lipid by Barne's Method
(Per cent composition) TL : Initial Value =7. 75 per cent

Diet	30 Days	60 Days	90 Days
A	7.65	7.15	5.35
B	7.50	6.80	5.00
C	7.10	5.75	4.50
D	6.05	5.30	4.25
E	5.80	4.50	3.30

The CP content in the flesh of *F. merguiensis* in diet A (40 per cent), diet B (45 per cent), diet C (50 per cent) and diet D (55 per cent), diet E (60 per cent) varied as 13.75 per cent, 14.20 per cent, 14.55 per cent, 15 per cent and 16.22 per cent respectively. The values of CP in the flesh were minimum on zero day of the experiment (5.25 per cent) and maximum on 90 days of experimental (16.22 per cent) (Table 12.4). Our results are also in agreement with those made by Hafedh (1999) in Nile tilapia, *Oreochromis niloticus* and Siddiqui and Howlader (1991) in *Oreochromis niloticus* who are reported that FCR decrease with increasing dietary protein level. In fact, FCR is also usually overestimated at a higher feeding rate as the experimental animals do not consume all the feed offered to them.

The CP content of the flesh prawn on any of the diet was minimum initially and high during the later culture period (Table 12.4). This is in accordance with Maynard and Loosli (1956) Sehgal and Thomas (1987) and Sehgal and Sharma (1991) who found that the body proteins in animal increased rapidly in young but remained constant in the aging animal. Moreover, a linear relationship was observed between dietary protein and body protein and is in accordance with Dabrowsi (1977) who have also reported that in grass carp fry as the dietary protein level increase, flesh protein also marks an increase. These results are however in contrast to those of Jayachandran and Paulraj (1975) and Sehgal and Sharma (1991) who recorded a negative correlation between body protein and dietary protein in *Cyprinus carpio* and *Cirrhina mrigala* respectively.

The total lipid content of the flesh of *F. merguiensis* in diet A (40 per cent), diet B (45 per cent), diet C (50 per cent), diet D (55 per cent) and diet E (60 per cent) varied from 5.35, 5.00, 4.50, 4.25 and 3.30 per cent respectively. The maximum and minimum total lipid content in all the treatments were observed on zero day and 90th day of the culture respectively (Table 12.5).

Thus, an inverse relationship has been observed between dietary protein and flesh lipid content. This relatively low flesh total lipid content in relation to increased dietary protein can be attributed to

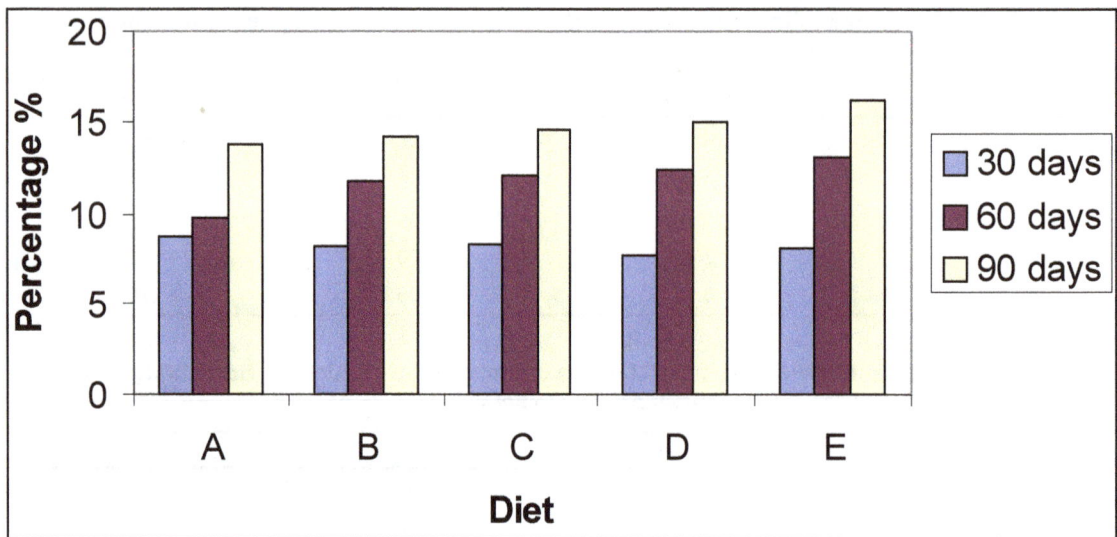

Figure 12.1: Quantitative Analysis of Muscles of Prawn Showing Crude Protein

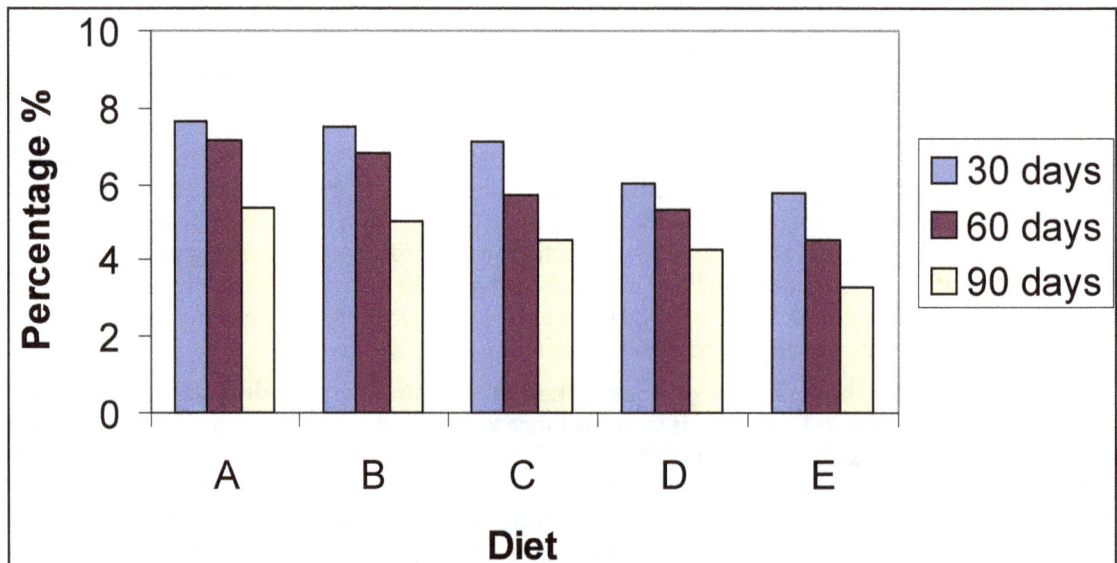

Figure 12.2: Quantitative Analysis of Muscles of Prawn Showing Total Lipid

high metabolic rate of experimental prawns in the orderE >D >C >B >A. This is supported by the higher growth of prawns in the referred order in various diets. The results get support from the observations made by Hafedh (1999) who also observed that Nile Tilapia, when fed low protein diets (25-30 per cent) had lower percentage of protein and higher lipid levels than fish fed 35 per cent and 40 per cent protein diets. The growth performance showed progressive increase in increasing dietary protein level. Among the various treatments, the most economic dietary protein level was found to be 60 per cent. In fact, FCR is also usually overestimated at a higher feeding rate as the experimental animal do not consume all the feed offered to them.

The dietary protein levels for maximal growth in different species of crustaceans have been worked out by several workers. The optimum level of dietary protein was determined to be within the range of 52-57 per cent for *P. japonicus* (Deshimaru and Yone, 1978) and later revised not to exceed 42 per cent (Koshio *et al.*, 1993), 40-44 per cent for *P. monodon* (Alava and Lim, 1983; Shiau *et al.*, 1991) and 36 per cent or higher for *P. vannamei* (Smith *et al.*, 1985) and later revised to 30 per cent (Cousin *et al.*, 1993) and 15 per cent (Araranyakananda, 1993). But in the present study the dietary protein level for the optimum growth of *F. merguiensis* (60 per cent) is almost similar to those worked out for *P. japonicus* (52-57 per cent) but higher than that of *P. vannamei* (36 per cent), *P. monodon* (40-44 per cent). This support the fact that postlarvae of shrimp require a higher dietary protein level than older shrimp. (Colvin and Brand, 1977, Chen *et al.*, 1985, Goddard, 1996).

Thus, the estimated protein level for optimum growth of *F. merguiensis* is 60 per cent and may be above (since higher levels have not yet been tried). This value is almost similar to those worked out in another closely allied species of the same genera. Since the nutritional requirement of *F. merguiensis* has not been much investigated, therefore, such studies are still needed and are expected to give a precise value for protein requirement in relation to the amino acid requirements and optimum levels of energy.

References

Akiyama, D.M., Dominy, W.G. and Lawrence, A.I., 1992. Penaeid shrimp nutrition. In: *Marine Shrimp Culture: Principles and Practices*. Elsevier Science Publishing Inc., New York, USA, p. 535–568.

Alva, V.R. and Lim, C., 1983. The quantitative protein requirement of *Penaeus monodon* juvenile in a controlled environment. *Aquaculture*, 30: 53–61.

Andrews, J.M., Sick, L.V. and Baptist, G.J., 1972. The influence of dietary and energy levels on growth and survival of penaeid shrimp. *Aquaculture*, 1(4): 341–347.

Aranyakananda, P., 1993. Dietary protein and energy requirements of *Penaeus vannamei* and the optimal protein to energy ratio. *Ph.D. Dissertation*, Texas A&M University, College Station, Texas, USA.

Balazs, G.H., Ross, E. and Brooks, C.C., 1973. Preliminary studies on the preparation and feeding of crustacean diets.

Barnes, H. and Stock, J. Black, 1973. Estimation of lipid in marine animal, detail investigation of the Sulphophosphovanilin method. *J. of Mar. Biol. Ecol.*, 1: 103–118.

Chen, H.Y., Zein-Eldin, P. and Aldrich, D.V., 1985. Combined effect of shrimp size and dietary protein source on the growth of *Penaeus setiferus* and *Penaeus vannamei. J. World Maricult. Soc.*, 6: 288–296.

Colvin, L.B., Brand, C.W., 1977. The protein requirement of penaeid shrimp at various life cycle stages in controlled environmental systems. *J. World Maricult. Soc.*, 8: 821–840.

Cousin, M., Cuzon, G., Blanchet, E., Ruelle, F. and Aquacop, 1993. Protein requirements following an optimum dietary energy to protein ratio for *Penaeus vannamei* juveniles. In: *Proceedings of the Aquaculture Feed Processing and Nutrition Workshop: Fish Nutrition in Practice*, (Eds.) S.J. Kaushik and P. Luquet. INRA, Paris, France, pp. 599–606.

Cowey, C.B. and Forster, J.R.M., 1971. The essential amino acid requirements of the prawn *Palaemon serratus*: The growth of prawns diets containing proteins of different amino acid compositions. *Mar. Biol.*, 10: 77–81.

Deshimaru, O. and Yone, Y., 1978. Optimum level of dietary protein for prawn. *Bull of the Jpn. Soc. of Sci. Fish.*, 44: 1395–1397.

Foster, J.R.M. and Beard, T.W., 1973. Growth experiments with the prawn, *Palaemon serratus* and *Pandalus platycerus. J. Mar. Biol. Assoc., U.K.*, 51: 943–961.

Foster, J.R.M., 1976. Studies on the development of compounded diets for prawns. In: *Proceedings of the First International Conference on Aquaculture Nutrition,* (Eds.) S.P. Kent, W.N. Shaw and K.S. Donberg. Lewes/Rehoboth, Delaware Univ. Delaware Coll.

Goddard, S., 1996. *Feed Management in Intensive Aquaculture.* Chapman and Hall, New York, USA.

Hafedh, Al.Y.S., 1999. Effects of dietary protein on growth and body composition of Nile tilapia, *Oreochromis niloticus. J. Food Resources Division Aquaculture Research,* 30: 385–393.

Jayachandran, P. and Paulraj, S., 1975. Experiments with artificial feeds on *Cyprinus carpio* fingerlings. *Journal of Inland Fisheries Society of India,* 8: 33–37.

Kanazawa, A., Shimaya, M., Kawasaki, M. and Kashiwada, K., 1970. Nutritional requirements, of prawn: I. Feeding and artificial diets. *Bull. Jpn. Soc. Sci. Fish.,* 36: 949–954.

Koshio, S., Teshima, S., Kanazawa, A. and Watase, T., 1993. The effect of dietary protein content on growth, digestion efficiency and nitrogen excretion of Juvenile Kuruma prawns, *Penaeus japonicus. Aquaculture,* 113: 101–114.

Lim, C. and Akiyama, D.M., 1997. Nutrient requirements of Penaeid shrimp. *Fishing Chimes,* 17(8): 31–35.

Lowry, O.H., Rosenbrough, N., Forr, A.L. and Ronball, R.J., 1951. Protein measurement with Folinphenol reagent. *J. Biol. Chem.,* 193: 265–275.

Maynard, L.A. and Loosli, J.K., 1956. In: *Animal Nutrition.* McGraw Hill, New York, p. 484.

New, M.B., 1976. A review of dietary studies with shrimps and prawns. *Aquaculture,* 9: 101–144.

Sehgal, H.S. and Thomas, J., 1987. Efficacy of two newly formulated diets for carp, *Cyprinus carpio,* var. *Communis* (Linn.): Effects of flesh composition. *Biol. Wastes,* 21: 179–187.

Sehgal, H.S. and Sharma, S., 1991. Efficacy of two newly formulated diets for carp, *Cyprinus carpio* var. *Communis* (Linn.): Effects of flesh composition. *Biol. Wastes,* 21: 179–187.

Shiau, S.Y., Kwok, C.C. and Chou, B.S., 1991. Optimal dietary protein level of *Penaeus monodon* reared in sea water and brackish water. *Nippon Suisan Gakkaishi,* 57: 711–716.

Sick, L.V., Andrews, J.W. and White, D.B., 1972. Preliminary studies of selected environmental and nutritional requirements for the culture of penaeid shrimp. *Fish. Bull.,* (70)1: 101–109.

Siddiqui, A.Q. and Howlader, M.S., 1991.Growth of Nile tilapia *Oreochromis niloticus,* in response to winter feeding. *Journal of Aquaculture in Tropics,* 6: 153–156.

Smith, L.L., Lee, P.L., Lawrence, A.L. and Strawn, K., 1985. Growth and digestibility by three sizes of *Penaeus vannamei* Boone: Effect of dietary protein level and protein source. *Aquaculture,* 46: 85–96.

Snedecor, G.W. and Cochran, W.G., 1967. *Statistical Methods,* 6th Edn. Oxford and IBH publishing Co., New Delhi, p. 593.

Wickens, J.F., 1976. Prawn biology and culture. *Oceanogr. Mar. Biol.,* 4: 435–507.

Chapter 13

Induced Toxicity of Imidacloprid on Protein Metabolism in Estuarine Clam, *Katelysia opima* (Gmelin)

☆ *A.S. Kulkarni and M.V. Tendulkar*

ABSTRACT

The indiscriminate use of insecticide has caused serious pollution problems of aquatic ecosystems. Imidacloprid is found in a variety of commercial insecticides like Admire, Confidor, Premier etc. and mainly used to control sucking insects such as Rice hoppers, Aphids, Thrips, Termites and some species of Beetles. Further, it is known to cause apathy, myatonia, tremor and myospasms in humans. Toxic effects of Imidacloprid were estimated by selecting *Katelysia opima* as an animal model. Effect of Imidacloprid on total protein content of gills, mantle, Hepatopancreas, foot, male gonad and female gonad of estuarine clam, *Katelysia opima* was studied. The clams were exposed to 86.6 ppm Imidacloprid for acute treatment; it was found that there was decrease in protein content in various tissues in LC_{50} as compared to control. In LC_{50} group protein content was decreased in mantle hepatopancreas, foot male and female gonad as compared to the control. This decrease was more in foot, male gonad and female gonad in LC_{50} group as compared to LC_0 group. Gills, mantle and hepatopancreas show increase in protein content in LC_{50} group as compared to LC_0 group. Decrease in protein content was more in foot, male gonad and female gonad in LC_{50} group due to the higher concentrations of Imidacloprid.

Keywords: Imidacloprid, Protein content, Katelysia opima.

Introduction

Imidacloprid ($C_9H_{10}ClN_5O_2$) is a systemic, chloronicotinyl insecticide used to control the sucking insects like rice hoppers, aphids, thrips, white flies, termites, turf insects, soil insects and some beetles.

It is most commonly used on rice, cereals, maize, potatoes, vegetables, sugar beets, fruits, cotton, hopes and turf. It acts on the nervous system of insect which causes a blockage in the neural pathway. This blockage leads to the accumulation of acetyl choline, an important neurotransmitter, resulting in the insect's paralysis and eventually death. It is effective on contact and via stomach action (Kidd and James, 1994). It is found in the variety of commercial insecticides (Meister, 1995). In market, it is available in the form of the products like Admire, Confidor, Premier, Provado, etc., with Imidacloprid as an active ingredient. In Ratnagiri, it is mainly used against mango hoppers, mango mealy bugs, aphids and other insect pests of mango. Due to this, Imidacloprid is getting concentrated in the aquatic bodies.

It is bioaccumulated in estuarine fauna like estuarine clams. Clams are abundant along the coast of Ratnagiri (Maharashtra) and are important with reference to the food value. Shells are mainly used as a raw material for lime factories along the coast. In spite of the fact, very less attention has been paid by researchers with regard to effect of pollution on estuarine clams. Clams are known to accumulate pesticides without getting killed easily and have relatively long life span. Toxic material can induce considerable change in protein content of bivalves. Balchandra Waykar and Lomte (2001) reported the depletion of total protein content level in all tissues of freshwater bivalves, *Parreysia cylindrica* after Cypermethrin exposure. Mane and Gokhale (2002) observed significant alterations in protein content of *M. meretrix* and *K. opima* after exposure to fluoride. Prabhupatkar (2004) also observed the depletion of total protein in estuarine clam, *Meretrix meretrix* after exposure to Cypermetrin.

Since biochemical assessment is a useful tool to measure environmental quality, the present work is aimed to study the effect of Imidacloprid on protein metabolism in estuarine clam, *Katelysia opima* (Gmelin).

Materials and Methods

The experimental clams (*Katelysia opima*) used for the present study were collected from Bhatye estuarine region, Ratnagiri coast, Maharashtra state. Clams of medium size (4.0 - 4.4cms.) were selected and brought to the laboratory and stocked in the plastic containers containing filtered, aerated estuarine water for 48 hrs. Well acclimatized clams to the laboratory condition were grouped in ten and kept into plastic containers containing five liters filtered estuarine water. Static bioassay tests were conducted for 96 hrs. by using Imidacloprid (17.8 per cent SL). For every experiment, a control group of clams was also run simultaneously.

The toxicity tests were repeated for three times and LC_0 and LC_{50} values were determined. After acute exposure to Imidacloprid the various tissues (gill, mantle, hepatopancreas, foot, male and female gonad) of live clams were pooled, weighed and dried in an oven at 70°C until a constant weight was obtained. Oven dried tissues were used for biochemical analysis. Live clams were analyzed for total proteins (Lowry *et. al.* 1951).

Results

Imidacloprid induced alterations in the total protein content in different organs of the estuarine clam, *Katelysia opima* are shown in Table 13.1 and Figure 13.1. In control group of *Katelysia opima* showed protein content in ascending order of, hepatopancreas < gills < mantle < foot < female gonad < male gonad, containing 13.732 ± 0.434524, 16.466 ± 0.379986, 22.799 ± 0.802759, 26.266 ± 0.641006, 32.799 ± 0.900512, 36.732 ± 0.641344 mg of protein/100mg dry wt. in respective organs.

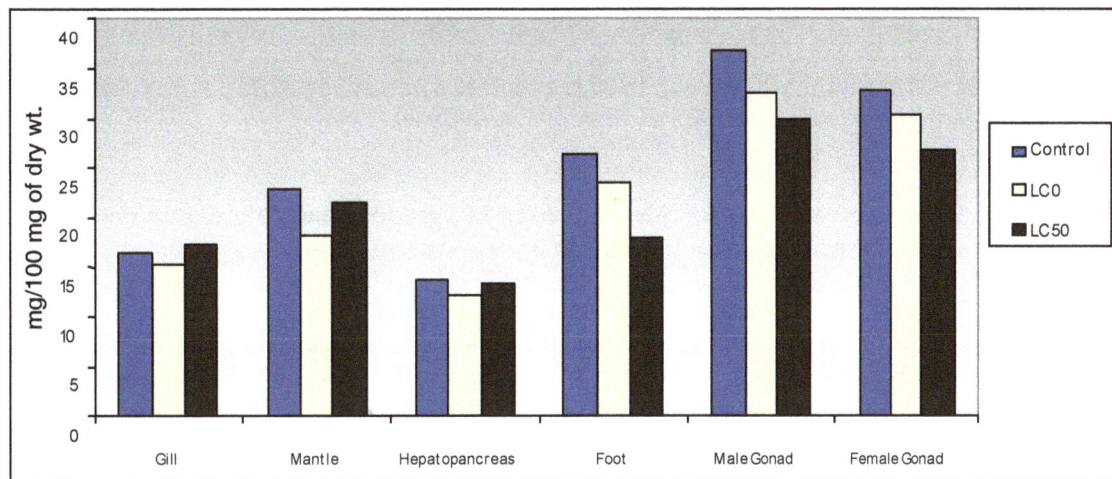

Figure 13.1: Imidacloprid Induced Changes in Protein Content of
***Katelysia opima* after Acute Exposure**

Table 13.1: Imidacloprid Induced Alterations in the Total Protein Content of
***K. opima* after Acute Exposure**

Tissue	Control	LC_0 Group	LC_{50} Group
Gills	16.466±0.379986	15.266±0.795843 (-7.28)*	17.266±0.86278 (5.66)**
Mantle	22.799±0.802759	18.266±0.924902 (-19.00)*	21.532±0.960318 (-7.89)*
H.P.	13.732±0.434524	12.332±0.849771 (-10.19)*	13.399±0.795781 (-2.42)*
Foot	26.266±0.641006	23.532±0.691319 (-12.18)*	17.999±0.505481 (-31.47)*
M.Gonad	36.732±0.641344	32.666±0.623654 (-11.06)*	29.932±0.98863 (-18.51)*
F.Gonad	32.799±0.900512	30.266±1.016301 (-6.19)*	26.866±0.900494 (-17.96)*

Values in parenthesis are percentage difference S.D. of five readings m.

In LC_0 (38.5 ppm) protein content was present in ascending order of hepatopancreas < gills < mantle < foot < female gonad < male gonad with 12.332 ± 0.849771, 15.266 ± 0.795843, 18.266 ± 0.924902, 23.532 ± 0.691319, 30.266 ± 1.016301, 32.666 ± 0.623654 mg of protein/100 mg dry wt. In LC_0 group as compared to control there was 19.88 per cent, 11.06 per cent, 10.40 per cent, 10.19 per cent, 7.72 per cent and 7.28 per cent decrease in mantle, male gonad, foot, hepatopancreas, female gonad and gills respectively.

In LC_{50} (86.6 ppm) group, protein content was present in ascending order of, hepatopancreas < gills < foot < mantle < female gonad < male gonad containing 13.399 ± 0.795781, 17.266 ± 0.86278,

17.999 ± 0.505481, 21.532 ± 0.960318, 26.866 ± 0.900494, 29.932 ± 0.98863 mg of protein/100mg dry wt. respectively. As compared to control group, there was increase in protein content of gills as 4.85 per cent and 31.98 per cent, 19.78 per cent, 18.08 per cent, 4.96 per cent and 2.42 per cent decrease in foot, male gonad, female gonad, mantle and hepatopancreas respectively.

Discussion

After acute exposure of Imidacloprid, clams from LC_0 group showed significant decrease in protein content in all target organs. In LC_{50} group protein content showed significant decrease in mantle, hepatopancreas, foot, male gonad and female gonad and increased in protein content was observed in gills. It may attributed to their primness, impairment of protein synthesis and increase in the rate of their degradation into amino acids, which may be fed to TCA cycle through amino transferase probably due to the high energy demand to cope with the stress condition. Jha (1988) supports the idea of consumption of amino acid for metabolic processes as energy source.

Decrease in protein level may be due to increased proteolytic activity or might be due to anaerobic conditions produced by pesticide (Sivaprasad Rao and Ramana Rao, 1980). Increase in protease activity also supported depletion of protein content (Srinivas and Purushottam Rao, 1987). Jadhav (1993) and Waykar (1998) also observed increased protease activity in molluscs after pesticide treatment. Depletion of protein content in animal tissue after exposure to various pollutants was reported by some workers. Lomte and Alam (1982) reported depletion of protein content in various body tissue of snail, *Bellamia bengalensis* after malathion treatment, Waykar and V. S. Lomte (2001) also observed the total protein content of all tissue of freshwater bivalve, *Parreysia cylindrical* after Cypermethrin exposure. Swami *et.al.* (1983) recorded significant decline in protein content in freshwater mussel *Lamellidens marginalis* exposed to flodit and metacid. Ramana Rao and Ramamurthi (1978) observed alteration in protein content in tissue of *Pila globosa* after sumithion exposure. Chaudhari and Kulkarni (1998) reported the significant decrease in the protein content due to the active use of proteins for overcoming to the molluscicidal stress.

The protein content was decreased in various tissues in LC_0 group as compared to control. In LC_{50} group protein content was decrease in mantle, hepatopancreas, foot, male gonad and female gonad as compared to control. This decrease was more in foot, male gonad and female gonad in LC_{50} group as compare to LC_0 group. Gills, mantle, hepatopancreas shows increase in protein content in LC_{50} group as compared to LC_0 group. Decrease in protein content was more in foot, male gonad and female gonad in LC_{50} group due to higher concentrations of Imidacloprid. Decrease in protein level might be due to increased proteolytic activity or due to anaerobic condition produced by pesticidal stress. Increased protein level may be due to enhanced protein synthesis induced by pesticide at transitional level.

References

Chaudhari, R.D. and Kulkarni, A.B., 1998. Alterations in protein contents in the reproductive organs of the snail, *Cerastus monssonianus* due to Monocrotophos intoxication. *Poll. Res.*, 17(4): 325–329.

Jadhav, S.M., 1993. Impact of pollutants on some physiological aspect of the freshwater bivalve, *Corbicula straitella*. *Ph.D. Thesis*, Marathwada University, Aurangabad, (M.S.), India.

Jha, B.S., 1988. Effect of lead nitrate on certain organs of air breathing teleost, *Channa punctatus*. *Ph.D. Thesis*, L.N. Mithila University, Darbhanga, India.

Kidd, H. and James, D. (Eds.) 1994. *Agrochemicals Handbook*, 3rd Edn. Royal Society of Chemistry, Cambridge, England.

Lomte, V.S. and Sabiha, Alam, 1982. Changes in the biochemical compounds of the *Prosobranch Bellamia* (V). *bengalensis* on exposure to malathion. In: *Proc. Symp. Phy. Resp. Anim. Pollutants*, Marathwada University, Aurangabad, p. 69–72.

Lowry, C.H., Rosenbrough, N.J., Farr, A.L. and Randall, R.J., 1951. Protein measurment with folin phenol reagent. *J. Biol. Chem.*, 193: 265–275.

Mane, U.H. and Gokhale, A.A., 2002. Fluoride toxicity to the estuarine bivalve's molluscs. *J. Mar. Biol., Assoc.*, India 40(1–2): 16–29.

Meister, R.T. (Ed), 1995. *Farm Chemicals Handbook*, 95. Meister Publishing Company, Willoughby, OH.

Prabhupatkar, M.M., 2004. Cypermetrin induced toxicity to the estuarine clam, *Meretirx meretrix* (Linnaeus). *M.Sc. Thesis*, Mumbai University, Mumbai.

Ramana Rao, M.V. and Ramamurthi, R., 1978. Studies on the metabolism of the Apple snail, *Pila globosa* (Swainson) in relation to pesticide impact. *Ind. J. Her.*, 11: 10.

Sivaprasad Rao, K. and Rao, K.V. Ramana, 1980. Effect of insecticide on freshwater snail, *Telugu Vydhanika Patrika*, 9: 27–32.

Srinivas, P. and Rao, Purushottam, 1987. Biochemical effects of phosphomidon on silk gland and other tissues of *Bombyx mori* (L). In: *Proc. Natl. Symp. on Env. Poll. and Pest. Toxicol. and 8th Annual Session of Academy of Environ. Biology, India*, University of Jammu (J&K) from Dec. 10–12, pp. 173–176.

Swami, K.S., Rao, J.K.S., Reddy, S.K., Jagannatha Rao, K.K.S., Moorthy, L.C., Chetty, C.S. and Indira, S.K., 1983. The possible metabolic diversions adapted by the freshwater mussel to control the toxic metabolic effect of selected pesticides. *Ind. J. Comp. Anim. Physiol.*, 1: 95–103.

Waykar, B.B., 1998. Effect of pesticides on some physiological activities of freshwater bivalve, *Parreysia cylindrica*. *Ph.D. Thesis*, Dr. Babasaheb Ambedkar Marathwada University, Aurangabad, M.S., India.

Waykar, B.B. and Lomte, V.S., 2001. Total protein alternation in different tissue of freshwater bivalve, *Parreysia cylindrica* after cypermethrin exposure. *Ecol. Env. and Cons.*, 7(4): 465–469.

Chapter 14

Population of *Lactobacillus* sp. in Two Freshwater Fishes, *Glossogobius giuris* (Ham) and *Labeo rohita* (Ham)

☆ *Krishna Ram H., Shivabasavaiah, R. Manjunath and*
M. Ramachandra Mohan

ABSTRACT

Fish are continuously exposed to a wide range of microorganisms present in the environment, and the microbiotas of fish have been the subject of several reports. In present study isolation and enumeration of heterotrophic bacteria from the intestine, gills and skin of *Glossogobius giuries* (Ham) and rohu, *Labeo rohita* (Hamilton) have been carried out to find out their importance in the nutrition of the host fish. The results of the investigations showed that highest *Lactobacillus* sp. load in intestine > gills > skin of *Glossogobius giuris*, while in *Labeo rohita* highest in skin> intestine > gills. The dominant numbers in the intestinal tract of endothermic animals, have beneficial effects in nutritional processes, stimulate the immune system and act as probiotics. Colony characteristics, biochemical and physiological characteristics, gram positive, coccobacilli, non motile, absence of the starch hydrolysis, lipase activity, and casease positive organisms. This harmless bacteriocin-producing strain may reduce the need to use antibiotics in future aquaculture.

Keywords: Freshwater fish, Lactobacillus sp., Probiotics, Intestine.

Introduction

Fish receive bacteria in the digestive tract from the aquatic environment through water and food that are populated with bacteria. Being rich in nutrient, the environment of the digestive tract of fish confers a favorable culture environment for the microorganisms. Lactic acid bacteria are characterized as gram - positive, usually non motile, non-sporulating bacteria that produce lactic acid as a major or sole product of fermentative metabolism. Members of this group containing both rods (*Lactobacilli* and carnobacteria) and cocci (streptococci), they are generally catalase - negative and they usually lack cytochromes.

Lactic acid bacteria are nutrdinally fastidious, requiring carbohydrates, amino acids, peptides, nucleic acid derivates and vitamins. Different species of Lactic acid bacteria have adopted to grow under widely different environmental conditions, and they are widespread in nature. It is known that several species of lactic acid bacteria are part of natural intestinal flora of healthy fish (Jankauskiene, 1995; Ringo and Gatesoupe, 1998; Jankauskine 2000a, 2000b), and it is determined that lacto acid bacilli often produce bacteriocins which may inhibit the growth of fish pathogens in the fish intestine (Austin and Al-Zahrani, 1998; Gatesoupe, 1994 and Gildbarg *et al.,* 1997).

Nowadays the intestinal tract ecosystem is considered as the protective system of the organism and the genus *Lactobacillus* bacteria play a very important role in it (Lenzer, 1973; Mitsuoka, 1992; McCartney *et al.,* 1996 and Tannock, 1998). On colonization of the microorganisms digestive tract *Lactobacilli* from a natural ecological barrier and prevent other pathogenic microbes from penetration and reproduction there (Hanson and Olafsen, 1999). The present investigation to understand the bacterial load with special reference to *Lactobacillus sp.* in the intestine, gills and skin of freshwater fishes *Glossogobius giuris* (Ham) and *Labeo rohita* (Hamilton).

Materials and Methods

Fish Samples

Samples of *Glossogobius giuris* (Ham) and *Labeo rohita* (Hamilton) were collected for a period of three years *i.e.,* January 2003 to December 2005 in lakes, Thalli lake (12°35"-02.5" N latitude and 077°39'–33.7" E longitude) in Krishnagiri district of Tamil Nadu and the Hebbal lake, (13°02' - 37.2" N latitude and 0.77°35'–25.7" E longitude) in Bangalore North city, Karnataka, using cast and gillnets (10mm) and fishes were brought to the laboratory in the live condition.

Physico-chemical Analysis

Temperature, dissolved oxygen and pH were measured on site using a digital thermometer, (OAKTON, Thermometer, China), dissolved oxygen (Aguasol) and digital pH meter (WP 12, Singapore).

Microbiological Samplings

The weight and length of the live fish were recorded, prior to the aseptical removal of the intestine, gills and skin. About 250 mg of the above tissue was crushed in 5 ml sterile distilled water and serial dilution of the samples was made with 9ml of distilled water to the level of 10^{-6}. 0.1 ml of the aliguots ranging from 10^{-3} to 10^{-6} dilution was added to the sterile Petridish containing 15-20 ml of the specific media at ear bearing temperature of 40°C. The media used are PNY media, the Petri dishes containing selective media for growth of *Lactobacillus* sp. incubated at 20°C for 24 hours. They were stored in refrigerator at 4°C for further analysis of the characteristics of the bacteria. Colony characteristics of *Lactobacillus* were studied apart from the gram character; biochemical characteristics of *Lactobacillus* were studied by the method (APHA, 1998) and physiological characteristics of the selected isolates were studied by the method (Harrington and McCance, 1972).

Media Composition

PNY medium (g/L^{-1}) : Piptic digest of animal tissue, 5.0; yeast extract, 5.0; Dextrose, 5.0; Monopotassium phosphate, 0.50; Dipotassium phosphate, 0.50; Magnesium sulphate, 0.25; Manganese sulphate, 0.01; Ferrous sulphate, 0.01; Sodium chloride, 0.01; Zinc sulphate, 0.001; Copper sulphate, 0.001; Cobalt sulphate, 0.001; and Agar, 15; pH (at 25°C) 6.0 ± 0.2.

Results

Physio-chemical Analysis

Data on physio-chemical characteristics of water at the time of sampling are presented in Table 14.1. There was no significant difference in physico-chemical values of rearing pools, which were with in the optimum range for *Glossogobius giuris* and *Labeo rohita* for rearing.

Lactobacillus Population in Fishes

The *Lactobacillus* population in *Glossogobius giuris* and *Labeo rohita* of intestine, gills and skin, the population showed maximum colonies in intestine 62.2×10^2 colonies/g wet weight, in gills showed about 41.2×10^2 colonies of/g wet weight and in skin showed about 38.8×10^2 colonies of/g wet weight at 10^{-3} and minimum was recorded at 10^{-6} dilution were showed 38.4×10^2 colonies/g wet weight of the intestine, 6.8×10^2 colonies of *Lactobacillus*/g wet weight of the gills and 24.4×10^2 colonies of *Lactobacillus*/g wet weight of the skin of *Glossogobius giuris* (Table 14.2) and on the other hand, on PNY media, *Lactobacillus* populations were maximum in intestine 39.2×10^2 colonies/g wet weight, with the gills showing about 26.4×10^2 colonies/g wet weight and the skin showing about 42×10^3 colonies/g wet weight at 10^{-3} dilution and the minimum *Lactobacillus* colony count was recorded at 10^{-6} dilution showed 17×10^3 colonies/g wet weight of the intestine, 12×10^3 colonies/g wet weight of the gills and 7.2×10^2 colonies/g wet weight of the skin of *Labeo rohita* (Table 14.2).

Table 14.1: Mean Physico-chemical Characteristics of Lakes at the Time of Samplings during 2003–2005

Months	Hebbal Lake			Thalli Lake		
	t°C	*pH*	*Dissolved Oxygen/ (mg/L)*	*t°C*	*pH*	*Dissolved Oxygen/ (mg/L)*
January	23.3	7.10	4.8	22.0	7.8	7.2
February	23.6	7.20	4.2	22.8	7.6	7.0
March	24.8	780	3.10	24.0	7.4	5.7
April	24.6	7.70	3.0	27.2	77	5.3
May	24.9	7.80	4.2	24.3	7.8	6.3
June	24.2	7.30	3.6	24.0	7.3	7.8
July	23.6	7.80	4.0	23.0	7.4	8.0
August	23.6	7.20	4.2	24.0	7.5	8.3
September	24.0	7.00	3.8	22.9	7.6	8.8
October	24.0	7.40	4.0	22.8	7.3	7.6
November	23.8	7.0	3.8	23.0	7.4	8.3
December	24.6	7.6	4.2	22.8	7.5	8.0

Comparative Variation of *Lactobacillus* in *Glossogobius giuris* and *Labeo rohita*

In *Lactobacillus* population in *Glossogobius giuris*, the intestine shows highest 62.2×10^2 colonies/g wet weight, the gills shows 41.2×10^2 colonies/g wet weight and the skin shows lowest

38.8 x 10^2 colonies/g wet weight. In *Labeo rohita*, the skin shows highest 42 x10^3 colonies/g wet weight, the intestine shows 39.2 x 10^2 colonies/g wet weight and the gills shows least 26.4 x 10^2 colonies/g wet at 10^{-3} dilution (Table 14.2).

Table 14.2: *Lactobacillus* Colony Count on PNY Media in Intestine, Gills and Skin from Freshwater Fishes of *Glossogobius giuris* (Ham) and *Labeo rohita* (Hamilton)

Media Used	Tissue	No. of Colonies/gm Wet Wt. of Tissue			
		10^{-3}	10^{-4}	10^{-5}	10^{-6}
PNY media	Intestine	62.2 x 10^2	59.2 x 10^2	45.6 x 10^2	38.4 x 10^2
(*Glossogobius giuries*)	Gills	41.2 x 10^2	39.6 x 10^2	13.8 x 10^2	6.8 x 10^2
	Skin	38.8 x 10^2	32.4 x 10^2	32 x 10^3	24.4 x 10^2
PNY media	Intestine	39.2 x 10^2	29.2 x 10^2	23.2 x 10^2	17 x 10^3
(*Labeo rohita*)	Gills	26.4 x 10^2	25.6 x 10^2	20.8 x 10^2	12 x 10^3
	Skin	42 x 10^3	39.6 x 10^2	9.4 x 10^2	7.2 x 10^2

Colony Characteristics of *Lactobacillus*

Colony characteristics of *Lactobacillus* sp. from intestine, gills and skin on PNY media, the isolated colonies were small, medium, round, entire, cream with white colour, convex, flat batyrous, translucent, which were gram positive coccobacilli and non motile (Table 14.3).

Biochemical Characteristics of *Lactobacillus*

All the biochemical characteristics which aid in the identification and classification of microorganisms that is morphologically identical. A series of biochemical tests can provide a microbial finger print. In the present investigation *Lactobacillus* sp, shows gram negative rods which biochemical characteristics was indicated in (Table 14.4).

Physiological Characteristics of *Lactobacillus*

The physiological characteristics of *Lactobacillus* sp. in intestine, gills and skin of *Glassogobbius giuris* and *Labeo rohita* were studied based on starch hydrolysis, Lipase activity and casein hydrolysis presented in (Figure 14.1).

Starch Hydrolysis

The colonies of *Lactobacillus* cultured from the intestine, gills and skin of the freshwater fishes, *Glossogobius giuris* and *Labeo rohita* were streaked on the starch hydrolysis medium. A week after streaking, iodine was poured into the plates; white transparent region was observed around the colonies of *Lactobacillus* while the rest of the media turned blue. In the absence of starch hydrolysis, it is noticed that the blue colour completely, surrounds the colonies (Figure 14.1a).

Lipase Activity

The plates with lipase activity media were streaked with *Lactobacillus* from *Glossogobius giuris* and *Labeo rohita*. A waxy material around the colonies of *Lactobacillus* was observed which indicated the liberation of insoluble oleic acid. However the extent of transparently around the bacterial colonies was more as compared to waxy material around them, thus indicating the *Lactobacillus* showed

Table 14.3: Colony Characteristics of *Lactobacillus* sp. from Intestine, Gills and Skin of *Glossogobius giuris* (Ham) and *Labeo rohita* (Hamilton) on PNY Media

Media Used	Tissue	Colour	Shape	Margin	Size	Edge	Elevation	Consistency	Transferency	Gram reaction	Motility
PNY Media	Intestine	Cream	round	smooth	small	entire	convex	bytyrous	translucent	Gram positive	Non motile
	Gills	Cream	round	smooth	medium	entire	flat	bytyrous	translucent	Gram positive	Non motile
	Skin	Cream	round	smooth	small	entire	convex	bytyrous	translucent	Gram positive	Non motile

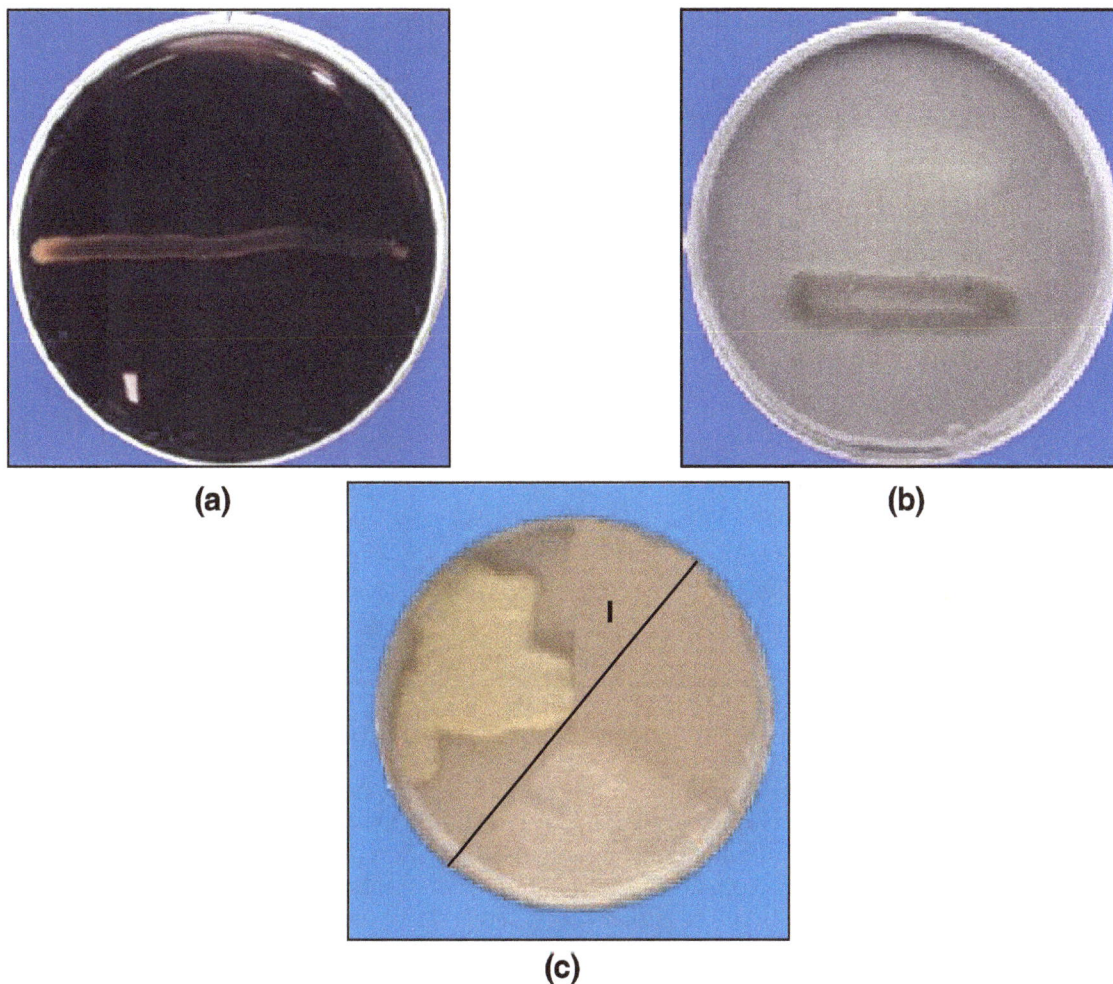

(a)

(b)

(c)

**Figure 14.1: Physiological Characteristics of *Lactobacillus* sp.
(a) Starch hydrolysis: Absence of starch hydrolysis notice there is the blue colour
completely surrounds the colonies of the organisms (b) Lipase activity: the absence of lipid
hydrolysis (c) Casein Hydrolysis: (I) represents casease positive of organisms.**

predominance of amylase activity than lipase activity, the colonies of Lactic acid bacteria indicated absence of Lipase activity (Figure 14.1b).

Casein Hydrolysis

Hydrolysis of Casein (Protein) by protease enzyme activity of *Lactobacillus* cultured from the intestine, gills and skin of *Glossogobius giuris* and *Labeo rohita* were studied. Generally the bacteria showed more affinity towards protein molecules then to the lipids. Casease is an exoenzyme that is produced by some bacteria in order to degrade casein. If an organism can produced casein, then there will be a zone of clearing around the bacterial growth. The colonies of *Lactobacillus* show casease positive organisms (Figure 14.1cI).

Table 14.4: Biochemical Characteristics of Identification of *Lactobacillus* sp. in *Glossogobius giuris* (Ham) and *Labeo rothia* (Hamilton) on PNY media

Name of the Test	*Lactobacillus* spp.
Gram stain	Cocco bacilli (+)
Indole Production	Not produced
M.R. Reaction	(-)
Vogas Proskauer	(-)
Citrat	(-)
H_2S production	Not produced
NO_3 reduction	do not reduce
Gelatin liquefaction	Not hydrolyzed
Motility	Non motile
Catalase activity	(-)
Oxidase activity	(-)
Fermentation test	Produced
Acid production	Produced
Sugar (Glucose and Maltose)	Formed
Arabinose	Not formed

Discussion

Lactobacillus genus bacteria are common members of the indigenous microflora of healthy fish (Ringo and Gatesoupe, 1998; Hansen and Olafsen, 1999; Jankauskiene, 2000 a, b). In the present study, as a result of *Lactobacillus* population in found in many different environments throughout the world. It is specially found in fish products and is found in the intestinal tract and along the lining of the mouth of help in food digestion. The earlier study, showed that fish contained lactic acid bacteria in the gastro intestinal tract which was isolated from the *Lactobacillus* and *Clupea harmgus*(L). The population density of the *Lactobacillus* in fishes *Glossogobius giuris* and *Labeo rohita*) usually varies quantitatively and qualitatively in different organs such as intestine, gills and skin. Highest *Lactobacillus* load in intestine > gills > skin of *Glossogobius giuris*, while in *Labeo rohita* highest in skin > intestine > gills, *Lactobacillus* sp. dominant numbers in the alimentary tract of endothermic animals, have a beneficial effect in nutritional processes, stimulate the immune system and act as probiotics (Conway, 1989; Fuller, 1989 and Sissons, 1989).

Nowadays the intestinal tract ecosystem is considered as the protective system of the organisms, and the genus *Lactobacillus* plays a very important role. It flows that the effective control of pathogenic bacteria in the fish intestine by using the active antibacterial lactobacilli strains may be successful in aquaculture; lactobacilli have complex nutritional requirements, needing to be supplied with carbohydrates, amino acids, peptides, fatly acids or fatly acid esters, nucleic acid derivates, and vitamins. Lactobacilli do not produce extra cellular derivative enzymes, and other numbers of the normal microbiota may act on complex molecules within the digestive tract before appropriate nutrients become available to the lactobacilli identification *Lactobacillus* sp. from fish intestine can often be successful. However, identification of from fish intestines can often be successful. However, identification of these bacteria with a large number of tests is tedious and time consuming. Moreover biochemical tests may not always be necessary because their differentiation can often rely on only a few tests. Physiological grouping of *Lactobacillus* in the intestine, gills and skin of *Glossogobius giuris* and *Labeo rohita*, indicate predominance of gram positive cocci in these fishes. Ability of the bacterial isolates to elaborate various hydrolytic enzymes indicates that majority of them are capable of utilizing various substrates such starch, gelatin and lipid, some investigations have also suggested that micro organisms exert a beneficial effect on the digestion process of fish (Danulat and Kausch, 1984 and Kono *et al.*, 1987). Lipolytic activity has been reported in bacterial isolates from gastrointestinal tract of grass carp (Trust *et al.*, 1979). Alex *et al.* (1993) observed maximum percentage of amylase and lipase producing bacteria associated with fishes. Our results confirmed the above findings, bacterial population play a significant role in the metabolism of fishes.

Conclusion

Lactic acid bacteria are not dominant in the normal intestinal microbiota of fish, at variance with homeotherms, but some strains can colonize the gut. On the one hand, it is possible to maintain artificially the lactic acid bacterial population at high level by regular intake with food. The applications of such treatments should be further considered to improve health and quality of fish in culture. One of the most promising and urgent application of harmless lactic acid bacteria is therefore to investigate their use both as probiotics against gram positive pathogens, and as sources of immune stimulants.

References

Alex Eapen, Maya, K., Dhevendran, K. and Natarajan, P., 1993. Phosphatase producing microbes associated with fish and shell fish in Trivendram coast. *Fish. Technology*, 30(1): 62–66.

APHA, AWWA, WPCF, 1998. Standard Methods for Examination of Water and Wastewater, 20th edn. Washington, D.C.

Austin, B., Al-Zahrani, A.M., 1988. The effect of antimicrobial compounds on the gastrointestinal microflora of rainbow trout, *Salmo gairdneri* Richardson. *J. Fish. Biol.*, 33: 1–14.

Conway, P., 1989. Lactobacilli fact and fiction. In: *The Regulatory and I Protective Role of the Normal Microflora*, (Ed.) R. Grubb, T. Middvedd and E. Norin. Stockton Press, New York, pp. 263–281.

Danulat, E. and Kaush, H., 1984. Microorganisms exert a beneficial effect of the digestive process of fish. *J. Fish. Biol.*, 24, 91.

Fuller, R., 1989. Probiotics in man and animal. *Journal of Applied Bacteriology*, 66: 365–378.

Gatesoupe, F.J., 1994. Lactic acid bacteria increase the resistance of turbot larvae *Scophthalmus maximus*, against pathogenic vibro. *Aquatic Living Resource*, 7: 277–282.

Gildberg, A., Mikkelsen, H., Sandaker, E. and Ringo, E., 1997. Probiotic effect of lactic acid bacteria in the feed on growth and survival of fry of Atlantic cod (*Gadus morhua*). *Hydrobiologia*, 352: 279–285.

Hansen, G.H. and Olafsen, J.A., 1989. Bacterial colonization of cod (*Gadus morhua* L.) and (*Hippoglossus hippoglossus*) eggs in marine aquaculture. *Appl. Environ. Microbiol.*, 55: 1435–1446.

Harringon, W.F. and McCance, 1972. *Laboratory Methods in Microbiology*. Academic Press, London, p. 362.

Jankauskiene, R., 1995. The lactoflora on the content of carps intestinal tract. *Ecology*, 1: 59–63.

Jankauskiene, R., 2000a. Defense mechanisms in fish: *Lactobacillus* genus bacteria of intestinal wall in nutriting and hibernating carps. *Ecology*, 1: 3–6.

Jankauskiene, R., 2000b. The dependence of the species composition of lactoflore in the intestinal tract of carps upon their age. *Acta Zoologica. Lituanica,* 10(3): 78–83.

Kono, M., Matsui, T. and Shimizu, C., 1987. The Micro-organisms exert a beneficial effect on the digestive process of fish. *Nippon Suissan Gakkashi,* 53: 305.

Lenzer, A.A., 1973. *Lactobacilli* in human microflora. *Doctoral Thesis,* Tartu University, Tartu, pp. 163.

McCartney, A.L., Wang, W. and Tannock, G.W., 1996. Molecular analysis of the composition of the bifidobacterial and *Lactobacillus* microflora of humans. *Appl. Env. Micro.,* 62: 4608–4613.

Mitsuoka, T., 1992. The human gastrointestinal tract. In: *The Lactic Acid Bacteria, Vol. 1: The Lactic Acid Bacteria in Health and Disease,* (Ed.) B.J.D. Wood. Elsevier Applied Science, London, pp. 69–114.

Ringo, E. and Gatesoupe, F.J., 1998. Lactic acid bacteria in fish: A review. *Aquaculture,* 160: 177–200.

Sissons, J.W., 1989. Potential of probiotic organisms to prevent diarrhea and promote digestion in farm animals. *A Review. J. Sci. Food Agric.,* 49: 1–13.

Tannock, G.W., 1988. The normal mocroflora: New concepts in health promotion. *Microbiol. Sci.,* 5: 4–8.

Trust, T.J., Bull, L.M., Curry, V.R. and Buckey, J.T., 1979. The lipolytic activity in bacterial isolates from gastrointestinal tract of grass carp. *J. Fish. Res. Bd.,* Canada, 36: 1174.

Chapter 15

Study of Food and Feeding Habit in *Nemipterus japonicus* along the Ratnagiri Coast of Maharashtra

☆ *P.S. Suresh Kumar and S.A. Mohite*

ABSTRACT

The food and feeding habits of the threadfin bream, *Nemipterus japonicus* was studied during the present work. The qualitative analysis of food of *N. japonicus* indicated that this species is a predaceous bottom feeder carnivore and feeds mainly on crustaceans and fishes. Molluscs, polychaetes, unidentified matter and semi digested matter were the other minor food items observed. The analysis of food with respect to size revealed variation in quantity of food consumed and also food preferences based on size groups. In size groups 40-60, 60-80mm, 100-120mm to 160-180 mm size group, crustaceans dominated over the other items of food. Increase in the feeding intensity was observed during February- March and August. This can be associated with the higher requirement of energy for the development of gonads and the spawning.

Keywords: Nemipterus japonicus, Feeding habits, Feeding intensity, Molluscs, Polychaetes.

Introduction

Nemipterus japonicus, the Japanese threadfin bream, has wide spread distribution throughout the Indian Ocean and is distributed in the tropical waters from 30°N-10°S latitude. It is benthic species, very abundant in coastal waters, found on mud or sand bottoms in 5 to 80 m, usually in schools. Major landings of *N. japonicus* locally called as "Rani masa" or "Bamni"are also reported from Ratnagiri coast of Maharashtra, India. Nemipterids are mostly consumed in fresh condition. They are also used for fish sausage and fishery by-products and have a great demand in surimi industry due to their low fat content, more whiteness and more myosin content in meat, as well as the availability of these species at low price in ample quantity. Threadfin breams are usually taken in multispecies catches and often three or more species of *Nemipterus* occur in the same trawl catch. To understand their distribution and exploitation pattern, study of food and feeding habits of these fishes was undertaken from February 2009 to January 2010.

Materials and Methods

Total 1000 specimens for this study were collected at random, at weekly intervals from Mirkarwada landing centre, Ratnagiri. Both qualitative and quantitative analysis was used to study the feeding of *N. japonicus*. It was done as per the procedure given by Qasim and Jacob (1972).

Qualitative Analysis

The identification of different organisms was usually done upto the generic level and whenever possible upto the species level, depending on the state of digestion. Identification of individual item was not always possible due to the semi digested condition of food inside the stomach and the advanced state of digestion. Hence the categories of crustaceans and fishes included both identifiable and unidentifiable specimens. Food items in the advanced stage of digestion were treated as semi digested matter; while pulpy mass as unidentified matter.

Quantitative Analysis

It was carried out by using volumetric method (Hynes, 1950; Pillay, 1952). The stomach contents were emptied into a petridish. The volume of individual food item was measured and later converted into percentage. From the volume obtained for individual fish, monthly averages and percentages were worked out. The volume index which is the percentage volume of each food item was calculated from the volumes displaced by all the food items over the whole period.

Results

Food and Feeding

About 19.82 per cent of the specimens examined were found with empty stomach while 9.02 per cent of the specimens examined were found with everted stomach showing regurgitation. The details of qualitative and quantitative analyses of stomach contents of *N. japonicus* during different months are presented in Table 15.1 and percentage composition of food items of *N. japonicus* during different months from February 2009 to January 2010 are shown in Figure 15.1.

The gut content analysis indicated that crustaceans and fishes formed the main food of the species. Molluscs, polychaetes, unidentified matter and semi digested matter also occurred in considerable quantity. The average proportions of the gut contents for the whole period of study were: Crustaceans 47.99 per cent, Fishes 31.07 per cent, Molluscs 6.66 per cent, polychaetes 1.13 per cent, unidentified matter 6.85 per cent, and semi digested matter 6.27 per cent.

Crustaceans were recorded in all the months with peak in December (55.69 per cent) with the lowest quantity during April (34.66 per cent). Among the crustaceans the prawns (20.03 per cent) were the most dominant item followed by squilla (14.96 per cent) and crabs (10.91 per cent). The prawns formed the major food during February (27.80 per cent) and April (27.95 per cent). The prawns were represented by *penaeus* sp, prawn mysis, other prawns and prawn remains. Among prawns as a group, the percentage of other prawns (8.64 per cent) was the highest followed by prawn remains (5.73 per cent). Other prawns were recorded in highest quality during April (16.67 per cent) and lowest in

December (0.87 per cent). Prawn remains were highest in May (7.92 per cent) and lowest in January (4.18 per cent). *Penaeus* sp. was recorded in all months of the study except December. Their quantity was highest in February (5.36 per cent) while lowest in May (1.24 per cent). Prawn mysis occurred in the all months of the study except December and May. Their quantity was highest in August (4.91 per cent) and lowest in April (1.38 per cent). Squilla was the second important food item among the crustaceans recorded. Squilla were recorded in highest quantities during December (31.98 per cent), and lowest during April (4.13 per cent). Crabs were the third important item among the crustaceans recorded. Crabs were occurring in highest quantity during December (16.57 per cent) and lowest in March (0.79 per cent). *Neptunus* sp. was recorded highest during August (7.90 per cent) followed by December (5.09 per cent), March (5.00 per cent), September (3.61 per cent), February (3.25 per cent) and in remaining months this was less than 1 per cent in the guts. It was absent during January and October.

**Table 15.1: Percentage Composition of Food Items in the Stomachs of
N. japonicus from February 2009 to January 2010**

Items	Feb	Mar	Apr	May	Aug	Sept	Oct	Nov	Dec	Jan	Annual Avg.
Crustaceans	55.33	49.21	34.66	47.93	51.76	48.92	39.71	48.68	55.96	47.81	47.99
Prawns	27.8	21.56	27.95	16.75	19.78	15.8	25.59	22.57	6.87	15.72	
Crabs	14.56	0.79	2.58	12.44	15.75	13.28	6.03	13.32	16.57	13.82	
Squilla	12.97	19.72	4.13	6.04	16.23	19.84	8.09	12.79	31.98		17.87
Pisces	24.04	25.06	35.61	22.69	38.26	26.14	35.24	33.16	36.2	33.92	31.07
Platycephalids	4.82	2.03	11.01	1.81	5.79	3.28	7.10	4.86	3.81	3.23	
Clupeids	–	1.28	7.33	–	–	3.41	8.01	5.44	–	–	
Nemipterids	0.62	–	–	–	–	–	–	–	–	2.03	
Other fishes	12.42	15.97	11.01	9.84	18.18	8.57	12.24	10.48	19.67	16.75	
Fish remains	6.18	5.77	6.26	9.45	14.37	10.88	7.89	12.38	12.73	11.92	
Molluscs	2.65	9.9	18.8	4.02	0.96	4.82	14.15	5.7	0.68	4.55	6.66
Cephalopods	2.09	9.66	18.45	2.93	0.71	3.61	3.61	4.2	0.09	3.63	
Molluscan shells	0.56	0.24	0.35	1.09	0.25	1.21	1.07	1.5	0.59	0.92	
Polychaetes	0.65	2.63	1.03	-	-	3.00	2.80	-	-	1.28	11.39
Unidentified matter	9.27	9.31	5.93	10.69	6.02	7.27	3.64	6.05	5.06	5.64	6.85
Semi digested matter	8.06	3.90	3.97	14.68	3.00	9.45	4.03	6.38	2.50	6.80	6.27

Fishes were recorded in good amount in all the months. Highest quantity was during August (38.26 per cent) and lowest in May (22.69 per cent). Among fishes, platycephalids occurred highest during April (11.01 per cent) and lowest during May (1.81 per cent). Clupeids occurred during April (7.33 per cent) and March (1.28 per cent). Nemipterids were recorded during January (2.03 per cent) and February (0.62 per cent) in very low percentage. The items grouped as other fishes and fish remains were seen during all the months. Other fishes (13.50 per cent) were recorded highest in December (19.67 per cent) and lower values seen during September (8.57 per cent). Fish remains (7.43 per cent) were recorded highest in August (14.37 per cent) and least in March (5.77 per cent).

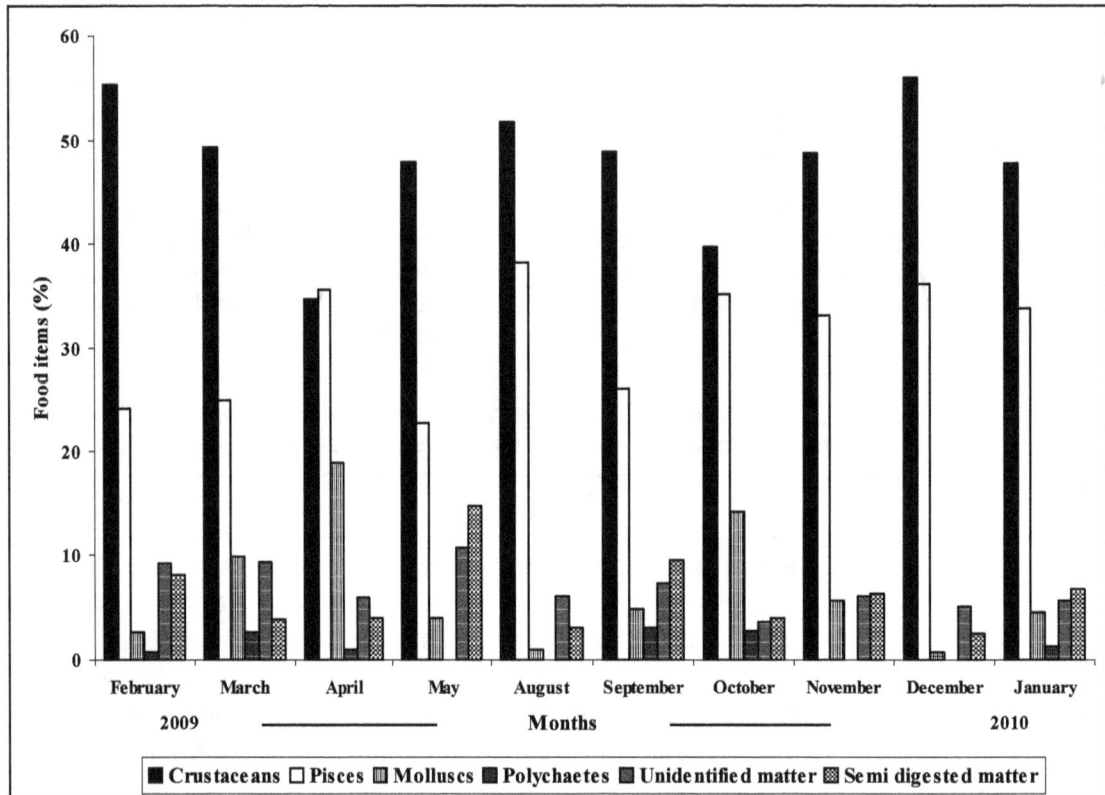

Figure 15.1: Percentage Composition of Food Items of *N. japonicus* during Different Months from February 2009 to January 2010

Unidentified matter was recorded in fairly good quantities in all the months. Its occurrence was maximum in May (10.69 per cent). The lowest was observed December (5.06 per cent).

Cephalopods and molluscan shells were also recorded under molluscs in all months. Cephalopods were highest during April (18.45 per cent) and lowest during March (9.66 per cent), while rest of the period they formed less than 5 per cent. Molluscan shell were recorded in low quantity with highest of 1.50 per cent in November.

Semi digested matter (6.27 per cent) was recorded in all the months with peak abundance in May (14.68 per cent). The lowest quantity was noticed during December (2.50 per cent).

Polycheates (0.92 per cent) were seen during all the months except December and May with highest during March (2.63 per cent) followed by January (1.28 per cent) and lowest in February (0.65 per cent).

Food in Relation to Size

The details of percentage composition of various food items in the stomach contents of *N. japonicus* in various size groups are given in the Table 15.2. It can be seen from the table that crustaceans were recorded in all the size groups except 40-60 mm and 60-80 mm, showing dominance in most size groups. Their percentage was highest (41.44 per cent) in the size group 240-260 and lowest in the size group 260-280. Prawns were not found in the gut of fishes in the size group 40-60 mm and 60-80 mm. size groups from 80-100 mm to 160-180 mm had prawns in their guts in appreciable quantity. The highest and lowest percentage of prawns was observed in the size groups 100-120 mm (21.82 per cent) and 260-280 mm (1.69 per cent) respectively.

Table 15.2: Percentage Composition of Food Items *N. japonicus* in Various Size Groups from February 2009–January 2010

Size Group in mm (TL)	Food Items							
	Fishes	Crustaceans			Molluscs	Polychaetes	Unidentified Matter	Semi Digested Matter
		Prawns	Crabs	Squilla				
40–60	–	–	–	–	–	–	–	4.61
60–80	–	–	–	–	–	–	32	28.2
80–100	–	16.00	–	–	3.76	–	5.12	22.5
100–120	11.24	21.82	9.47	12.54	0.30	0.17	5.58	3.04
120–140	14.60	13.03	6.07	11.17	4.83	0.58	2.88	2.67
140–160	19.37	11.02	6.24	11.30	4.47	0.16	2.55	2.50
160–180	24.79	9.56	6.57	12.58	5.10	0.82	1.36	2.39
180–200	26.13	3.63	4.06	11.19	6.60	0.97	1.63	2.43
200–220	31.46	4.36	4.19	15.24	7.77	0.19	1.40	1.95
220–240	28.47	5.50	1.00	16.68	1.23	2.36	–	3.42
240–260	51.33	3.14	7.09	41.44	15.23	1.69	–	–
260–280	49.45	1.69	2.03	–	–	–	–	–
Total	**21.40333**	**7.479167**	**3.893333**	**11.01167**	**4.1075**	**0.578333**	**4.774545**	**6.1425**

Crabs formed food for all the size groups except 40-60 mm and 60-80 mm. Highest percentage (9.47 per cent) of crabs were recorded in 100-120 mm size group and the lowest (1.00 per cent) in 220-240 mm size group.

Squilla as the food item was observed to form important item for all the size groups except 40-60, 60-80 mm and 260-280. The highest percentage (41.44 per cent) in the size group 240-260 mm and lowest was recorded in the size group 180-200 (11.19 per cent).

Fishes were observed to form important food items in most of the size groups, although they occurred in different proportions. The fishes were recorded from the size group 100-120 mm to 260-280 mm, forming major food for the size groups 180-200 mm to 260-280 mm. The size group 240-260 mm recorded highest percentage (51.33 per cent) of fish, while the lowest (11.24 per cent) was recorded in 100-120 mm.

Molluscs were found to occur in size groups 80-100 mm to 240-260 mm. the highest percentage (15.23 per cent) was observed in the size group 240-260 mm and lowest (0.30 per cent) in the size group 100-120 mm.

Unidentified matter was observed in different proportions in size groups 60-80 mm to 200-220 mm. The highest percentage (32 per cent) was recorded in the size group 60-80 mm and the lowest

(1.36 per cent) in 160-180 mm size group. Semi digested matter was recorded in variable percentages in different size groups from 40-60 mm to 220-240 mm size group. The maximum percentage (28.2 per cent) in the size group 60-80 mm. The minimum percentage was found in the size group 200-220 mm (1.95 per cent).

Discussion

The qualitative analysis of food of *N. japonicus* indicated that this species feeds mainly on crustaceans and fishes. Molluscs, polychaetes, unidentified matter and semi digested matter were the other minor food items observed. The present study therefore shows that *N. japonicus* is carnivorous. The main food items indicate that *N. japonicus* is predaceous carnivore, and a bottom feeder. All the earlier works (Anon, 1960; Kuthalingam, 1965; Rao, 1967; Krishnamoorthi, 1971; Eggleston, 1972; Quasim and Jacob, 1972; Muthiah and Pillai, 1979; Vinci 1982, Rao, 1989, Rao and Rao, 1989, Bakhsh, 1994, Raje 2002) also indicate that threadfin breams are bottom feeders and mainly feed on crustaceans.

The percentage occurrence of crustaceans as food item was recorded in different proportions in all the months. Highest quantity of crustaceans occurred during December (55.69 per cent) and the lowest quantity was found during April (34.66 per cent), with prawns consumed in lowest quantity during December (6.87 per cent) and highest during February (27.80 per cent). Among prawns *Penaeus* sp. was recorded in all months of the study except December. Their quantity was highest in February (5.36 per cent) while lowest in May (1.24 per cent). Prawn mysis occurred in the all months of the study except December and May. Other prawns and prawn appendages occurred during all the months. Squilla dominated during December and March. Crabs occurred in high quantity during December followed by August but they occurred during all the months of sampling. Among crabs *Neptunus* sp. was observed during all the months except October and January. Other crabs were seen in the gut during all the months, but their proportion of occurrence was different. Crab appendages dominated in March and the lowest was in April forming variable proportion in other months.

Studies on the feeding behaviour of *N. japonicus* off Mangalore waters (Kuthalingam, 1965) indicated that the fish in 10-20 m. depth showed preference for crustaceans mainly on *Metapenaeus dobsonii* and *Parapenaeopsis stylifera* of which *M. dobsonii* dominated constituting 35 per cent. Other crustacean items were represented by *Penaeus* species, appendages of copepods and crabs. Rao (1967) found that crustaceans form a major part of gut content in *N. japonicus*. According to Krishnamoorthi (1971) *N. japonicus* off Visakhapatnam coast was actively predacious and feeding substantially on crustaceans; squilla, crabs, caridean prawns were the major components. According to Vinci (1982) *N. japonicus* is a carnivorous fish voraciously feeding on the crustaceans. The major items were crustaceans with *Acetes indicus* as the main individual item followed by squilla and crabs.

Fishes formed the second important food item of *N. japonicus* with highest percentages during August (38.26 per cent) and lowest in May (22.69 per cent). This indicates that *N. japonicus* feeds on fishes during all the months. According to Kuthalingam (1965), fish formed the bulk of the diet of *N. japonicus* caught from 30-40 mt depth, constituting 40 per cent, that included *Clupea* sp, *Engrauli* sp, *Thrissocles* sp, *Dussumieria* sp, *Leiognathus* sp and *Cynoglassus* sp. He also found that fishes of *N. japonicus* caught from 40-50 mt. depth fed mainly on teleostean constituting 65 per cent. According to Bakhsh (1994), small amounts of *Saurida sp* and *N. japonicus* were found in the gut of nemipterids along the Jizan Region of the Red Sea. Raje (2002) reported that 20.32 per cent volume of teleosts are seen in the stomachs of *N japonicus* along Veraval coast represented by *Scianids, N. mesoprion, N. japonicus, Lactarius lactarius, Apogon* sp., *Trichirus* species, *Cynoglossus* sp, *Saurida* sp, *Bregmaceros macclellandi, Leiognathus* sp and fish larva in decreasing order of abundance. Occurrence of juveniles of *N. japonicus* in January and February reveals the cannibalistic tendency of this fish. Occurrence of juvenile in the gut during these months can be directly related to the availability of juveniles in the feeding ground and can be an indication of spawning period. Kuthalingam (1965) reported cannibalism in *N. japonicus* from 40-50 m. depth off Mangalore. Bakhsh (1994) also reported the cannibalistic nature of *N. japonicus* along the Jizan Region of the Red Sea.

Unidentified matter was found to be next in abundance among the food items. Due to its nature as grayish pulpy mass, it was not possible to separate them. Its occurrence was highest in May (10.69 per cent) and lowest in December (5.06 per cent).

The occurrence of cephalopods in all the months except April and molluscan shell in all the months in small quantities clearly indicate that molluscs are not of primary importance as a food item in *N. japonicus*. Rao (1967) while studying the gut contents of *N. japonicus*, observed gastropods forming 6.3 per cent of total. According to Krishnamoorthi (1971) molluscs were contributed by Squids mainly.

Semi digested matter was in variable proportion in all the months. It mainly consisted of fish and prawn remains in an advanced stage of digestion. Since, it was difficult to put it under the category fish or prawn due to its advanced state of digestion, the actual quantity of fishes and prawns would have been more than that recorded. Polychaetes were found to be least in abundance among the food items. They were seen during all the months except May, August, November and December with highest during September (3.00 per cent) and lowest in February (0.65 per cent).

The analysis of food with respect to size revealed variation in quantity of food consumed and also food preferences based on size groups. It may be seen that only semi digested matter in the size range of 40-60 mm TL. The food items of the size group 60-80 mm was unidentified matter and semi digested matter were found. A close observation of the Table 15.2 reveals that in size groups 40-60, 60-80 mm and from size group 100-120 mm to 160-180 mm size group, crustaceans dominated over the other items of food. These size groups also include molluscs, polychaetes, unidentified matter and semi digested matter in considerable amount. In fishes of size group 80-100 mm semi digested matter was found dominating the food items followed by crustaceans, unidentified matter and molluscs. From the Table 2, it is also seen that from the size group 100-120 mm onwards, the proportion of fishes in the diet also increased reaching to the maximum (51.33 per cent) in the size group 240-260 mm. The above observation indicated that there is a change in the composition of food with increase in size of the fish. Fishes of less than 180 mm TL seem to prefer crustaceans (prawns, crabs, squilla) while the adults (180 mm onwards) upto 260-280 mm size group preferred to feed on fishes. The occurrence of molluscs from 80-100 mm size group to 240-260 mm size group in increasing proportion indicates preference for molluscs as food item in adult fishes. However, proportion of unidentified and semi digested matter observed decreasing with increase of size. Increase in the feeding intensity was observed during February- March and August. This can be associated with the higher requirement of energy for the development of gonads and the spawning.

References

Anon., 1960. Government of India, Central Marine Fisheries Research Station, Marine Fisheries, Mandapam camp, South India–Annual Report of the Chief Research Officer. *Indian J. Fish.*, 7(1 and 2): 496–552.

Bakhsh, A.A., 1994. The biology of thread bream, *Nemipterus japonicus* (Bloch) from the Jizan region of the Red Sea. Special Issue: Symp. on Red sea Mar. Environ., Jeddah. *J. KAU, Mar. Sci.*, 7: 179–189.

Eggleston, D., 1972. Patterns of biology in Nemipteridae. *J. Mar. Biol. Assoc., India*, 14 (1): 357–364.

Hynes, H.B.N., 1950. The food of the freshwater sticklebacks (*Gastrosteus aculeatus* and *Pygosteus pungitius*) with a review of methods used in studies of the food of fishes. *J. Anim. Ecol.*, 19: 36–58.

Krishnamoorti, B., 1971. Biology of threadfin bream, *Nemipterus japonicus* (Bloch). *Indian J. Fish.*, 18(1 and 2): 1–21.

Kuthalingam, M.D.K., 1965. Notes on some aspects of the fishery and biology of *Nemipterus japonicus* (Bloch) with special references to the feeding behaviour. *Indian J. Fish.*, 12(2A): 500–506.

Muthiah, C. and Pillai, S.K., 1979. A new distributional record of *N. delagoae* (Smith) from Bombay waters with notes on its biology. *J. Mar. Biol. Assoc., India*, 21(1–2): 174–178.

Pillay, T.V.R., 1952. A critique of the methods of study of food of fishes. *J. Zool. Soc., India*, 4(2): 185–195.

Qasim, S.Z. and Jacob, P.G., 1972. The estimation of organic carbon in the stomach contents of some marine fishes. *Indian J. Fish.*, 19(1–2): 29–34.

Raje, S.G., 2002. Observations on the biology of *Nemipterus japonicus* (Bloch) from Veraval. *Indian J. Fish.*, 49(4): 433–440.

Rao, K.V.N., 1967. The flat fishes. *Souvenir, 20th Anniversary, CMFRI*, pp. 62–66.

Rao, T.A., 1989. Fishery of threadfin breams of Waltair with notes on some aspects of biology of *Nemipterus mesoprion* (Bleeker). *J. Mar. Biol. Assoc., India*, 31(1–2): 103–109.

Rao, D.M. and Rao, K.S., 1989. Studies on the age determination and growth of *Nemipterus japonicus* (Bloch) off Visakhapatnam. *Indian J. Fish.*, 33(4): 426–439.

Vinci, G. K., 1982. Threadfin bream (*Nemipterus*) resources along the Kerala coast with notes on biology of *Nemipterus japonicus* (Bloch). *Indian J. Fish.*, 29: 37–49.

Chapter 16

Studies of Monthly Variations in DO, BOD and COD Parameters of Gangapur Dam Water at Nashik

☆ *R.S. Bhadane*

ABSTRACT

In the present investigation the study was conducted to determine the monthly mean variation in dissolved oxygen, biological oxygen demand and chemical oxygen demand parameters of Gangapur Dam water at three sampling Stations namely Station 'A', Station 'B' and Station 'C' during January 2004 to December 2005. The maximum dissolved oxygen values are found in winter and minimum recorded in summer months. Similarly the biological oxygen demand and chemical oxygen demand values were observed more during monsoon than winter and summer.

Keywords: *Gangapur dam, Dissolved oxygen, Biological oxygen demand, Chemical oxygen demand.*

Introduction

Water quality of any water body is measured in terms of parameters like dissolved oxygen, biological oxygen demand and chemical oxygen demand. Dissolved oxygen is an important parameter which can be used as an index of water quality, primary production and pollution. Biological oxygen demand is a measure of water pollution based on the organic matter it contains. Similarly chemical oxygen demand test is useful in pinpointing the toxic condition and presence of biological resistant substances in water. Higher values of chemical oxygen demand indicate higher microbial activity, similar findings were reported by Khulbe and Durgapal (1993).

Materials and Methods

Sampling Sites

Three sampling sites namely Station 'A' (Centre of Dam), Station 'B' (Near Filtration Plant) and Station 'C' (Opposite station 'A') were selected and samples were collected at an interval of one month for 2 years *i.e.* January 2004 – December 2004 and January 2005 – December 2005. Station 'C' receives polluted water from various sources.

Sample Analysis

Water samples were analysed with the help of Winkler's Iodometric method in the laboratory for dissolved oxygen and biological oxygen demand. Similarly chemical oxygen demand was measured by using Dichromate digestion method, samples were collected in dark bottles, incubated at $20°$ C for five days.

Results and Discussion

Dissolved Oxygen

In the present investigation the dissolved oxygen values ranged from $3.0 – 7.8$ mg/lit. Maximum dissolved oxygen was noted in winter and minimum was recorded and in summer months. Similar findings were observed by Hancock (1973), Mishra and Yadav (1978) Adebisi (1981) and Mitra (1982).

Table 16.1: Monthly Mean Value of Dissolved Oxygen (DO) (mg/lit.) of Water Samples at Various Sampling Stations during the Monitoring Period (2004 - 2005)

Year	2004			2005		
Month	A	B	C	A	B	C
January	6.8	5.9	4.7	6.2	6	5.4
February	6.1	4.9	4.9	6.8	5.9	5.2
March	6.9	5.8	5.4	5.3	5	4.3
April	4.4	4.2	3.6	4.7	4.4	3.7
May	4.2	4.1	3.6	3.8	3.7	3
June	4.1	4	3.8	4.2	4	3.8
July	5.5	4.6	4.2	5.6	5	4.8
August	6.2	5.6	4.8	6	5.8	4.4
September	6.8	6.7	4.7	6.4	6.2	4.8
October	7.5	7	6.8	7.6	7.1	6.2
November	7.2	7.1	6.6	7.7	7.4	6.1
December	7.7	7.4	7.2	7.8	7	6.7

Biological Oxygen Demand

It is an indicator parameter which indicates the presence of biodegradable matter in the water and express degree of contamination. In the present investigation the range of biological oxygen demand was 3.5 - 24.7 mg/lit. The values were found more during monsoon and low in winter may be due to lesser quality of solids and microbial population. Similar trends were also observed by Singhai *et al.* (1990) and Patki (2002).

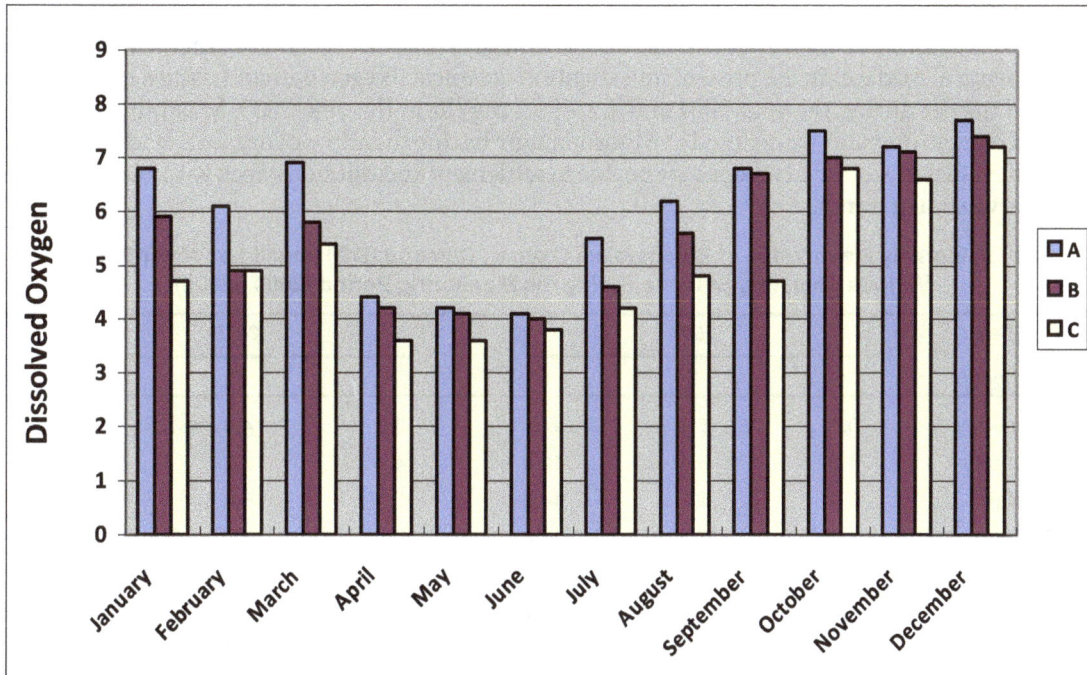

Figure 16.1: Average Monthly Mean Value of Dissolved Oxygen at Various Sampling Stations during the Period 2004

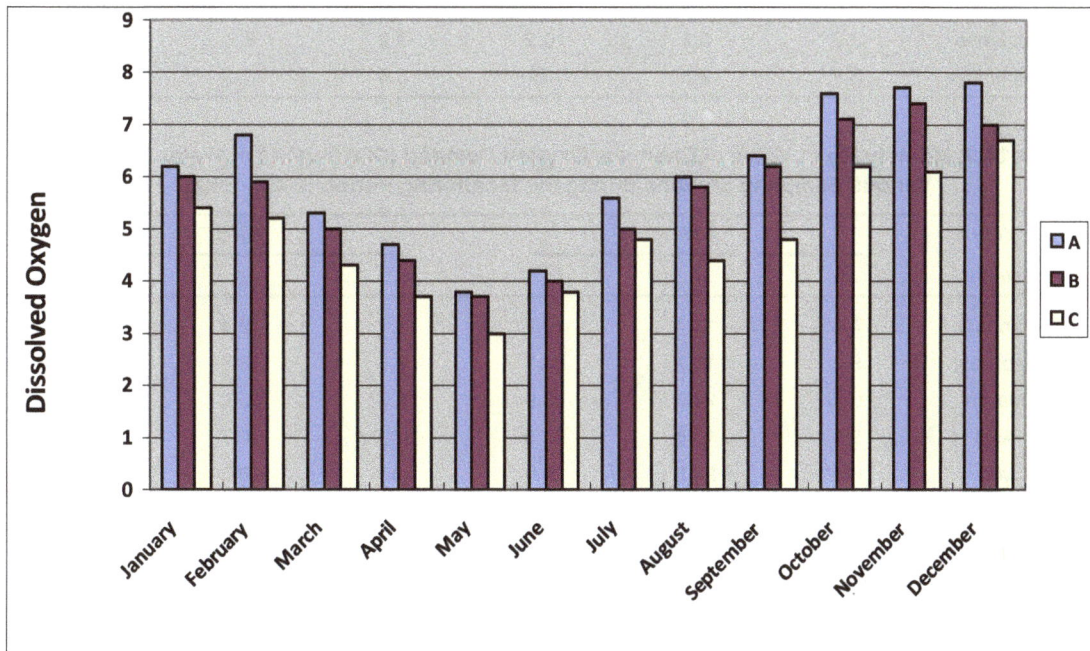

Figure 16.2: Average Monthly Mean Value of Dissolved Oxygen at Various Sampling Stations during the Period 2005

Advances in Aquatic Ecology Volume 7

Chemical Oxygen Demand

This test determines the oxygen required for chemical oxidation of organic matter with the help of strong chemical oxidant. In the present investigation chemical oxygen demand values ranged from 4.4 – 54.8 mg/lit during the year 2004 and 4.7 – 56.7 mg/lit in the year 2005. Maximum values of chemical oxygen demand were noted in Monsoon may be due to inflow of organic dead matter and minimum values are found in winter may be due to settlement and dilution effect. Kudesia *et.al.* (1986) also observe similar trends.

Table 16.2: Monthly Mean Value of Biochemical Oxygen Demand (DO) (mg/lit.) of Water Samples at Various Sampling Stations during the Monitoring Period (2004 - 2005)

Year	2004			2005		
Month	A	B	C	A	B	C
January	3.5	5.0	8.8	3.5	4.6	9.0
February	4.9	5.4	12.4	4.4	5.2	12.8
March	5.8	6.2	14.8	5.6	6.0	14.0
April	4.1	6.0	12.8	4.9	5.8	12.9
May	4.2	6.6	12.5	4.6	6.2	12.2
June	7.4	7.9	22.2	6.2	7.8	20.3
July	10.7	12.4	24.7	10.4	11.0	22.4
August	10.4	12.2	22.8	10.0	10.3	20.1
September	8.2	10.6	16.6	7.4	9.8	14.7
October	3.8	9.4	12.4	4.0	8.6	10.8
November	3.6	8.4	10.2	3.9	8.2	9.7
December	3.8	7.8	9.3	3.6	7.0	9.6

Table 16.3: Monthly Mean Value of Chemical Oxygen Demand (DO) (mg/lit.) of Water Samples at Various Sampling Stations during the Monitoring Period (2004 - 2005)

Year	2004			2005		
Month	A	B	C	A	B	C
January	4.4	10.4	20.2	4.7	12.4	16.9
February	9.2	16.7	28.7	6.8	18.9	27.8
March	10.0	20.0	30.9	9.7	20.2	29.2
April	9.7	18.8	35.8	10.9	16.7	28.4
May	11.7	24.8	39.9	11.4	18.9	37.2
June	14.9	28.7	52.4	13.7	26.7	42.8
July	24.7	32.4	54.8	27.8	30.8	56.7
August	20.2	30.1	52.2	19.9	30.2	50.4
September	10.4	29.7	49.4	18	27.8	43.4
October	7.2	20.2	30.2	9.8	20.8	32.2
November	6.8	14.8	29.8	6.7	14.3	24.4
December	5.7	10.7	20.9	5.2	12.0	18.2

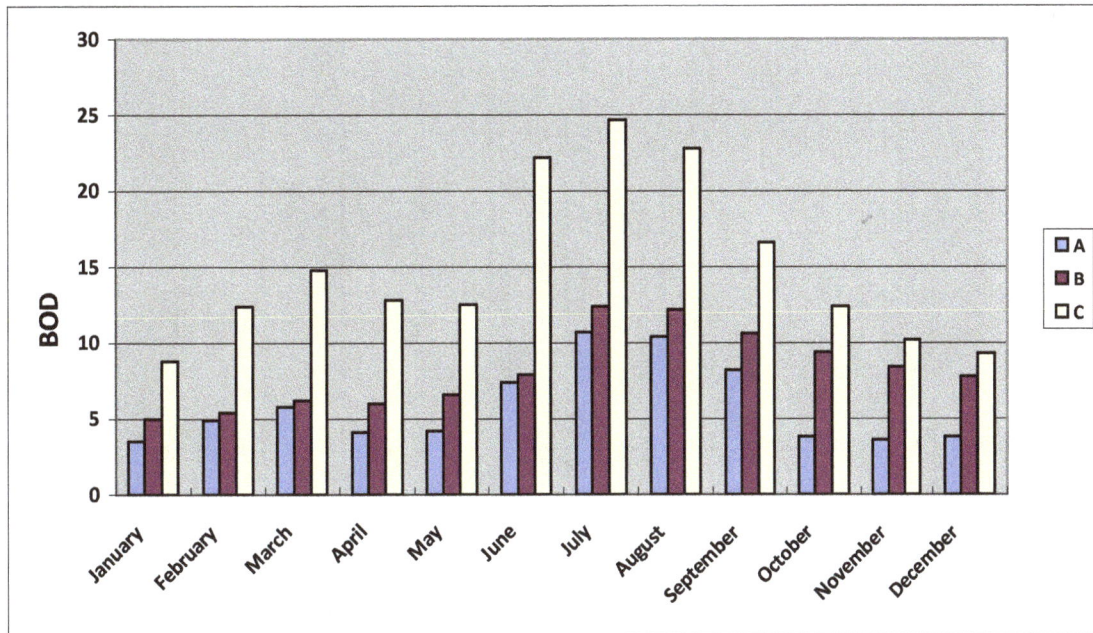

Figure 16.3: Average Monthly Mean Value of BOD (mg/lit) of Water Samples at Various Sampling Stations during the Period 2004

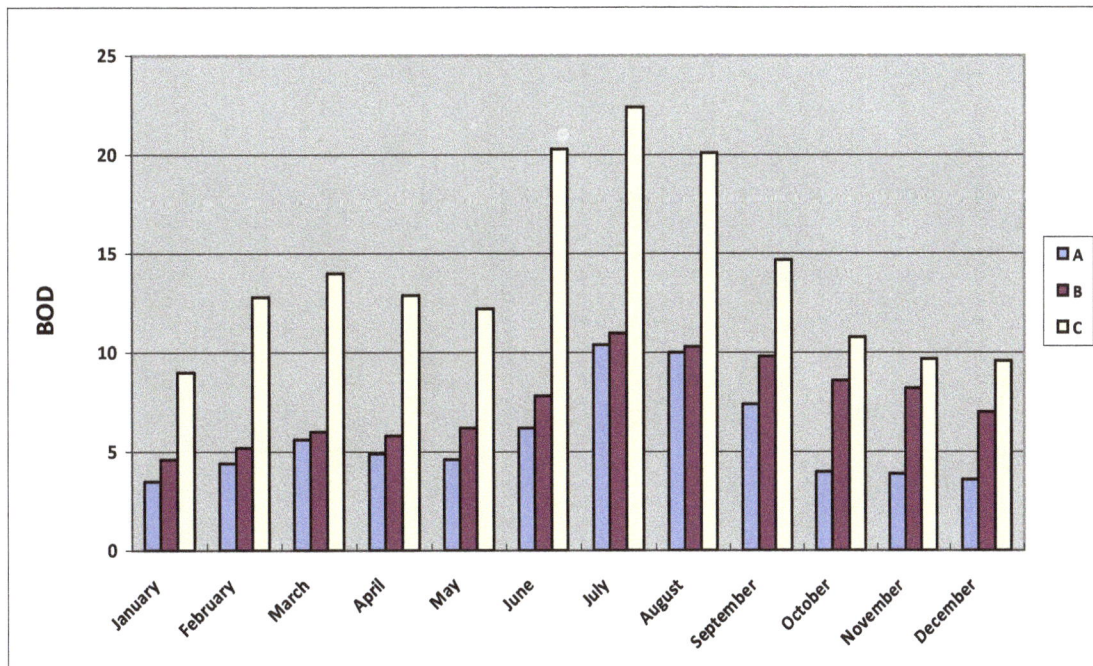

Figure 16.4: Average Monthly Mean Value of BOD (mg/lit) of Water Samples at Various Sampling Stations during the Period 2005

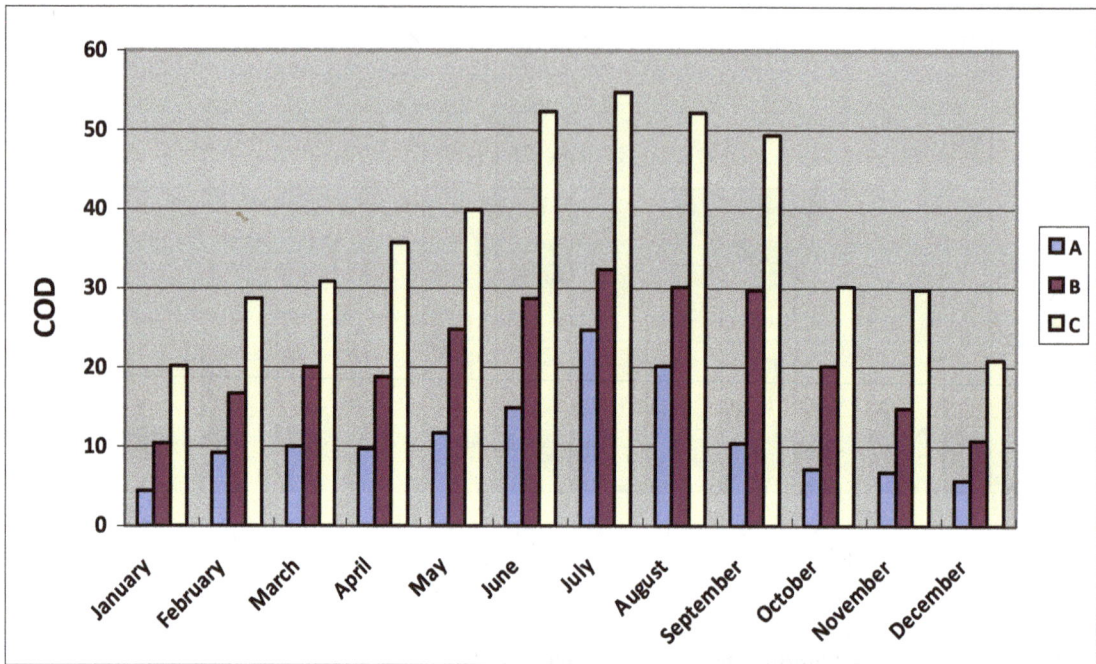

Figure 16.5: Average Monthly Mean Value of Chemical Oxygen Demand (mg/lit) of Water Samples at Various Sampling Stations during the Period 2004

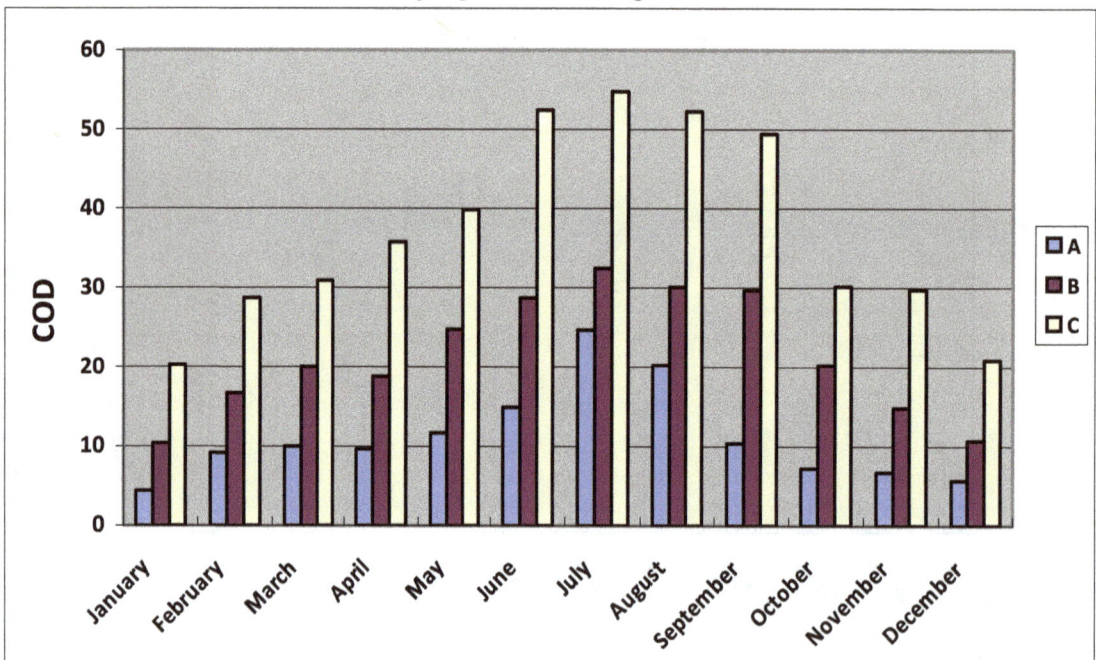

Figure 16.6: Average Monthly Mean Value of Chemical Oxygen Demand (mg/lit) of Water Samples at Various Sampling Stations during the Period 2005

Conclusion

Thus present study concludes that the Gangapur dam water was not polluted, all results are within permissible limits when compared with WHO. The water of the dam is good for drinking after normal processing. Similarly the water can also be used for the purpose of irrigation as well as for aquaculture.

Acknowledgements

The author is thankful to the Coordinator M.G. Vidyamandir and Principal L.V.H. Mahavidyalaya for providing necessary research laboratory facilities.

References

Adebisi, A.A., 1981. The physico-chemical hydrology of a tropical seasonal river upper Ogun river Nigeria. *Hydrobiologia*, 79(2): 157–165.

Hancock, F.D., 1973. Algal ecology of a stream polluted through gold mining in wint water strand. *Hydrobiol.*, 43: 189–229.

Khulbe, R.D. and Durgapal, A., 1993. Evaluation of drinking water quality at Bhimtal, Nainital, Uttar Pradesh. *Poll. Res.*, 21(2): 109–111.

Kudesia, V.P., and Verma S.P., 1986. Physico-chemical studies on industrial pollution of Kalinadi due to combined effluents of canesugar, chemical industry, distillery and rubber industries in Merrut region. *Indian J. Env. Agric.*, 1(1): 1–11.

Mishra, G.P. and Yadav, A.K., 1978. A comparative study of physico-chemical characteristics of river and lake water in Central India. *Hydrobiologia*, 59(30): 275–278.

Mitra, A.K., 1982. Chemical characteristics of surface water at a selected gauging station in the river Godavari, Krishna and Tungbhadra. *Ind. J. Environ. Health*, 24(2): 165–179.

Patki, Saroj, 2002. Hydrobiological studies of Banshelki dam at Udgir. *Ph.D. Thesis*, Dr. Babasaheb Ambedkar Marathwada University.

Singhai, S., Ramani, G.M.A. and Gupta, U.S., 1990. Seasonal variations and relationship of different physico-chemical characteristics in the newly made Tawa Reservoir. *Limnologia (Berlin)*, 21(1): 293–301.

Chapter 17

Ecological and Protein Estimation Studies in Genus *Cassia* L. from Kolhapur District

☆ *Sagar Anant Deshmukh*

ABSTRACT
Five *Cassia* L. species form Kolhapur district were analyzed to reveal their ecological aspects and protein estimation of fresh seeds by Biuret method.

Keywords: Biuret, Cassia, Kolhapur, Maharashtra.

Introduction

The genus *Cassia* L. ranks among the 25 largest genera of the dicotyledonous plants (Irwin and Turner, 1960). According to Irwin and Barneby 1981, 1982; and Willis, 1988, it includes about 650 species of trees, shrubs and herbs and has a pantropical distribution, but very few in Asia. From Kolhapur district of Maharashtra state 19 species of *Cassia* L. including one variety have been recorded (Yadav and Sardesai, 2002). Ecological studies has great significance in understanding plant's status while proteins occupy a central position in the architecture and functioning of living matter (Jain *et al.*, 2005) hence their estimation to determine the quantity is important.

Materials and Methods

Cassia L. species were collected form Kolhapur district and identified with the help of "Flora of Kolhapur" (Yadav and Sardesai, 2002). These species were analyzed to reveal their ecological aspects (Gupta, V. P. and Yadav, A. S.) and fresh seed protein percentage by Biuret method as per standard protocols.

Results and Discussion

Ecological Studies

Cassia obtusifolia L.

Morphology

Annual herb, common in wastelands, frequent along the road sides away from the habitations. Occurs mixed with the populations of *Cassia tora* L. and *Cassia uniflora* Mill. Easily get propagated through seeds. Generally plant height, number of leaves/plant, number of pods/plant and number of seeds/plant varies from 57.3±17.10, 63.4±12.42, 53.4±12.23, 1110±260.12 respectively.

Phenology

At the beginning of the monsoon seeds get germinated. Vegetative and reproductive growth takes place from June to November.

Population Density

Average population density of *Cassia obtusifolia* L. at various places in Kolhapur district especially around road sides varies from 35 to 46 plants m^{-2}.

Cassia occidentalis L.

Morphology

Perennial herb or undershrub. Frequently occurs singly or in groups of 1-5 along road side and in wastelands. Easily get propagated through seeds. Generally plant height, number of leaves/plant, number of pods/plant and number of seeds/plant varies from 108.9±36.12, 103.2±22.48, 86.2±34.97, 1511±792.56 respectively.

Phenology

At the beginning of the monsoon seeds get germinated. Vegetative growth takes place from June to November. Flowering and fruiting takes place throughout the year, more profuse during September-October.

Population Density

Average population density of *Cassia occidentalis* L. at various places in Kolhapur district especially around road sides varies from 01 to 05 plants m^{-2}.

Cassia sophera L.

Morphology

Perennial herb or undershrub. Frequently occurs singly or in groups of 1-6 along road side and in wastelands. Easily get propagated through seeds. Generally plant height, number of leaves/plant, number of pods/plant, number of seeds/plant varies from 123.1±43.18, 238.9±143.23, 153.5±65.24, 2017.87±1235.95 respectively.

Phenology

At the beginning of the monsoon seeds get germinated. Vegetative growth takes place from June to November. Flowering and fruiting takes place from September–December.

Population Density

Average population density of *Cassia sophera* L. at various places in Kolhapur district especially around road sides varies from 01 to 06 plants m^{-2}.

Cassia tora L.

Morphology

Annual herb, common in wastelands, frequent along the road sides away from the habitations. Occurs mixed with the populations of *Cassia obtusifolia* L. and *Cassia uniflora* Mill. Easily get propagated through seeds. Generally plant height, number of leaves/plant, number of pods/plant, number of seeds/plant varies from 61.10±18.79, 71.80±19.73, 56.2±15.71, 942.77±185 respectively.

Phenology

At the beginning of the monsoon seeds get germinated. Vegetative and reproductive growth takes place from June to December.

Population Density

Average population density of *Cassia tora* L. at various places in Kolhapur district especially around road sides varies from 15 to 45 plants m^{-2}.

Cassia uniflora Mill.

Morphology

Annual herb, common in wastelands, frequent along the road sides away from the habitations. Occurs mixed with the populations of *Cassia tora* L. and *Cassia obtusifolia* L. Easily get propagated through seeds. Generally plant height, number of leaves/plant, number of pods/plant, number of seeds/plant varies from 94.9±31.38, 77.6±17.69, 136.37±33.36, 991.22±210 respectively.

Phenology

At the beginning of the monsoon seeds get germinated. Vegetative and reproductive growth takes place from June to January.

Population Density

Average population density of *Cassia uniflora* Mill. at various places in Kolhapur district especially around road sides varies from 35 to 45 plants m^{-2}.

Protein Estimation by Biuret Method

Protein percentage was determined in fresh, mature seeds of *Cassia* L. showing following results.

Table 17.1: Protein Percentage in Fresh Mature Seeds of *Cassis* L.

Sl.No.	Name of the Plant	Protein Percentage (g)
1.	*Cassia obtusifolia* L.	10.72
2.	*Cassia occidentalis* L.	7.59
3.	*Cassia sophera* L.	8.12
4.	*Cassia tora* L.	12.46
5.	*Cassia uniflora* Mill.	8.75

It was observed that protein content was high in seeds of *Cassia tora* L. than the other studied species.

Acknowledgements

Author is thankful to UGC-WRO, Pune for financial assistance and Head of the Botany Department and the Principal, the New College, Kolhapur.

References

Gupta, V.P. and Yadav, A.S., 2007. Ecology of *Cassia tora* L. in the Sarika Tiger Reserve Forest in Rajasthan. *J. Phytol. Res.*, 20(2): 265–269.

Irwin, H.S. and Turner, B.L., 1960. Chromosomal relationship and taxonomic consideration in the genus *Cassia*. *American Journal of Botany*, 47: 309–318.

Irwin, H.S. and Barneby, R.C., 1981. Cassinae. In: *Advances in Legume Systematics*, Part I, (Ed.) R.M. Polhill and P.H. Raven. Royal Botanic Gardens, Kew, p. 97–106.

Irwin, H.S. and Barneby, R. C., 1982. The American Cassiinae: A synoptic revision of Leguminosae tribe Cassieae subtribe Cassiinae in the New World. Mem. *New York Bot. Gard.*, 35(1): 1–918.

Jain, J.L., Jain, Sanjay and Jain, Nitin, 2005. *Fundamentals of Biochemistry*. S. Chand and Company Ltd., New Delhi.

Willis, J.C., 1988. *A Dictionary of the Flowering Plants and Ferns*, 8th Edn. Revised by H.K. Airy Shaw, Cambridge University Press, Cambridge.

Yadav, S.R. and Sardesai, M.M., 2002. *Flora of Kolhapur District*. Shivaji University, Kolhapur.

Chapter 18

A Checklist of Avifauna in the Mangrove Areas of Aghanashini Estuarine Complex, Uttara Kannada, Karnataka, West Coast of India

☆ *T. Roshmon and V.N. Nayak*

ABSTRACT

The Aghnashini River originates in the Western Ghats of Sirsi taluk in Uttara Kannada district and winds its way through a rugged terrain for about 121 kms before its confluence with the Arabian Sea. Towards the coast, in between the towns of Kumta and Gokarna, is the estuarine region which has a spread of about 48 sq. km. The passage of the river through the forest clad Western Ghats and through an agricultural landscape has made it rich in biodiversity. Plenty of organic matter from the forests gets deposited in this estuary making it rich in nutrients. Mangroves that flourish in the estuary also provide ideal habitats for fish and avifauna and confer on them relative safe habitat from predators and fishing by humans. The avifauna of these mangroves has received little attention from researchers. The lack of baseline information is therefore a major limiting factor in the preparation of management plans. Here we present comprehensive checklist on the species composition of the avifauna in the mangrove areas of Aghanashini estuarine complex based on the studies conducted for two years from January 2007 to December 2009. A total of 119 species belonging to 45 families were recorded, with increased number of species and individuals around the peak migration period *i.e.* winter season.

Keywords: Avifauna, Mangroves, Aghanashini estuary, Migration, Winter.

Introduction

Mangrove ecosystems are highly complex water and land interactive systems and are supposed to be the most fertile and productive ecosystems in the world. This wetland constitutes a treasury of biodiversity. The social demand and dependence on these wetlands provide an unaccountable economic value to such habitats. Due to inadequate attention and ignorance of common man wetlands are referred as waste lands in the past leading to its disappearance in the process of urbanization and development. Therefore there is an urgent need to develop the conservation strategies and management plan by inventorying, monitoring and documenting the diversity and density of biodiversity. The loss of mangrove community has affected the lives of many birds and other related organisms in the marine food chain. The slide began with the removal of crowns from the mangroves for fodder and other household purposes. The thick foliage and crowns are crucial for mangrove-dependent species such as sandpipers, storks, herons, egrets, bitterns, godwits, curlews, plovers, oystercatchers, terns, gulls and snipes. They not only take refuge and roost amid the foliage but also build nests. They enrich the mangrove habitats with their nitrogen-rich droppings. Waterbirds, being generally at or near the top of most wetland food chains are highly susceptible to habitat disturbances and are therefore good indicators of the general condition of wetland habitats (Kushlan, 1992). They also play a crucial role in mass and energy fluxes between terrestrial and aquatic food chains (Moreira, 1997).

Mangroves play a vital role in the environmental and ecological processes of Aghanashini estuarine ecosystem. The natural ecological processes of this fragile habitat are degraded day by day due to urbanization and development. In order to draw global attention towards this fragile and unique ecosystem, many research works have been made mainly linked on fishery aspects and mangrove flora. As far as avifauna of this estuary is concerned, there is hardly any document is available so far. Therefore, it is pertinent to study the ecology and avian community structure from a totalitarian scientific perspective so as to plan a proper management structure for the conservation of wetland ecosystems. The present paper aims at providing baseline data on the avifaunal community structure of one of the most fertile estuarine ecosystems in the state.

Materials and Methods

Study Area Description

River Aghanishini is originating in the Western Ghats running westwards forming major estuarine complex with sizable mangrove spread, sustaining estuarine fishery playing important role in the economy of thousands of dependents. The Unique feature of the Aghanashini estuary is that, it opens into the sea after taking a U turn. This turn causes the flowing water to halt for a while in this region before joining the sea. The second interesting fact is that the area around the estuary is low lying causing a vast backwater spread. Therefore, the total backwater area of Aghanashini is much more than the total of all the remaining estuaries of rivers and rivulets of Uttara Kannada. Also, the Aghanashini has barely any anthropogenic damage in the flow pattern as against Kali, Sharavati and Gangavali (Bedti) where dams are constructed. The dominant mangrove vegetation comprised of dense stands of mixed mangroves of *Rhizophora* sps, *Sonneratia* sps, *Avicennia* sps, *Kandelia* sps, *Excoecarea* sps and *Acanthus* sps.

The climate of Uttara Kannada is strongly influenced by the Southwest monsoon occurring from June to September. However, since the present study is focused on avifaunal ecology, the seasons have been categorized as summer extending from March to May, Monsoon from June to September, Post-Monsoon from October to November and winter from December to February.

Monthly observations were carried out from January 2007 to December 2008 in the Mangrove areas of Aghanashini estuary. The method of total count was employed for the estimation of population of birds in the mangrove and adjacent mud flat areas of the study sites, in which direct counting was made with the help of binocular and spotting scope. Fieldwork was carried out mainly by trekking along the dykes of mangroves and also traversing the wetland by boat which is not accessible by walk. The avifauna sighted were identified using relevant field guides (Ali.S., 1996. and Grimmet *et.al.* 1998). Checklist of birds was sequenced according to orders, families, season of occurrence and possible nesting in mangroves was noted.

Results and Discussion

From the investigation, 119 species of birds belonging to 45 families that use Aghanashini Mangrove areas for feeding, nesting, roosting or other activities (Table 18.1) were listed. The maximum species were represented by the families Scolopacidae (14), Ardeidae (9), Laridae (8), Accipitridae (8), Charadriidae (8) and Anatidae (7). From the present investigation we found that the mangrove avifauna of Aghanashini estuary can be divided into six groups based on similarities in methods of procuring food. These groups are the large wading birds, probing shore birds, floating and diving water birds, aerially-searching birds, birds of prey and arboreal birds. However, arboreal bird group feed and or nest in the mangrove canopy.

Large Wading Birds

Herons, egrets, ibises, open bills and spoonbills are the most conspicuous groups of birds that are found in mangroves (Table 18.1). Eleven species were recorded, eight species are year round residents, seven species breed in Aghanashini estuary and the three are winter visitors. The dominant species were intermediate egret, Indian pond heron followed by little egret and the least one observed was black-headed ibis. Mangrove swamps function as the feeding ground for wading birds, since two thirds of these species feed almost exclusively on fishes (Odum *et al.*, 1982). Ibis feed predominantly on crabs of the genus Uca from mangroves (Kushlan and Kushlan, 1975) and spoonbills prefer mollusks and invertebrates of the sediments (Allen, 1942) as has been recorded earlier. Mangroves serve as a breeding ground for large breeding colonies of herons and egrets. Feeding activity of herons and egrets are observed throughout the year. Successful breeding of all these mangrove nesters is undoubtedly correlated with the abundant supply of fishes associated with mangroves which is also supported by the reports of Odum *et al.* (1982) in Florida Mangroves. Wading birds play an important role in nutrient cycling in the coastal mangrove zone.

Probing Shore Birds

Birds of this group are commonly found associated with intertidal and shallow water habitats. Plovers, curlews, whimbrel, sand pipers, godwits, stint, Ruff, sanderling, shanks, waterhen, jacanas, lapwings, stilts, moorhens, pratincole and curlews come under this group. Plovers and sand pipers are opportunistic feeders taking the most abundant, proper sized invertebrates present in whatever habitat the birds happen to occupy (Wolff, 1969; Schneider, 1978). Of the 28 species of probing shore birds, (Table 18.1) nine species are year-round residents, three species are found to breed in mangroves and the others are migrants or winter visitors. Among those density wise the dominant species were little ringed plovers followed by red shank and Kentish plover and the least was yellow-wattled lapwing. Baker and Baker (1973) indicated that winter was the most crucial time for the birds in terms of survival but coincidentally in tropical regions winter is the time when most of the shore birds use mangrove areas (Odum *et al.*, 1982). The invertebrate fauna (mollusks, crustaceans and worms) which occur on sediments under intertidal mangroves forms the principal diets of these species.

Table 18.1: A Complete Checklist of Avifauna in the Mangrove Areas of Aghanashini Estuarine Complex Observed during 2007 and 2008

Sl.No.	Family/Species	Scientific Name	Migratory Status	Abundance	Nesting
I	**WADING BIRDS**				
	Ciconiiformes/Ardeidae				
1.	Black-crowned Night-Heron	*Nycticorax nycticorax*	RLM	C	Y
2.	Cattle egret	*Bubulcus ibis*	RLM	C	Y
3.	Indian Pond heron	*Ardeola grayii*	RLM	C	Y
4.	Intermediate egret	*Mesophoyx intermedia*	RLM	C	Y
5.	Great egret	*Bubulcus ibis*	RLM	C	Y
6.	Grey Heron	*Ardea cinerea*	RLM	C	Y
7.	Western reef egret	*Egretta gularis*	M	C	-
8.	Little egret	*Egretta garzetta*	RLM	C	Y
9.	Asian Openbill	*Anastomus oscitans*	RLM	C	-
	Ciconiiformes/Threskiornithidae				
10.	Black-headed Ibis	*Theskiornis melanocephalus*	M	FC	-
11.	Eurasian Spoonbill	*Platalea leucorodia*	M	C	-
II	**PROBING SHORE BIRDS**				
	Gruiformes/Rallidae				
12.	Whitebreasted waterhen	*Amourornis waterhen*	R	C	Y
	Charadriiformes/Jacanidae				
13.	Bronze-winged Jacana	*Metopidius indicus*	R	FC	-
	Charadriiformes/Charadriidae				
14.	Little-ringed plover	*Charadrius dubius*	R	C	-
15.	Common ringed Plover	*Charadrius hiaticula*	R	C	-
16.	Lesser Sand Plover	*Charadrius mongolus*	M	C	-
17.	Greater Sand Plover	*Charadrius leschenaultii*	M	C	-
18.	Yellow-wattled Lapwing	*Vanellus malabaricus*	R	R	Y
19.	Red-wattled Lapwing	*Vanellus indicus*	R	C	-
20.	Kentish plover	*Charadrius alexandrinus*	M	C	-
21.	Pacific Golden-Plover	*Pluvialis fulva*	M	C	-
	Charadriiformes/Scolopacinae				
22.	Black-tailed godwit	*Limosa limosa*	M	C	-
23.	Bar-tailed godwit	*Limosa lapponica*	M	C	-
24.	Eurasian Curlew	*Numenius arquata*	M	C	-
25.	Curlew Sandpiper	*Calidris ferruginea*	M	C	-
26.	Broad-billed Sandpiper	*Limicola falcinellus*	M	C	-
27.	Whimbrel	*Numenius phaeopus*	M	C	-

Contd...

Table 18.1–*Contd...*

Sl.No.	Family/Species	Scientific Name	Migratory Status	Abundance	Nesting
28.	Common Redshank	*Tringa totanus*	M	C	-
29.	Common Greenshank	*Tringa nebularia*	M	C	-
30.	Green Sandpiper	*Tringa ochropus*	M	C	-
31.	Common sandpiper	*Tringa hypoleucos*	R	C	-
32.	Marsh sandpiper	*Tringa stagnatilis*	R	C	Y
33.	Sanderling	*Calidris alba*	M	C	-
34.	Wood sandpiper	*Tringa glareola*	R	C	-
35.	Little Stint	*Calidris minuta*	M	C	-
	Charadriiformes/Glareolidae				
36.	Small Prantincole	*Glareola lactea*	M	C	-
37.	Oriental Pratincole	*Glareola maldivarum*	M	C	-
	Charadriiformes/Recurvirostridae				
38.	Black-winged Stilt	*Himantopus himantopus*	M	C	-
	Charadriiformes/Burhinidae				
39.	Stone-Curlew	*Burhinus oedicnemus*	M	C	-
III	**FLOATING AND DIVING WATER BIRDS**				
	Pelecaniformes/Podicipitidae				
40.	Little Grebe	*Tachybaptus ruficollis*	RLM	C	-
	Anseriformes/Anatidae				
41.	Little cormorant	*Phalacrocorax niger*	RLM	C	Y
42.	Greater cormorant	*Phalocrocorax carbo*	RLM	C	-
43.	Common Teal	*Anas crecca*	M	C	-
44.	Cotton Teal	*Nettapus coromandelianus*	M	C	-
45.	Northern Pintail	*Anas acuta*	M	C	-
46.	Northern Shoveller	*Anas clypeata*	M	C	-
47.	Spot-billed Duck	*Anas poecilorhyncha*	M	C	-
IV	**AERIALLY SEARCHING WATER DEPENDENT BIRDS**				
	Charadriiformes/Laridae				
48.	Black-headed Gull	*Larus ridibundus*	M	C	-
49.	Brown-headed Gull	*Larus brunnicephalus*	M	C	-
50.	Gull-billed Tern	*Gelochelidon nilotica*	M	C	-
51.	River Tern	*Sterna aurantia*	RLM	C	-
52.	Common Tern	*Sterna hirundo*	RLM	C	-
53.	Caspian Tern	*Sterna caspia*	M	C	-
54.	Whiskered Tern	*Chlidonias hybridus*	M	C	-
55.	Little tern	*Sterna albifrons*	RLM	C	-

Contd...

Table 18.1–*Contd...*

Sl.No.	Family/Species	Scientific Name	Migratory Status	Abundance	Nesting
	Coraciiformes/Halcyonidae				
56.	Whitebreasted kingfisher	*Halcyon smyrnensis*	R	C	Y
57.	Storkbilled kingfisher	*Pelargopsis capensis*	R	C	Y
58.	Black-capped kingfisher	*Halcyon pileata*	R	R	Y
	Coraciiformes/Halcyonidae				
59.	Small blue kingfisher	*Alcedo atthis*	R	C	Y
60.	Blue-eared Kingfisher	*Alcedo meninting*	R	C	Y
61.	Oriental Dwarf kingfisher	*Ceyx erithacus*	R	C	Y
	Coraciiformes/Halcyonidae				
62.	Pied kingfisher	*Ceryle rudis*	RLM	C	-
V	**BIRDS OF PREY**				
	Falconiformes/Accipitridae				
63.	Black Kite	*Milvus migrans*	R	C	Y
64.	Black-shouldered Kite	*Elanus caeruleus*	R	UC	-
65.	Brahminy kite	*Haliastur indus*	R	C	Y
66.	Osprey	*Pandion haliaetus*	M	UC	-
67.	Pallid Harrier	*Circus macrourus*	M	R	-
68.	Pariah kite	*Milvus migrans*	R	UC	-
69.	White breasted Eagle	*Haliaeetus leucogaster*	R	C	Y
70.	Western Marsh-Harrier	*Circus aeruginosus*	M	UC	-
	Falconiformes/Falconidae				
71.	Common Kestrel	*Falco tinnunculus*	RLM	UC	-
VI	**ARBOREAL BIRDS**				
	Passeriformes/Corvidae				
72.	House crow	*Corvus splendens*	R	C	Y
73.	Jungle crow	*Corvus macrorhynchos*	R	C	Y
74.	Rufous Treepie	*Dendrocitta vagabunda*	R	C	-
	Apodiformes/Apodidae				
75.	Asian Palm-swift	*Cypsiurus balasiensis*	R	C	Y
76.	House swift	*Apus affinis*	R	C	Y
	Columbiformes/Columbidae				
77.	Blue Rock Pigeon	*Columba livia*	R	C	Y
	Psittaciformes/Psittacidae				
78.	Blossomheaded parakeet	*Psittacula cyanocephala*	R	R	Y
79.	Roseringed parakeet	*Psittacula krameri*	R	FC	-

Contd...

Table 18.1–*Contd...*

Sl.No.	Family/Species	Scientific Name	Migratory Status	Abundance	Nesting
	Cuculiformes/Cuculidae				
80.	Asian Koel	*Eudynamys scolopacea*	R	C	-
	Strigiformes/Tytonidae				
81.	Barn Owl	*Tyto alba*	R	UC	Y
	Strigiforme/Strigidae				
82.	Spotted owlet	*Athene brama*	R	UC	Y
	Caprimulgiformes/Caprimulgidae				
83.	Indian Nightjar	*Caprimulgus affinis*	R	FC	-
	Coraciiformes/Meropidae				
84.	Small green bee-eater	*Merops orientalis*	R	C	-
	Coraciiformes/Coraciidae				
85.	Indian Roller	*Coracias benghalensis*	RLM	FC	-
	Coraciiformes/Upupidae				
86.	Common Hoopoe	*Upupa epops*	RLM	R	-
	Piciformes/Picidae				
87.	Goldenbacked woodpecker	*Dinopium benghalense*	R	C	Y
88.	Black- Rumped flameback	*Dinopium javanense*	R	FC	-
	Passeriformes/Alaudidae				
89.	Crested lark	*Galerida cristata*	R	R	-
	Passeriformes/Hirundinidae				
90.	Red-rumped swallow	*Hirundo daurica*	M	C	-
91.	Barn swallow	*Hirundao rustica*	R	C	-
	Passeriformes/Laniidae				
92.	Brown shrike	*Lanius cristatus*	RLM	C	-
	Passeriformes/Dicruridae				
93.	Black Drongo	*Dicrurus macrocercus*	R	C	-
94.	White -bellied Drongo	*Dicrurus caerulescens*	R	R	-
95.	Greater racket-tailed drongo	*Dicrurus paradiseus*	R	FC	-
	Passeriformes/Sturnidae				
96.	Common Myna	*Acridotheres tristis*	R	C	Y
97.	Jungle Myna	*Acridotheres fuscus*	R	UC	-
	Passeriformes/Campephagidae				
98.	Scarlet minivet	*Pericrocotus cinnamomeus*	R	UC	-
	Passeriformes/Irenidae				
99.	Iora	*Aegithina tiphia*	R	R	Y

Contd...

Table 18.1–*Contd...*

Sl.No.	Family/Species	Scientific Name	Migratory Status	Abundance	Nesting
	Passeriformes/Pycnonotidae				
100.	Redwhiskered bulbul	*Pycnonotus jocosus*	R	C	Y
101.	Red-vented bulbull	*Pycnonotus cafer*	R	C	Y
	Passeriformes/Timaliinae				
102.	Jungle babbler	*Turdoides striatus*	R	FC	-
103.	Blackheaded babbler	*Rhopocichla atriceps*	R	FC	-
	Passeriformes/Sylviinae				
104.	Blyth's reed warbler	*Acrocephalus dumetorum*	M	C	-
105.	Greenish warbler	*Phylloscopus trochiloides*	M	C	-
106.	Common Tailorbird	*Orthotomus cuculatus*	R	C	Y
107.	Malabar whistling thrush	*Myiophonus horsfieldii*	R	UC	-
108.	Magpie robin	*Copsychus saularis*	R	C	-
109.	Indian robin	*Saxicoloides fulicata*	R	C	-
	Passeriformes/Parinae				
110.	Grey tit	*Parus major*	R	C	-
	Passeriformes/Motacillidae				
111.	Large pied wagtail	*Motacilla maderaspatensis*	R	R	-
112.	White wagtail	*Motacilla alba*	M	FC	-
113.	Paddyfield pipit	*Anthus rufulus*	R	C	Y
	Passeriformes/Dicaeidae				
114.	Tickell's flowerpecker	*Dicaeum erythrorhynchos*	R	C	-
	Passeriformes/Nectariniidae				
115.	Purple sunbird	*Nectarinia asiatica*	R	C	Y
116.	Purple-rumped Sunbird	*Nectarinia zeylonica*	R	C	Y
	Passeriformes/Passerinae				
117.	House sparrow	*Passer domesticus*	R	C	Y
	Passeriformes/Estrildinae				
118.	Indian Silverbill	*Lonchura malabarica*	R	C	-
119.	Black-headed munia	*Lonchura malacca*	R	C	-

C: Common (Total popn >25); FC: Fairly Common (<25-10); UC: Uncommon (<10-5); R: Rare (<5); RD: Resident; RLM: Resident with Local Movements; M: Migrant.

Floating and Diving Water Birds

Darter, Ducks, grebes, cormorants, shovellers, and Teals fall under this group. Eight species of floating and diving water birds were identified (Table 18.1). Three species are the year-round residents, while others are migrants or winter visitors. Among those, density wise the dominant species were Northern Pintail followed by cotton teal and Northern Shoveller and the least was darter. From the

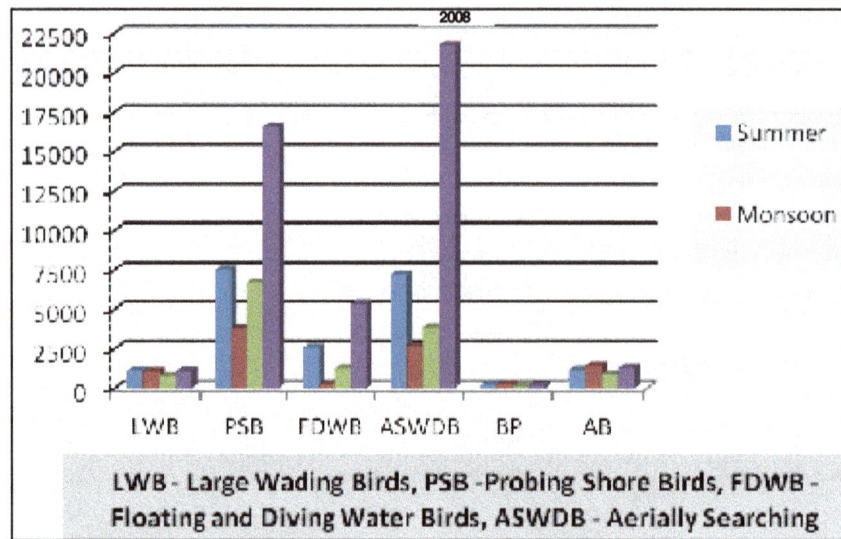

Figure 18.1: Season-wise Population of Avifauna Observed in the Mangrove Areas of Aghanashini Estuary during 2007 and 2008

standpoint of feeding, members of this group are highly heterogeneous. Cormorants and Darters are exclusively piscivorous, some members exclusively feeds on benthic invertebrates and others like Pintails, ducks and teals consumes a significant portion of plant materials.

Aerially Searching Water Dependent Birds

Gulls, terns, and kingfishers, comprise this group of omnivorous and piscivorous species (Table 18.1). These birds hunt in pools, creeks and waterways adjacent to mangrove stands. Many fishes and invertebrates which they feed upon are from mangrove based ecosystems. Fifteen species were identified, and 10 species are year - round residents and remaining are winter visitors. The dominant species was Black-headed Gull followed by Brown-headed and Gull-billed Tern and the least species in number observed was Black-capped kingfisher. Gulls and terns prefer open sandy areas for nesting (Kushlan and White, 1977) and use mangrove ecosystem for feeding. The birds belonging to this group are observed from a variety of coastal and inland wet-land habitats.

Birds of Prey

Kites, eagles, shikra, osprey, harrier, owls and kestrel fall under this group. Out of 9 species identified, 6 species are permanent residents and remaining are winter visitors (Table 18.1). The dominant species observed were Black kite followed by Brahminy kite and white breasted sea eagle, and the least was Common kestrel. Their use of mangrove areas varies greatly from feeding, roosting and nesting. The prey species are the common inhabitants of mangrove areas. Kites are observed in flocks near the fish landings. Owls and harriers feed on the chicks of parakeets, reptiles, disabled birds and young mammals.

Arboreal Birds

This group is the largest and the most diverse group inhabiting the mangroves. 48 species (Table 18.1) identified includes Crows, swifts, doves, cuckoos, robins, warblers, woodpeckers, flycatchers, parakeet, swallow, thrushes, oriole, sparrow, flowerpeckers, bush larks, shrikes, mynas, babblers, bulbuls, pipits, sunbirds, munias, coucal, wagtails, hoopoe, and larks. Among these, 44 species are year round residents and 4 species are winter visitors. The dominant species was House Crow followed by Greenish warbler and purple sunbird, the least in numbers observed was blossom-headed parakeet. The divers groups living together indicates that these birds utilize mangrove ecosystem in similar ways. Significant portion of the diets of these birds includes invertebrates particularly insects, except for a few birds like doves and pigeons, which eat upon the reeds, berries and fruits. Different types of searching patterns were identified (Odum *et al.*, 1982). Hawking of insects is the primary mode of feeding by cuckoos and flycatchers. Gleaning is employed by most of the warblers. Woodpeckers are classic probers. Rose-ringed Parakeets are fond of feeding on the ripe fruits of mangroves at estuary. Flocks of parakeets nest on the *Sonneratia, Kandelia* and *Avicennia* stands. However, most of the birds from this group have expanded their range from mangrove to non-mangrove habitats also with irregular occurrence.

Species, like White-breasted sea eagle, Darter, Asian openbill stork, and Black-headed ibis, that are listed by Birdlife International (2001) as migratory waterbird species of special conservation interest in the Asia-Pacific region. The factors, which usually influence the selection of sites for roosting, are proximity to the food source and water (Haneda *et al.*, 1966) and isolation from human activity (Ross, 1973) suggested the hypothesis of protection against predators as the function of communal roosting. Protection from natural predators and increased feeding efficiency due to the proximity to the food source may be the two factors, which favour the Aghanashini as a preferred roosting site for the avifauna. Siltation at Aghanashini has created a vast landscape with shallow muddy zones of micro delta formations, where many species of wading and probing water birds feed. Cormorants and kingfishers hunting fish all the time in the lagoon is a scenic beauty. During the onset of monsoon, the winter migrants and visitors dominate the bird population in lagoon. Ibises, spoonbills, egrets, kites and storks roost in the canopy of the dense *Avicennia and Sonneratia* stands creating the scenario of immaculate cloud resting on the canopy till the onset of February. Even though the mangroves are protected ecosystem, poachers hunt the teals, gargeny ducks, geese, innumerable herons and egrets. Birds are accurate indicators of over-all imbalances in a habitat. The diminished health and disappearance of bird species worldwide indicates an environment at risk. The welfare of this avifauna community is inextricably tied to ours. Conversion of tidal swamps may severely reduce the bird population and the concentration of migratory and winter visitors.

References

Ali, S., 1996. *The Book of Indian Birds.* Bombay Natural History Society and Oxford University Press Mumbai.

Allen, R.P., 1942. *The Roseate Spoonbill.* Res. Rep. 2. National Audubon Society, New York.

Baker, M.C. and Baker, E.M., 1973. Niche relationships among six species of shorebirds on their wintering and breeding ranges. *Ecol. Monogr.*, 43: 193–212.

Birdlife, International, 2001. *Threatened Birds of the World.* Barcelona and Cambridge,U.K., Lynx Editions and Bird Life International, 3026 pp.

Grimmet, R., Inskipp, T. and Inskipp, C., 1998. *Birds of the Indian Subcontinent.* Oxford University Press. Delhi, 888 pp.

Haneda, K., Lida, V., Kagw, T., Moti, T. and Vamasingh, S., 1966. Roosting flock area of crows in North Nagano prefecture. *Japanese Journal of Ecology*, 16: 213–215.

Kushlan, J.A. and White, D.A., 1977. Nesting wading wild population in southern Florida. *Fla. Sci.*, 40: 65–72.

Kushlan, J.A. and Kushlan, M.S., 1975. Food of the white Ibis in southern Florida. *Fla Sci.*, 38: 31–38.

Naskar, K.R. and Guhabakshi, D.N., 1987. *Mangrove Swamps of the Sunderbans: An Ecological Perspective.* Naya Prokash, Calcutta.

Odum, W.E., McIvor, C.C. and Smith III, T.J., 1982. The ecology of the mangroves of south Flordia: A community profile. U.S. Fish and Wildlife Service, Office of Biological Services, Washington, D.C.FWS/OBC–81/24: 61–73.

Ross, P.X., 1973. Notes on wintering Great Cormorant in Nova Scotia. *Canadian Field Naturalist*, 88: 493–494.

Samant, J.S., 1985. Avifauna of the mangroves around Rathnagiri, Maharashtra. *Proc. Nat. Symp. Biol. Util. Cons. Mangroves*, p. 456–466.

Schneider, D., 1978. Equalization of prey numbers by migratory shorebirds. *Nature*, 271: 353–354.

Wolff, W.J., 1969. Distribution of non-breeding waders in an estuarine area in relation to the distribution of their food organisms. *Ardea*, 57: 1–28.

Chapter 19

Identification of *Vibrio harveyi* in Sea Water Sample by PCR Targeted to *vhh* Gene

☆ *Sreenath Pillai and Leena Muralidharan*

ABSTRACT

Today, marine water is broadly suffering from different types of aquatic bacterial species or the aqua technological organisms. One of the major risks involves the consumption of raw or under-cooked sea food that may be naturally contaminated by food borne pathogens present in the marine environment. Such risk is further increased if the food is mishandled during processing where pathogens could multiply exponentially under favorable conditions. In contrast to most other food borne pathogens, *Vibrio* sp. has the aquatic habitat as their natural niche. As a result, *Vibrios* are most commonly associated with seafood as natural contaminants. Food borne infections with *Vibrio* sp. are common in Asia. *Vibrio* infections usually occur in fish from marine and estuarine environments, and have been reported throughout the world. Occasionally, Vibriosis is reported in freshwater fish. The disease can cause significant mortality in fish culture facilities once an outbreak is in progress. The standard selective medium method used for *Vibrio* is Thiosulphate citrate bile salt sucrose agar (TCBS). *Vibrios* are highly abundant in aquatic environments, including estuaries, marine coastal waters and sediments, and aqua -culture settings worldwide. Sea water sample was collected and investigated which revealed that *Vibrio harveyi* species appear at particularly high densities in marine water in turn affecting the various aquatic organisms, *e.g.*, corals, fishes, molluscs, sea-grasses, sponges, shrimps and zooplanktons. Specific-PCR for *vhh* gene detection gave positive results in which bands with 235bp size appeared on gel confirming the presence of *Vibrio harveyi*.

Keywords: Vibrio harveyi, TCBS, vhh gene, Sea food, Pathogens.

Introduction

Vibrio harveyi is a ubiquitous, gram-negative bioluminescent marine bacterium, occurring either as a free-living form or in association with intestinal microbiota of marine animals (Makemson and Hermosa 1999) in aquatic environment. It lives in brackish saltwater and causes gastro- intestinal illness in humans. It is a halophile, or salt-requiring organism. Most people become infected by eating raw or undercooked shellfish, particularly oyster and cockles. It is a pathogen of fishes, eels, frogs as well as other vertebrates and invertebrates (Todar, 2005). As a pathogen, *V. harveyi* manifests itself as luminous vibriosis in hatchery reared and commercially farmed *penaeid* shrimps resulting in severe economic losses to shrimp industry in Asia (Lavilla-Pitogo *et al.*, 1990; Karunasagar *et al.*, 1994; Conejero and Hedreyda 2003). Vibriosis is well recognized as significant of disease and mortality. At least 12, *Vibrio* spp. are classified as pathogenic strains and become major factor for food- borne diseases.

There are many methods used in the detection of *Vibrio harveyi*. The standard selective medium method used for *Vibrio* is Thiosulphate citrate bile salt sucrose agar (TCBS). *Vibrio harveyi* colonies are green or blue green on the agar due to sucrose fermentation. The latest technique would be PCR that can be used for detection of *Vibrio harveyi* in various samples including sea- foods or other samples, and this method is faster, easier and more reliable. This technique can be applied because of the presence of *vhh* gene (was first discovered as the regulatory gene of the haemolysin operon), that appears to be well conserved among *Vibrio harveyi*. This gene can be used to develop a PCR method for identification of *Vibrio harveyi* (Kim *et al.*, 1999).

Vibrios are among the most important bacterial pathogens of various aquatic organisms responsible for a number of diseases, and mortalities up to 100 per cent have been reported due to vibriosis. The present study was designed to isolate identify *Vibrio harveyi* in sea water collected from Cochin Shipyard area of Kerala during May 2010.

Materials and Methods

Sampling Method

Sea water sample was collected from a jetty located at the Cochin Shipyard area of Kerala. Water sample was collected by submerging an autoclaved dark bottle under water and then opening its cap. The bottle was sealed under water and then kept in dark at 4°C until processed for isolation of bacterial strains.

Sample Preparation

300ìL of the water sample was spread onto four thiosulphate citrate bile salt sucrose agar (TCBS) (Hi-Media, India) plates for the selective growth of *Vibrio* sp. and incubated overnight at 25°C. 1-3 well isolated green colonies were inoculated on Luria Bertani broth for overnight (12-16 hours). A total of 4 isolates of bacteria were obtained from sample. These isolates were purified by re-streaking onto TCBS agar plates and gram stained for cellular characterization. The selected colonies from TCBS agar were tested using PCR method to detect the presence of *vhh* gene which appeared highly conserved in *Vibrio harveyi* isolates. Purified *Vibrio* cultures were then coded S_1, S_2, S_3 and S_4 and employed for further experimentation.

Extraction of Genomic DNA

Extraction of genomic DNA was accomplished by standard protocol of chloroform: iso-amyl alcohol method. After incubation, 1ml of the overnight culture was transferred to sterile micro-centrifuge tube and the cells were harvested by centrifugation at 12,000rpm for 10 minutes. Supernatant was discarded and the pellet was re-suspended in 1ml of 0.85 per cent (w/v) NaCl solution and centrifuged as above. The supernatant was discarded and added 600ìl lysis buffer along with 7ìl of *proteinase-k*. The mixture was mixed using vortex machine and incubated at 65°C for 1 hour. Equal volume of chloroform: iso-amyl alcohol (24:1) was added with gentle mixing by inverting the tubes for 2-5 minutes. The samples were centrifuged for 15 minutes at 12,000 rpm and 40°C. Aqueous phase was collected in another micro-centrifuge tube without disturbing the interface and lower phase. The steps of chloroform: iso-amyl alcohol extraction was repeated. Again the aqueous phase was collected and added 1/10th volume of 3M sodium acetate (pH 5.2) followed by equal quantity of ice cold iso-propanol, so that the DNA gets precipitated and centrifuged it again at 40°C for 5 minutes at 12,000rpm. The supernatant was discarded and rinsed the pellet twice with 70 per cent iso-propanol, followed by maintaining the tubes for one hour in vacuum desiccators. The desiccated DNA sample was completely re-suspended in 200ìl of DNA dissolving buffer (TE buffer) and stored at -20°C.

The supernatant contained genomic DNA and was ready to be used as a template for *vhh* gene detection in *Vibrio harveyi*. Specific primers (R: 5'-GAGTTTGATCCTGGCTCAG- 3' and F: 5' - GATATTACCGCGGCGGCTG- 3') correlated to *Vibrio harveyi vhh* region with GC contents 52.6 per cent and 66.7 per cent respectively. The melting temperature T_m was found to be 56.7 °C and 60.5 °C for forward and reverse primers respectively. The primers were synthesized by MWG Biotech Private Ltd, Bangalore, India.

PCR-Protocol

PCR (Perkin Elmer) was carried out using 0.5 ml microfuge tubes. The total volume of reaction mixtures was 25.0 ìl which consisted of 17.7 ìl sterile distilled water, 1x assay buffer (100mM Tris, 500mM KCl, 0.1 per cent gelatin, pH 9.0), 2.5 ìl 10x T_{aq} buffer with 1.0 ìl 15 mM MgCl$_2$ (Genei, Bangalore, India), 0.5 ìl 20 mM deoxyribonucleotide phosphate (Genei, Bangalore, India), 0.5 ìl 20 pm/ìl of each primer, 0.3 ìl of 5 units T_{aq} DNA polymerase and 1.0 ìl 20 ng/ìl template DNA. The cycling conditions were as follows; pre- denaturation at 96°C for 5 min, denaturation at 94°C for 30 sec, annealing at 58°C for 30 sec, and extension at 72°C for 45 sec, with a final extension at 72°C for 10 min at the end of 29 cycles. To check the DNA contamination, a negative control was set up omitting template DNA from the reaction mixture.

Agarose Gel Electrophoresis

The PCR products were run on 1.2 per cent agarose gel (Sigma) in 1x Tris-Borate- EDTA (TBE). 15-20 ìl PCR products were loaded into sample wells and the voltage used was 100 volt for 1 h. The gel was then stained in Ethidium bromide (0.5 ìg/ml) solution for 1 min and destained in distilled water for 30 min. Then the gel was visualized and photographed under UV trans- illuminator.

Results and Discussion

The marine water sample was collected from Cochin Shipyard area of Kerala, tested on the TCBS agar for the detection of *Vibrio harveyi*. When the samples were streaked on the TCBS agar, green colonies were observed which were of *Vibrio harveyi*. Various reports state that three species of *Vibrio* usually associated with *Vibrio harveyi* in aquatic environmental and sea food were *Vibrio vulnificus*, *Vibrio carchariae* and *Vibrio cholerae*. The selection of the isolates was based on the strong colony appearance purely green in color. The TCBS agar medium, which contains a chromogenic substrate as substrates for beta galactosidase, was used instead of sugar fermentation (sucrose) for the detection of *Vibrio harveyi* by Green colony (Kasthuri, V., *et al.*, 1998). Identifying *Vibrio harveyi* strains through

PCR based method which targets the conserved region of *Vibrio harveyi* such as *vhh* gene is more efficient, reliable and faster compared to biochemical tests. After the identification on the TCBS agar, the PCR assay for *vhh* detection was performed to confirm the presence of *Vibrio harveyi IJPBA, Nov - Dec, 2011, Vol. 2, Issue, 6*. After the identification on the TCBS agar, the PCR assay for *vhh* detection was performed to confirm the presence of *Vibrio harveyi*. Dileep *et al.* (2003) and Sujeewa *et al.* (2009) reported that the *vhh* gene fragment (~235bp) which is specific for *Vibrio harveyi* was successfully amplified and results are presented in (Figure 19.1). In this study, the primer used for the amplification of *vhh* gene is specific for *Vibrio harveyi* on 1.2 per cent agarose (Lanes 2 - 5) as the sample DNA of *Vibrio harveyi* and (Lane-1) as the 100 bp DNA ladder.

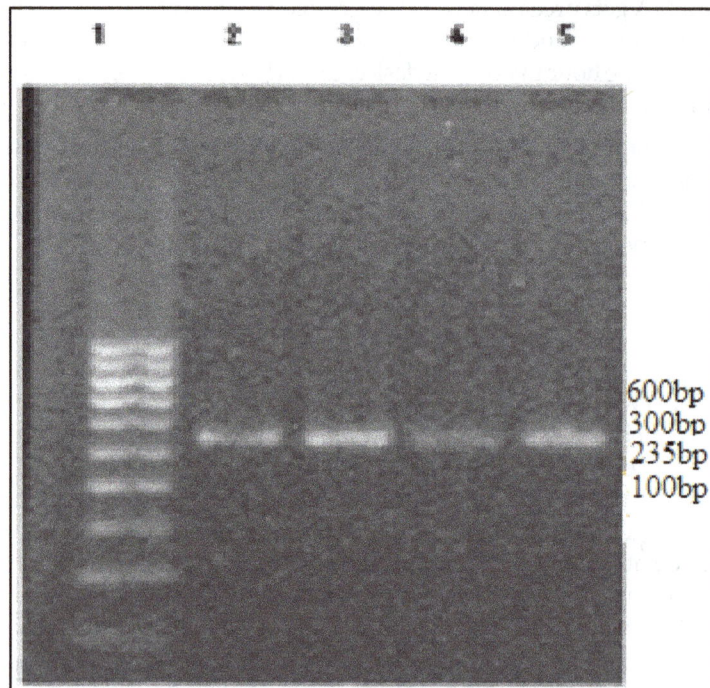

Figure 19.1: PCR Profile Using *vhh* Gene and DNA Templates from 4 *Vibrio harveyi* isolates (a) Lanes 2 – 5 contain amplified 235bp *vhh* gene of DNA templates from S$_1$, S$_2$, S$_3$ and S$_4$ respectively. (b) Lane 1 contains 100bp DNA ladder.

PCR for *vhh* Gene

All 4 isolates amplified the expected 253 bp of the *vhh* gene (Figure 19.1) indicating the presence of a hemolysin gene, thereby, further confirming the identification of *V. harveyi* used in this study.

PCR products were purified and the purified DNA was air dried and commercially sequenced by MWG Biotech Private Ltd, Bangalore, India. Sequence data obtained were analyzed using BLAST algorithm (http://www.ncbi.nlmnih.gov/blast/Blast.cgi).

The partial 16S rRNA gene sequence data was analyzed using NCBI BLAST algorithm search and the *Vibrio* isolates S$_1$ (Accession # EU 937751), S$_2$ (Accession # GO 907034), S$_3$ (Accession # GO 907023) and S$_4$ (Accession # FJ 227119) were identified as *Vibrio harveyi* species.

Conclusion

This approach targets the haemolysin toxin which is the conserved region of *V. harveyi* is more efficient, reliable and faster compared to biochemical tests. Various factors have been related to the apparent increased incidences of diseases such as poor water quality, sometimes resulting from increased self-pollution due to effluent discharge and pathogen transfer via movements of aquatic organisms appearing to be an important underlying cause of such epizootics Thus, this method can be applied in the coastal areas, shrimp farms, etc. for diagnosis work. Furthermore, this method is very inexpensive and needs less equipment compared to the other methods. It can be concluded that *V. harveyi* isolated during the present study exhibited high degree of very less degree of strain variation. 16S rDNA - PCR was indeed a very useful tool to reveal the degree of variation among the same strain of bacteria. Extent of growth as in the case of *V. harveyi* should be controlled through suitable measures so as to sustain the marine aquaculture along the coast of Kerala.

References

Baumann, P. and Schubert, R.H.W., 1984. Family II. Vibrionaceae. *Bergey's Manual of Systematic Bacteriology*, (Eds.) N.R. Krieg, and J.G. Holt. Williams and Wilkins Co., Baltimore/London, p. 516–550.

Dileep, V., Kumar, H.S., Kumar, Y., Nishibuchi, M., Karunasagar, I. and Karunasagar, I., 2003. Application of polymerase chain reaction for detection of *Vibrio parahaemolyticus* associated with tropical seafoods and coastal environment. *Letters in Applied Microbiology*, 36: 423–427.

Farmer, J.J., 1992. The Family *Vibrionaceae*. In: *The Prokaryotes: A Handbook on the Biology of Bacteria: Eco-Physiology, Isolation, Identification, and Applications*, 2nd Edn., (Eds.) A. Balows, H.G. Truper, M. Dworkin, W. Harder, and K.H. Schleifer. Springer-Verlag, KG, Berlin, Germany, p. 2938–2951.

Food and Drug Administration (FDA), 1992. *Bacteriological Analytical Manual* 7th edn, USA, pp. 111–140.

Feldhusen, F., 2000. The role of seafood in bacterial food-borne diseases *Microbes and Infection*, 2: 1651–1660.

Kasthuri, V., Nobuhiko, D. and Shigeaki, M., 1998. Cloning and nucleotide sequence of the *gyrB* gene of *Vibrio harveyi* and its application in detection of this pathogen in shrimp. *Applied and Environmental Microbiology*, 64: 681–687.

Kim, Y.B., Okuda, J., Matsumoto, C., Takahashi, N., Hashimoto, S. and Nishibuchi, M., 1999. Identification of *Vibrio harveyi* strains at the species level by PCR targeted to the *toxR* gene. *Journal of Clinical Microbiology*, 37: 1173–1177.

Lavilla-Pitago, C.R., Leano, E.M. and Paner, M.G., 1998. Mortalities of pond-cultured juvenile shrimp, *Penaeus monodon*, associated with dominance luminescent *Vibrios* in the rearing environment. *Aquaculture*, 164: 337–349.

Makemson, J.C. and Hermosa, Jr. G.V., 1999 Luminous bacteria cultured from fish guts in the Gulf of Oman. *Luminescence*, 14: 161–168.

Thompson, J., Gibson, T., Plewniak, I., Jeanmougin, F. and Higgin, D., 1997. The clustal X windows interface: Flexible strategies for multiple sequence alignment aided by quality analysis tools. *Nucleic Acids Res.*, 24: 4876–4882.

Todar, 2005. *Todars Online Textbooks of Bacteriology*. Viewed 18th October 2007.

Chapter 20

Chronic Effects of Organophosphorus insecticide 'Fenthion' in Melanophore Pigments of *Cyprinus carpio* (Linn)

☆ *Leena Muralidharan and Sreenath Pillai*

ABSTRACT

Fish skin, directly exposed to the ambient toxicants; is used extensively as a potential indicator of contaminated aquatic environment. Chronic effect of *fenthion* on melanophore of *Cyprinus carpio* was investigated to understand its toxicity on melanophore morphology. Observed toxicological alterations include significant variations in size, shape and in number of melanophores. *Fenthion*-induced morphological changes in melanophore seemed to be an attempt to protect the epidermis from toxic medium. The density of melanophore was found to a maximum after 60 days. Destructive changes that resulted could be due to accumulation of *fenthion* in nervous tissues. Due to the lysis of the melanophores, the melanin contents are poured into the surrounding matrix of the connective tissue between 20 and 30 days of exposure. The regenerated pigment cells were smaller in size and more in number as compared to the normal melanophores. Present study also showed the melanizing effect of *fenthion* toxicity.

Keywords: Organophosphorus, Fenthion, Melanophore, Cyprinus carpio, Toxicity.

Introduction

The body covering of multi- cellular animals is usually pigmented and it is a matter of general observation as well as careful experimentation that such coloration is frequently cryptic and often provides help in indicating the extent of pollution in the environment (Gopi, 1994). The pollutants like insecticides are widely used in agriculture and are drained into the nearby rivers or ponds which are the natural habitat of fishes. Hence, it is important to study the insecticide effect on melanophore cells to evaluate the extent of stress.

Very few records are available on melanophore studies. Work has been carried out on fish scale melanophore which showed that there is isolation of peripheral cytoplasm during the melanin dispersion (Marshland, 1944). Melanophore was cultured and found that melanophore contracts in sodium- free potassium medium and this even occurred in presence of MSH (Novales and Novales, 1961). The effect of drug and chemicals on aggreviation or dispersion of pigment due to effect of DDT on MSH was also studied in anuran tadpole (Fuji, 1969). The effect of organophosphorus insecticide on melanophore of cichlid fish *Serotherodon mossambica* was also observed (Novales and Novales, 1961).

The present study deals with the effect of insecticide *fenthion* on melanophore cells of *Cyprinus carpio* to determine the extent of pesticide stress on melanophore pigment.

Materials and Methods

The animals were weighed and acclimatized under laboratory conditions (APHA, 1975). At the end of acclimatization, the fishes were exposed to three different sub- lethal concentrations of *fenthion* (0.38, 0.193, 0.096 mg/l) respectively. To determine the changes in the state of melanophore, scales were removed from the tail region of the control and treated fish bodies at the end of the 60 days exposure period and were compared with the control for further studies.

Results and Discussion

The examination of melanophore pigment exposed to three different sub- lethal concentrations of *fenthion* revealed that scales exposed to low sub- lethal concentration of *fenthion* (0.096 mg/l) showed a much notable difference in melanophore cells whereas severe damage to melanophore cells was observed in fishes exposed to 0.38 and 0.193 mg/l of *fenthion* when compared to the control.

Melanophore Control (Figures 20.1–20.3)

Typical teleost scale consists of five types of melanophore pigment cells:

Punctate Melanophore
Very few small sized round structures without any dendritic processes

Puntostellate Melanophore
Small round structures with 5- 7 unbranched dendritic processes

Stellate Melanophore
Single branch at the tip region of every dendritic process

Reticulostellate Melanophore
Reticulate branching of every dendritic process

Reticulate Melanophore
The largest melanophore with numerously branched thick and long dendritic processes

Melanophore Chronic (Figures 20.4–20.6)

Punctate Melanophore
No particular change in punctate melanophore state was seen. The number of punctate melanophore increased significantly in 0.096, 0.193 and 0.38 mg/l *fenthion* exposed scales. This was proportional to the concentration of insecticide exposed.

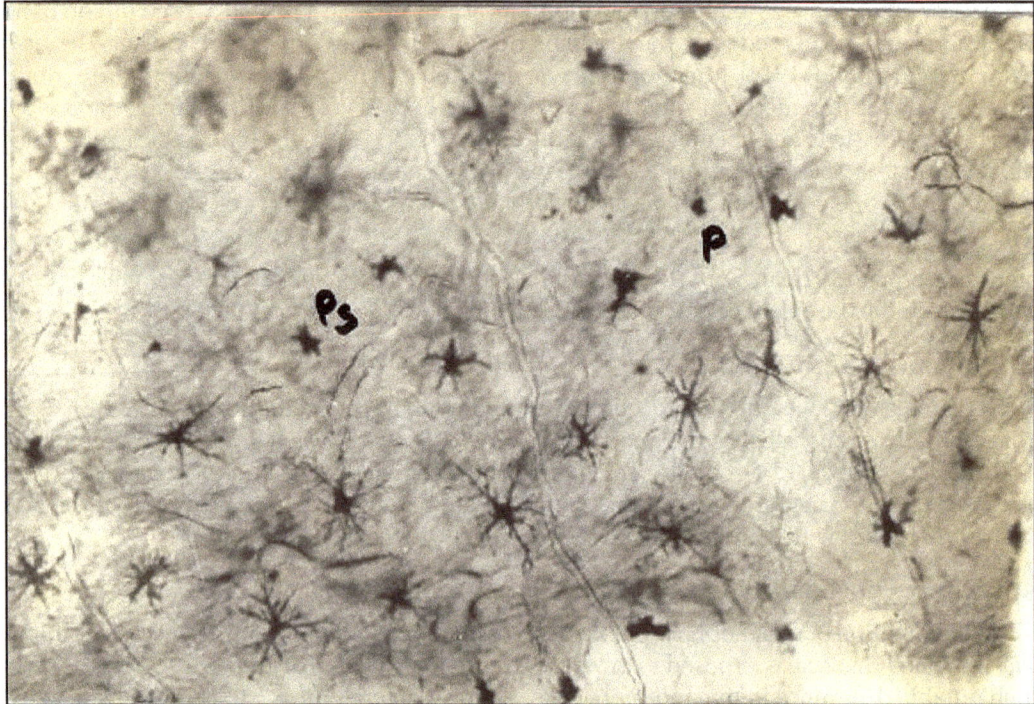

Figure 20.1: Melanophore Cells Exposed to 0.096 mg/l

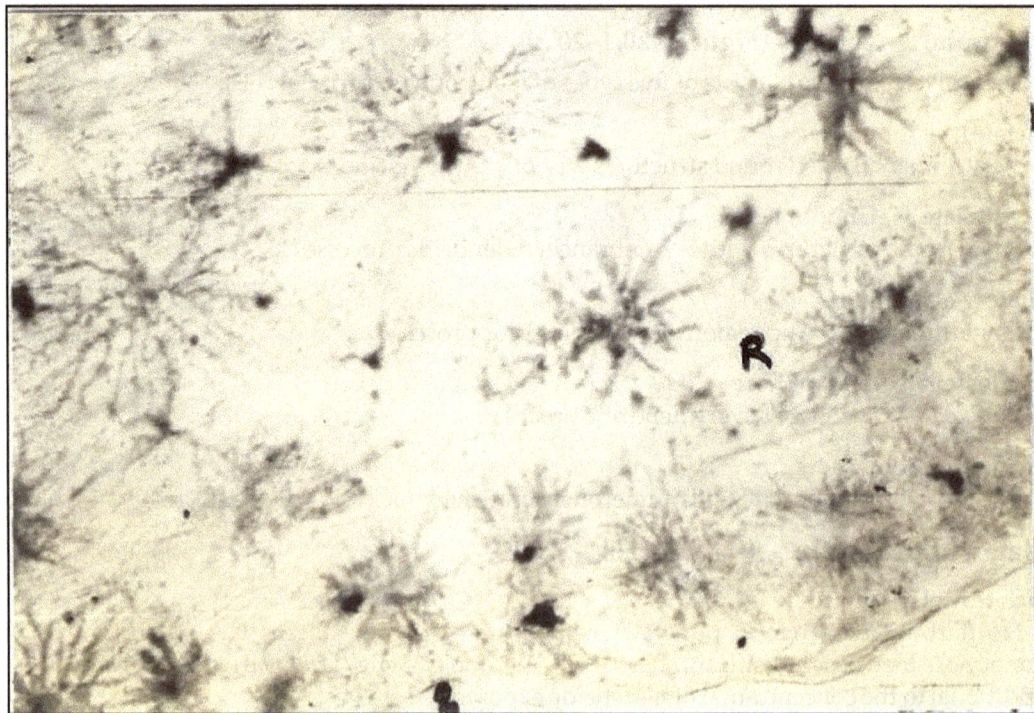

Figure 20.2: Melanophore Cells Exposed to 0.38 mg/l

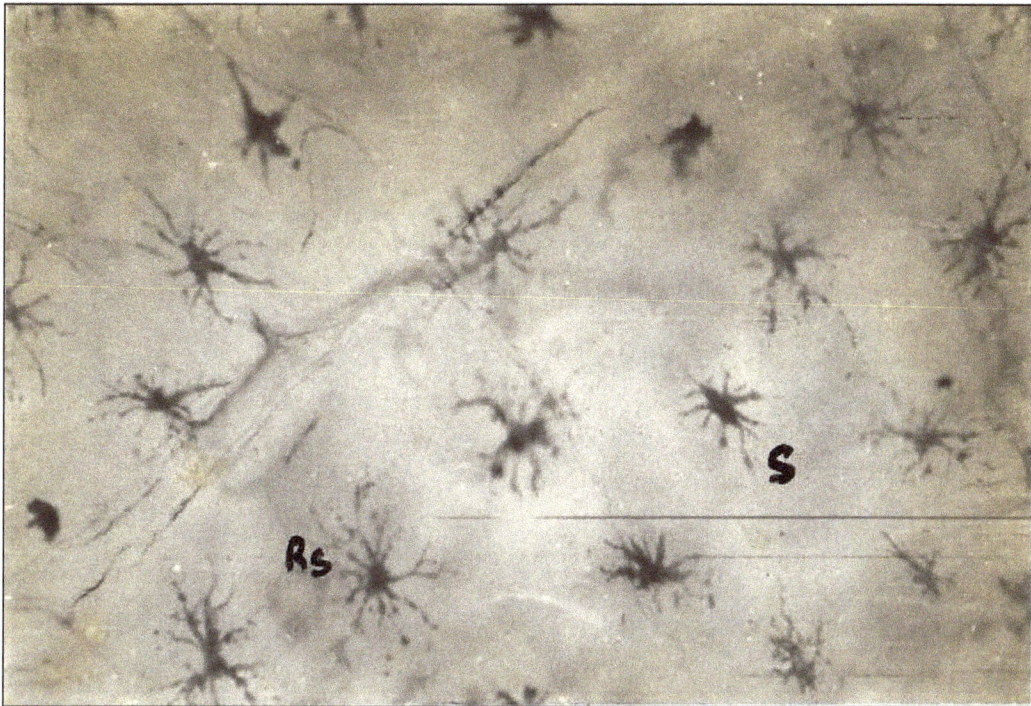

Figure 20.3: Melanophore Cells Exposed to 0.38 mg/l

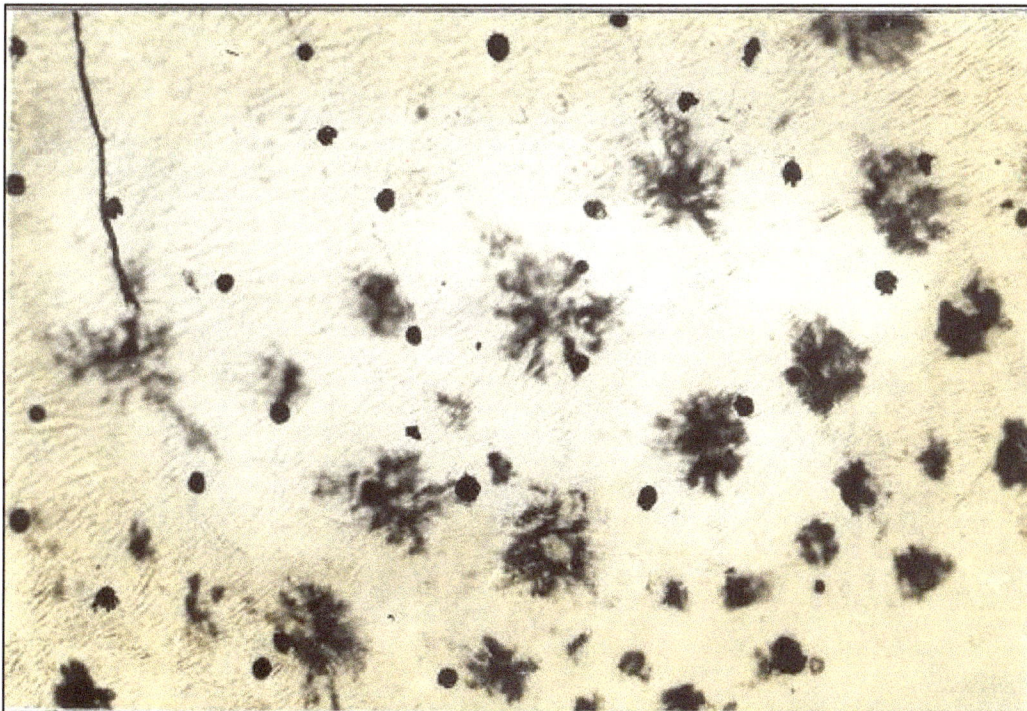

Figure 20.4: Melanophore Cells Exposed to 0.096 mg/l

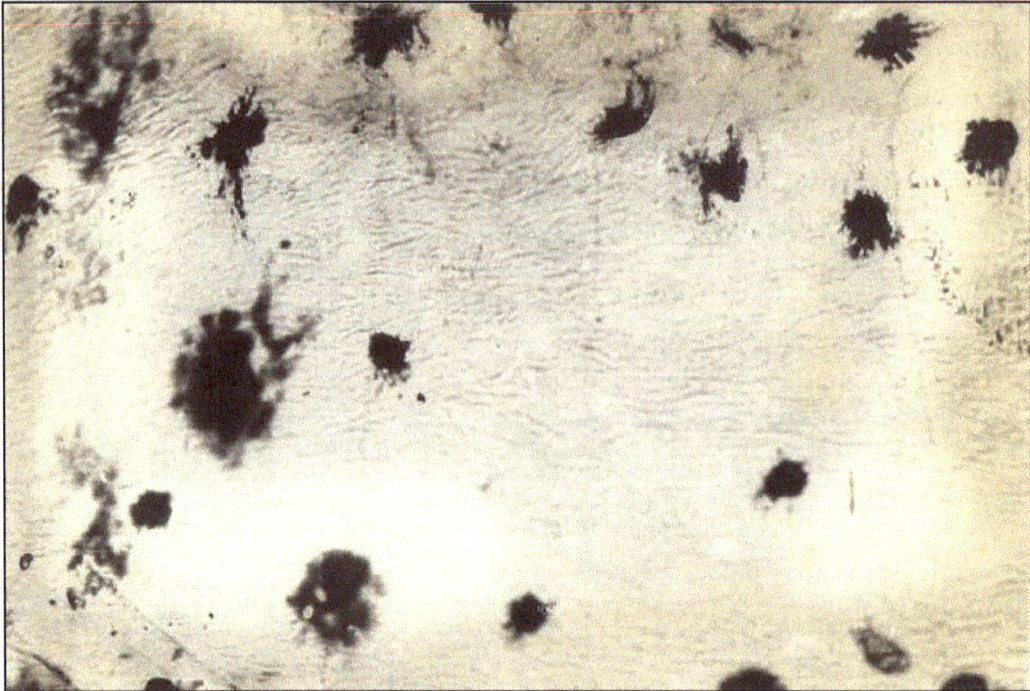

Figure 20.5: Melanophore Cells Exposed to 0.38 mg/l

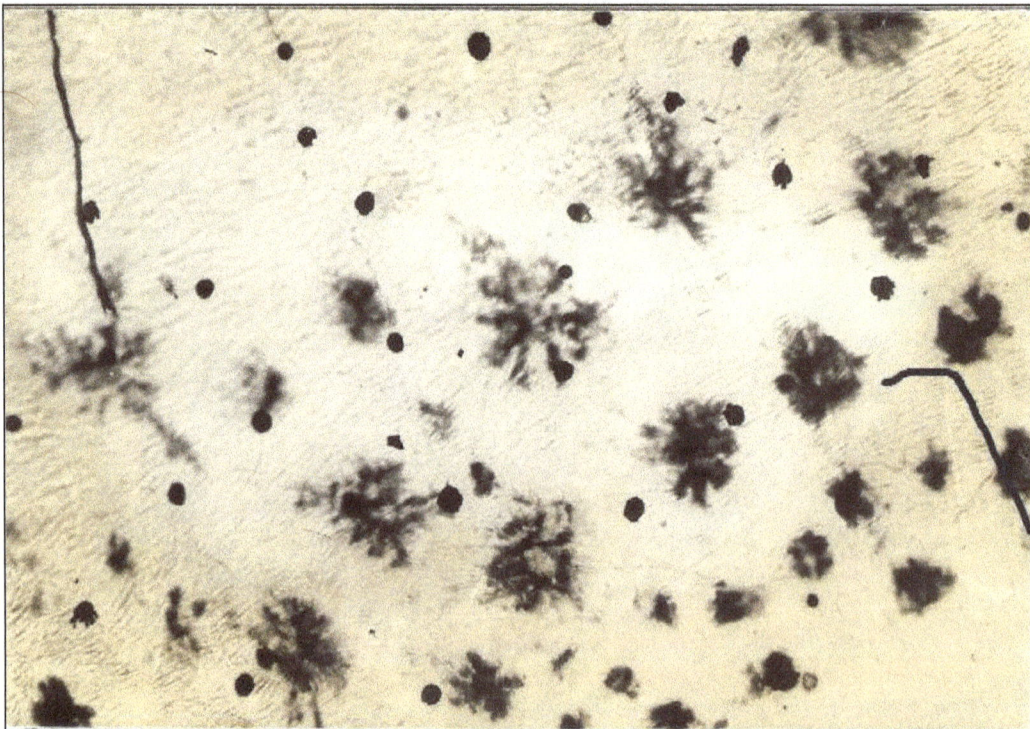

Figure 20.6: Melanophore Cells Exposed to 0.38 mg/l

Punctostellate Melanophore

Dendritic processes of punctostellate melanophore exposed to 0.096, 0.193 and 0.38 mg/l *fenthion* were broken severely and fused with each other. Damage was severe in those exposed to the highest concentration (Pandey *et al.*, 1981).

Stellate Melanophore and Reticulostellate Melanophore

Dendritic processes were broken and the shape was significantly deformed. In highly dosed fish scales, the dendritic processes were broken completely and fused with each other giving it an irregular appearance (Pandey *et al.*, 1981).

Reticulate Melanophore

The dendrites were severely damaged in highly exposed fish scales and the shape of the melanophore was significantly deformed.

In the scales of *Cyprinus carpio* exposed to 0.38 mg/l *fenthion* aggregation of broken melanophore cells was observed. The melanin pigment seemed to contract centrally. This could be the reason for darkened appearance of fish.

Melanophore changes are controlled by various factors such as neuro-humors, hormones, neuro-secretion and other nervous control of endocrine function. The ionic mechanism involved in MSH might be producing cytoplasmic solution by changing the permeability of melanophore membrane to ions (Jochle, 1958; Lerner and Takashashi, 1956). Finally, the real possibility that MSH acts on melanophore by an ionic and osmotic mechanism indicates that there is a similarity between melanophores and neurons. In support of this work, destruction observed in melanophore pigment in exposed fish during the present study may be its adaptation to stressful condition and could be due to high accumulation of *fenthion* in nervous system (Brain) as in Table 20.1.

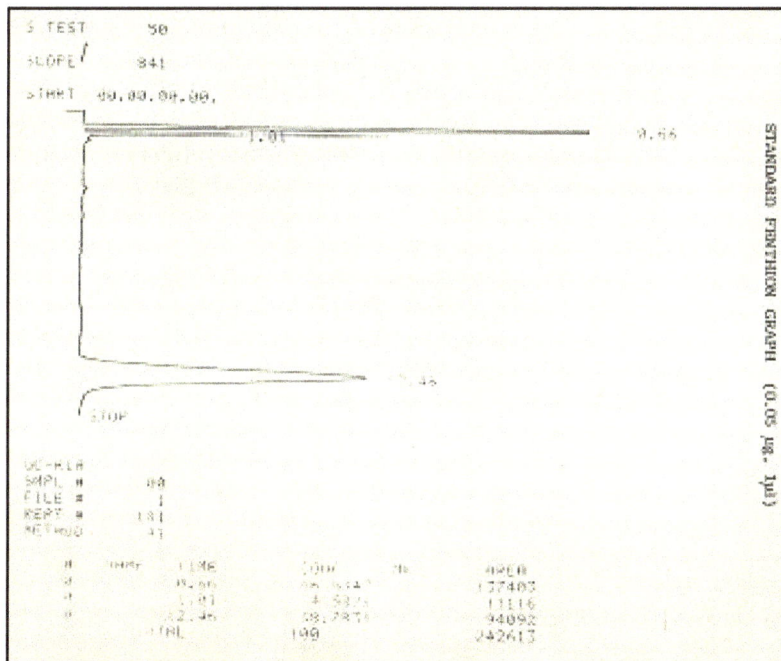

Figure 20.7: Fenthion Residue in Brain of *Cyprinus carpio*

Table 20.1: Residue Level of Fenthion in mg/l in the Aquarium Freshwater in which the Test Fish *i.e.* ***Cyprinus carpio*** **was Kept**

Exposure Period	Control	Concentrations of Fenthion in mg/l		
Initial	Nil	0.096	0.193	0.38
24 hours	Nil	0.02 ± 0.009	0.09 ± 0.004	0.19 ± 0.002
48 hours	Nil	0.08 ± 0.003	0.04 ± 0.02	0.03 ± 0.001

Conclusion

Destructive changes observed in melanophores of exposed fish are directly related to the strength of dose induced. Extensive damage observed in melanophore could be due to the accumulation of *fenthion* in nervous tissues and also maybe due to the changes in ionic regulation. The melanophore changes noted is a significant abnormality providing a symptomatic index of toxicity.

References

APHA, 1975. *Standard Method for Examination of Water and Wastewater* 14th Edn., Washington.

Fuji, 1969. *Chromatophore and Pigments in Fish Physiology.*

Gopi, L., 1994. Melanophore in *Cyprinus carpio. Ph.D. Thesis,* University of Mumbai.

Jochle, W., 1958. Melanophore trope virstoffe in saugetier organisms aundihre mogliche bedeutung.

Lerner, A.B. and Takashashi, Y., 1956.A hormonal control of melanin pigmentation. *Recent Progr. Hormone Res.,* 12: 303– 320.

Marshland, D.A., 1944. Mechanism of pigment displacement in unicellular chromatophore. *Biol. Bull.,* 87: 252–261.

Novales, R.R. and Novales, B.J., 1961. Sodium dependence intermedian action on melanophore in tissue culture. *Gen. Comp. Endocrinology,* 1: 134–144.

Pandey, A.K., Shukla, L., Tuji, R. and Miyashita, Y., 1981. Effect of sub-lethal malathion exposure of cichlid *Serotherodon mossambica. J. Lib. Ants and Sci. Sappro. Med.,* 22: 77.

Chapter 21

A Study on Organochlorine Pesticide Accumulation and its Effect on Nutrient Value of Edible Fish *Catla catla* (Ham) Sold in Local Market of Mumbai

☆ *Leena Muralidharan*

ABSTRACT

Organochlorine insecticides in use for more than three decades all over the world, have caused one of the most serious environment problems, and is being detected easily in food chain. They occur in higher concentration at higher tropic levels due to their great chemical stability. It reaches human bodies through daily diet and are deposited in tissues. The objective of the present study is to determine the concentration of certain organochlorine pesticides accumulated in edible fish *Catla catla* sold in Mumbai local market and its effect on nutrient value on three different season (rainy, winter and summer).

Keywords: Catla catla, Organochlorine pesticides, Concentration, Nutrient value.

Introduction

Our uses of pesticides have resulted in increased food production and other benefits but it has also raised concerns about potential adverse effects on human health. The persistence of organochlorine in aquatic ecosystem has special significance as they are picked up by aquatic organisms. Pesticide residues may bring about several changes in pathological and physiological conditions of fish. In the present communication a systematic survey was carried out to determine the extent and magnitude of organochlorine residual level in major carp *Catla catla* (Ham) collected from Mumbai local market. The fish sample was examined for residual level of organochlorine pesticide along with the nutrient value on three different seasons. The data would help in assessing the risk of human exposure to pesticides and in implementing interpreted pest management.

Materials and Methods

Fishes close to 46 cm long and 1500 g were selected from Mumbai local Market on three different seasons Rainy (June to September) Winter(December to February) and Summer (March to May) 2003-04. Liver, Muscles, Kidney and brain were removed and weighed every month. It was stored at -20°C. The lipid content of muscles and liver was determined by the method of Folch *et al.* (1957) and ash was measured by weighing samples after heating at 600°C for 12 hrs. Glycogen was determined by Seifter *et al.*, 1950, measuring the optical density at 620 nm. Protein was measured using Lowry's method. Water content of the sample was estimated by drying weighed samples at 90–110°C for five days and reweighed. Multi-residue methods of Venant *et al.* (1982) is used for residual analysis. Working standard solutions were prepared in n hexane of concentration ranging from 10-100 pg/µl. Standards were obtained from M/s. Hindustan Ciba Giegy Ltd., Bayer India Ltd., Raillis India Ltd. The reproducibility of results of all the pesticide was 89-92 per cent. Primary sample of liver, muscle, kidney and brain was kept in aluminium foil pre rinsed several times with n hexane and stored at 35°C until analysis. The solvents used were of pesticide residue grade. Perkin Elmer Gas chromatograph- Model Sigma 2000 equipped with an electron capture detector Ni 63 ECD was used with 2 to 3 packed glass columns.

1. 2.0 x ¼" OD x 2 mm ID containing 1.5 per cent SP – 2250 + 1.95 per cent SP – 2401 on 100/120 supel coport.
2. 3 per cent dexsil 300 on supel coport 100/120
3. 2 x ¼" OD x 2 mm ID glass containing 3 per cent SP – 2100 on 100/120 Supel Coport.

Operating Conditions

Temp.	Injector 270°C
Column	Oven 200°C
Detector	300°C
Flavorate	30ml/min.

Identification of peaks was done by comparison of relative retention times of the unknowns to those of standards for confirmation two or more columns where used. The reproducibility of results of all the pesticide was 89.92 per cent.

Results and Discussion

Present studies on *Catla catla* fish distributed in local market showed pesticide residue. It is especially important that this omnivorous fish might be expected to accumulate contaminants. The facts that they did so moderately suggest that available pesticide concentrations are moderately higher. This does mean that there is increase in entry of organochlorine pesticides. It is necessary to ensure that no determination occurs in future.

During winter season wet weight, dry weight, lipid, protein and glycogen content all attained peak values while in summer season all attained low values. Increase was also observed in Rainy season. The weight of the liver was peak in November and December month. The reduction in liver weight was unexpected. Highest concentration of pesticide observed in the month of March could be one of the reasons for depletion in lipid, protein and glycogen.

Table 21.1: Concentration mg/g Wet Wt. of Organochlorine insecticides in the Muscle, Liver, Brain and Kidney of *Catla catla* Sold in Mumbai Market 2003-04

Season		α BHC	β BHC	v BHC	HCB	Aldrin	o-PDDT	p-PDDT	Hepta-chlor	Dieldrin
Winter Π (14)	Muscle	ND	ND	ND	ND	0.005+ 0.002	0.007+ 0.003	0.001+ 0.002	0.002+ 0.001	0.005± 0.0043
	Brain	ND	ND	0.001	ND	0.0024	0.001	ND	ND	0.0002
	Kidney	ND	ND	0.0001	ND	ND	ND	ND	ND	Trace
	Liver	ND	ND	0.004	ND	0.002	0.002	0.023	0.013	0.002±0.001
Summer Π (15)	Muscle	ND	ND	ND	ND	0.014	0.008	0.001	0.021	0.009±0.002
	Brain	ND	ND	ND	ND	0.0014	0.001	ND	ND	Trace
	Kidney	ND	ND	ND	ND	ND	Trace	ND	ND	ND
	Liver	ND	ND	0.05	ND	0.063	0.026	0.110	0.03	0.009±0.012
Rainy Π (20)	Muscle	ND	ND	ND	ND	0.001	ND	ND	ND	0.003±0.002
	Brain	ND	ND	ND	ND	ND	0.007±0.003	ND	ND	Trace
	Kidney	ND	ND	ND	ND	ND	ND	ND	ND	ND
	Liver	ND	ND	0.01	ND	0.006	0.008	0.003	ND	ND

ND: Not Detected.

Table 21.2: Proximate Composition of the Liver and Muscle in *Catla catla* (Ham) Sold in Mumbai Market during the Year 2003-04

Season	Body wt.	Tissue	g/Wet Wt Glycogen	g/Wet Wt Lipids	g/Wet Wt Protein	g/Wet Wt Ash
Winter Π (13)	1448+49	Muscle	1.29±0.2	44.6±2.7	198.7±4.8	43.34±1.1
		Liver	14.48±1090	6.59±0.13	4.81±0.07	0.10±0.06
Summer Π (15)	1400+14	Muscle	0.96±0.1	21.3±3.8	124.7±4.8	30.10±0.8
		Liver	14.10±1.43	2.71±0.18	2.92±0.10	0.348±0.16
Rainy Π (20)	1422+13	Muscle	1.70±0.12	45.24±1.7	136.5±1.4	34.83±0.6
		Liver	14.22±1.3	6.5±0.11	4.58±0.02	0.26±0.09

± = Standard deviation

It is not possible to obtain the exact information regarding the quantity and type of pesticides exposed in Mumbai water. There is trend that apart from the persistent compound, the substance that can be metabolized are broken down easily in water. Therefore there is an unexpected increase in different spectrum of chemical substances entering into Mumbai lakes.

The present study revealed occurrence of organochlorine pesticide HCH, DDT, Aldrin, Heptachlor, Dieldrin even through their agricultural uses were discontinued. This reflects the persistence of organochlorine compound in Mumbai lakes ever a long period of time and their extensive use in the past while comparing the total DDT concentration with EPA water quality criteria of freshwater aquatic life. It appears that residue values in fish samples exceeded the standard laid down by ISEPA (1997, 1995). The highest level of pesticides recoded in summer may be due to its high use.

The finding collected from the study will provide information for making effective conservation programme and also to investigate the mechanism and path way of pesticide contamination in Mumbai aquatic biota.

Acknowledgements

This project was supported by a research grant from the University of Mumbai.

References

Kannan, K., Tanabe, S. and Tatsukawa, K., 1995. Geographical distribution and accumulation feature of organochlorine residues fish in tropical Asia and Accania. *Environ. Sci. Technol.*, 29: 2673–2683.

Seifter, S. Dayton, Novic, B. and Munt Wyler, E., 1950. The estimation of glycogen with the Antrone reagent. *Archs Biochem.*, 25: 191–200.

Senthilkumar, K., Kannan, K., Sinha, R.K., Tanabe, S. and Giesy, J.P., 1999. Bioaccumulation profiles of polychlorinated biphenyl co-geners and organochlorine pesticides in Ganga river Dolphins. *Environ. Toxical. Chem.*, 18: 1511–1520.

Venant, A., Borrel, S. and Richou-Bac, L., 1982. Method rapide pour la determination des residues de composes organochlores dans les produits laitiers et les graisses animals. *Analysis*, 10: 333–335.

U.S. Environmental Protection Agency, 1979. *Quality Criteria for Water*, Washinton D.C.

U.S. Environmental Protection Agency, 1995. Guidance for assessing chemical contaminant data for use in fish advisories.V. 1. Fish sampling and analysis, 2nd edition, EPA 823–R–97–006.

Chapter 22

In vivo Studies on the Therapeutic Values of Marine Macroalgae against the Fish Pathogen *Aeromonas hydrophila*

☆ *M.Parimala Cella and A.P. Lipton[1]*

ABSTRACT

This study is an integrated systemic approach to develop immunostimulation or disease resistant property in ornamental gold fish *Carassius auratus* that results in longer life span as well as increased body weight in fishes. Macroalgae were reported to possess a variety of therapeutic property. Hence the present study has made an attempt to explore therapeutic compounds from selected macroalgae. This study investigates the potential therapeutic property of the methanolic extracts of macroalgae *viz., Ulva fasciata, Gracilaria corticata, Enteromorpha compressa, Hypnea musciformis* and *Sargassam wightii*. The gold fish *Carassius auratus* in 10 numbers were acclimatized in the fish tank for experimental purpose and were fed with normal fish feed. The LD$_{50}$ value was calculated and the 50 per cent death was noted at 10^6 dilution and the bacterial titre was calculated as 12×10^6 CFU/ml. After acclimatization, macroalgae extract top coated fish feeds in three different concentrations (1mg, 10mg and 20mg) were provided to each set of ten fishes for 30 days and normal fish feed was provided to the control fishes. After thirty days, macroalgae extract topcoated fish feed fed gold fishes were injected with the test pathogen *Aeromonas hydrophila* by intradermal injection.The test fishes were observed for the development of disease symptoms or mortality. The control fishes were injected with saline and were observed for disease symptoms or mortality. Similarly the body weight of the fishes has also been observed after feeding with medicated (macroalgal extract topcoated) fish feed. 100 per cent survival rate was observed in test fishes fed with the extracts of

Sargassam wightii, better survival was noted with *Gracilaria corticata* and *Ulva fasciata.* All the selected macroalgae showed good survival percentage with 10mg concentration of extract top-coated fish feed consumed fishes. Considerable weight gain has been noted with macroalgae extract fed fishes than control fishes.

Keywords: Fish pathogen, *Aeromonas hydrophila, Therapeutic value, Macroalgae–Carassius auratus, Ulva fasciata, Gracilaria corticata, Sargassam wightii, Enteromorpha compressa, Hypnea musciformis.*

Introduction

Fish diseases may occur due to many reasons such as nature of water, type of feed, salt content and external infectious agents such as parasites, bacteria, fungus and viruses. The oxygen content and pH plays a major role.Similarly the parasites attached to the gills and the body of the fish creates a lot of problems to the fish farms. Many bacteria cause infections in fish and cause ulcers, blood spots or misty eyes on susceptible fishes. Similarly fungal infections can also be noted on the skin of infected fishes. Viral infection may cause sudden death of fish masses. The bioactive compounds isolated from the macroalgae showed remedial measures against fish infections caused by *Vibrio parahaemolyticus, Aeromonas hydrophila,* etc and gradually increased the resistant property and prolonged the survival rate of the fishes. Macro algae are the floating submerged plants of shallow marine meadow having salt tolerance. Macroalgae is the common term used for the gelatinous or leathery, brown green or red plant like organism found in the sea shore and in well illuminated sub tidal waters. They are commonly called as "Algae". Like those of bacteria, fungi, plants and animals, the algae are the photosynthetic members of "Protista", according to the five kingdom concept. Macroalgae are distinguished by their biochemistry, appearance and life cycle. The simplest macroalgae is made up of a holdfast and a thallus. The holdfast holds the the algae to the rock. Absorption of mineral nutrients takes place directly through the walls of the thallus. The macroalgae are the primary producers of rocky seashores. The macroalgae are found to contain vitamins and variety of trace elements. Many macroalgae are edible and are useful in the production of salads, soups, jellies and in vinegar production. Macroalgae associated foods are more popular in many countries such as Malaysia, Indonesia, Korea, Australia, Singapore and Japan. Variety of algae contain 10 times more calcium than milk. Macroalgae are rich source of iodine needed by the thyroid gland to produce the hormone thyroxine. Macroalgae contain high amount of calcium and phosphorous. They have high calorific value. Important chemical substances obtained from macroalgae are alginic acid, mannitol, laminarin, fucoidin, iodine, agar, agarose and carageenan. In Asian maritime areas macroalgae extracts were used as curative or preventive agents for various maladies and can serves as antibiotics, anthelmintics, antihypertensive, antipyretic, antitumour, antidiarrhea, wound healings, upper respiratory tract infections, post operative infections (Ravikumar *et al.,* 2002), treatment for gallstones and goiter (Chengkui and Chang, 1984). Macroalgae serve as a potent reservoir of many bioactive metabolites. These bioactive compounds have the potential to treat many plant, animal, human and fish pathogens. Some macroalgae have antifungal activities also (Chesters and Stott, 1956). Hepatoprotective nature of macroalgae has been reported early. Fattyacids, terpenes, carbonyls and bromo phenol compounds in macroalgae are responsible for antibiotic activity. Many types of macroalgae are useful in the treatment for the degenerative changes of nerves, chronic bronchitis and also in the edema of the foot. It is also useful in treatment of cancer, tumour, dissolving fat deposits; eliminate heavy metal contamination from the body, arthritis and Rheumatics.

Materials and Methods

Collection, Identification and Processing of Nacroalgae

The 5 different types of marine macroalgae such as *Ulva fasciata, Sargassam wightii,Gracilaria corticata, Enteromorpha compressa* and *Hypnea musciformis* were collected from the rocky shores of Colachel, Muttam and Kanyakumari, Tamil Nadu, India. Only healthy full- grown plants still submerged under water during low tides were collected. The sampled algae were cleaned repeatedly with freshwater to remove epiphytes, shells etc attached to the algal biomass. Then the samples were brought to the laboratory and then thoroughly washed with freshwater fo llowed by distilled water to remove the salt remains on the surface of the samples. After draining the water, macroalgal samples were spread out and were shade dried for nearly 10 days. After thorough drying, dried samples were then powdered, sieved and were stored separately in clean plastic containers for further use.

Extraction of Macroalgae Using Soxhlet Apparatus

Nearly 125g of finely powdered macroalgae was added to the soxhlet apparatus and was extracted with the standard solvent methanol (125ml) and the extraction process was carried out at 30-40 °C. After the complete extraction of macroalgae powder the extract was collected in the pre-weighed glass beaker and the solvent was allowed to evaporate in a rotary evaporator so that a semisolid mass of macroalgal extract was obtained. From this stock, macroalgae extracts in different concentrations such as 1mg, 10mg and 20 mg were prepared by adding solvent methanol to the extracts of macroalgae.

Preparation of Macroalgal Extract Top-Coated Fish Feed

The selected 5 macroalgae were extracted individually and the methanol extract of individual macroalgae in three different concentrations such as 1mg, 10mg and 20mg quantities were added separately into 100 g of fish feed. The fish feed used belong to the company "**TRIO**". It is found to contain as follows (Table 22.1).

Table 22.1: Different Concentration of Ingredients and their Percentage Composition

Sl.No.	Ingredients	Percentage
1.	Crude protein min	32 per cent
2.	Crude ash max	10 per cent
3.	Crude fat min	4 per cent
4.	Moisture max	9 per cent
5.	Crude fibre max	5 per cent
6.	Nitrogen free extract min	31 per cent

Mainly it contains carotenoid pigment to enhance colour in most tropical fishes. The individual macroalgal extracts in 3 different concentrations such as 1mg, 10mg and 20mg were topcoated over the fish feed by spraying using 4 per cent gelatin as a binding agent.

Testing the Therapeutic Efficacy of Macroalgae against the Fish *Carassius auratus*

Acclimatization of *C. auratus* for Experimental Trial

The gold fish *C. auratus* in 10 numbers were reared in the fish tank with a length of 58cm, breadth of 30cm.The water was added to the fish tank upto the 3/4th the volume of fish tank with aerators and filters for better survival of gold fishes.

Lethal Dose Determination in *C. auratus* (LD$_{50}$ Determination)

The fish pathogen, *A. hydrophila* stored in slant culture was activated in nutrient broth. From this, 1ml of 18 hrs culture was inoculated in nutrient broth and was kept overnight at 30±2úC in the mechanical shaker at 80 rpm. Then the culture was centrifuged at 10,000rpm for 5 minutes and was washed twice in normal sterile saline (0.85 per cent NaCl). The pellet was serially diluted and plated on nutrient agar and 0.1ml quantities of individual dilutions (10^{-3} to 10^{-7}) were injected separately into a set of 4 fishes and the appearance of diseased symptoms and mortality rate was found out. The highest dilution of the culture which produced death in 50 per cent of the fishes *i.e.* LD$_{50}$ was found out which was used for injection purpose in test fishes.

Testing the Therapeutic Potential of Macroalgae in Test Fishes

A set of *C. auratus* in 10 numbers were fed with the macroalgae extract top-coated fish feed for 30 days. Then they were injected with the fish pathogen *A. hydrophila* in 0.1ml quantity (LD50) by intrdermal inoculation using tuberculin syringe and the control fishes (10 numbers) were injected with 0.1ml saline. The rate of mortality and disease symptoms were observed for 30 days. Any death occurring within 24hrs should be considered nonspecific. Then specific signs and disease symptoms of *Aeromonas* infection were observed. Total number of specific death in each dilution and the bacterial titre was calculated by using Reed and Muench (1938).

Similarly the macroalgal extracts of *Sargassam, Gracilaria, Enteromorpha, Hypnea* and *Ulva* in different concentrations such as 1mg, 10mg and 20mg were topcoated over fish feed Trio in 100gram quantity and were applied to test fishes. Each set of fishes after 30 days of feeding(medicated fish feed), they were injected with the test pathogen *A. hydrophila* in 0.1ml quantities by intra-dermal inoculation and then they were observed for any disease symptoms or mortality for 30 days. After 30 days, the survival percentage was calculated in each set of fishes using the following formula:

$$\text{Survival percentage} = \frac{\text{Total number of fishes on day 30}}{\text{Initial number of fishes}} \times 100$$

Similarly absolute growth rate (AGR) can be calculated using the following formula:

$$\text{Absolute growth rate (AGR)} = \frac{\text{Final body weight} - \text{Initial body weight}}{\text{Total number of days}}$$

The mortality rate of fishes were calculated using the following formula:

$$\text{Percentage of Mortality} = \frac{\text{Number of dead fish}}{\text{Total number of infected fish}} \times 100$$

Results

The methanolic extracts of macroalgae such as *U. fasciata, S. wightii, G. corticata, E. compressa* and *H. musciformis* were prepared and used as immunomodulators in test fishes and were challenged with the test pathogen *A. hydrophila* (0.1ml) by intradermal inoculation.

The LD$_{50}$ value was calculated by Reed and Muench and the 50 per cent death was noted at 10^6 dilutions and the bacterial titre was calculated as 12×10^6 CFU/ml (Table 22.2). In case of *G. corticata* extract top-coated fish feed consumed fishes, the lowest concentration (1mg/100g of fish feed) of macroalgal extract when incorporated into the fish feed for 30 days has considerably raised the survival of gold fishes. Out of 10, 8 were alive and after injecting with the test pathogen *A. hydrophila* only 4 out of 8 remained alive whereas in the control set of fishes only 2 were alive. In the control set of fishes, the fishes were dead due to the development of various disease symptoms such as white spot, dermatitis, fungal infection etc. Whereas in case of 10mg/100g of macroalgal extract top-coated fish feed fed fishes they showed a better survival *i.e.,* out of 10, nine were alive. In this set of fishes even after injection with the test pathogen *A. hydrophila* remained healthy without only disease symptoms.

Table 22.2: Computation of LD$_{50}$ Titre by Reed and Muench (1938)

Sl.No.	Aeromonas Culture per cent Dilution	Number of Gold Fishes		Mortality Ratio	Bacterial Titre
		Died	Survived		
1.	10^{-3}	4	0	100 per cent	TNTC
2.	10^{-4}	4	0	100 per cent	282
3.	10^{-5}	3	1	75 per cent	40
4.	10^{-6}	2	2	50 per cent	12
5.	10^{-7}	1	3	25 per cent	–

Table 22.3: Survival Studies of *C. auratus* by Top-coating the Fish Feeds with Selected Macroalgal Extracts in Different Concentrations

Seaweed	Seaweed Concentration (mg)	Survival of Carassius auratus			Survival (per cent)
		During Acclimatization	After Feeding with Seaweed Top-coated Fish Feed	After injecting with A. hydrophila	
S.wightii	1	10	9	4	44
	10	10	10	9	90
	20	10	5	3	60
G. corticata	1	10	8	4	50
	10	10	9	8	89
	20	10	4	Nil	0
U. fasciata	1	10	6	2	33
	10	10	8	3	38
	20	10	3	Nil	0
Enteromorpha compressa	1	10	7	2	29
	10	10	7	3	43
	20	10	5	Nil	0
Hypnea	1	10	6	1	17
	10	10	8	2	25
	20	10	5	Nil	0
Control	-	10	Normal fish feed Trio	2	20

In (20mg/100g) of macroalgal extract top-coated fish feed fed fishes, the survival has been gradually decreased to 4. In such cases, after injection with the test pathogen *A. hydrophila* nothing remained live (Table 22.3).

Among the 3 concentrations of *Gracilaria* extract incorporated fish feed consumed fishes the highest survival percentage(89 per cent) was found in 10mg concentration of algal extract. 20mg concentration showed highest mortality rate (100 per cent.).The absolute growth rate of fishes (AGR) fed with the extract of *Gracilaria* was found to be 0.017g/day (Table 22.4).

Table 22.4: Absolute Growth Rate of *Carassius auratus*

Seaweed	Weight of Carassius auratus			
	During Acclimatization	Feeding with Medicated Fish Feed	After injecting with A. hydrophila	Absolute Growth Rate (AGR)
Sargassam wightii	7.0g	7.5g	8g	0.017g
Gracilaria corticata	5.5g	6.0g	6.5g	0.017g
Ulva fasciata	4.0g	4.5g	5g	0.017g
Enteromorpha compressa	7.5	7.9	8.2g	0.013g
Hypnea	7.5	7.9	8.3g	0.013g
Control fish (without seaweed)	5.0	5.2	5.4	0.007g

In case of *U. fasciata* extract top-coated fish feed consumed fishes, the lowest concentration (1mg/100g of fish feed) of macroalgal extract when incorporated into the fish feed for 30 days. Out of 10, 6were alive and after injection with the test pathogen *A. hydrophila* only two out of 6 remained alive where as in the control set of fishes 2 were alive. In case of (10mg/100g) of macroalgal extract (Ulva) top-coated fish feed fed fishes, they showed a better survival rate *i.e.*, out of 10, 8 were alive and after injection with the test pathogen *A. hydrophila* only 3 were alive. In 20mg/100g of macroalgal extract top-coated fish-feed fed fishes, the survival rate has been gradually decreased to 3 out of 10. In such cases, after injection with the test pathogen (*A. hydrophila)* all were died.

Among the three concentrations of *U. fasciata* extract top-coated fish-feed consumed fishes, the highest survival percentage (38 per cent) was noted with 10 mg concentration of macroalgae. 20mg concentration showed 100 per cent mortality rate. The absolute growth rate of the fishes fed with the extract of *U. fasciata* was found to be 0.017g/day (Figure 22.1).

In case of *S. wightii* extract top-coated fish feed consumed fishes, the lowest concentration (1mg/100g of fish feed) of macroalgal extract, when incorporated into the fish feed for 30 days Out of 10, nine were alive and after injection with the test pathogen *A. hydrophila* only 4 out of 9 remained alive whereas in the control set of fishes only 2 were alive. In case of (10mg/100g) of macroalgal extract(*Sargassam)* incorporated fish feed fed fishes they showed a better survival rate *i.e.*, out of 10,all are alive. In this set of fishes, after injection with the test pathogen, 9 remained healthy without any disease symptoms.

In (20mg/100g) of macroalgae incorporated fish feed fed fishes, the survival has been gradually decreased to 5 out of 10. In such cases, after injection with the test pathogen only 3 remained healthy. Among the 3 concentrations of *Sargassam* extract incorporated fish feed consumed fishes, the highest survival percentage(90 per cent) was noted with 10mg concentration of macroalgae and 1mg showed

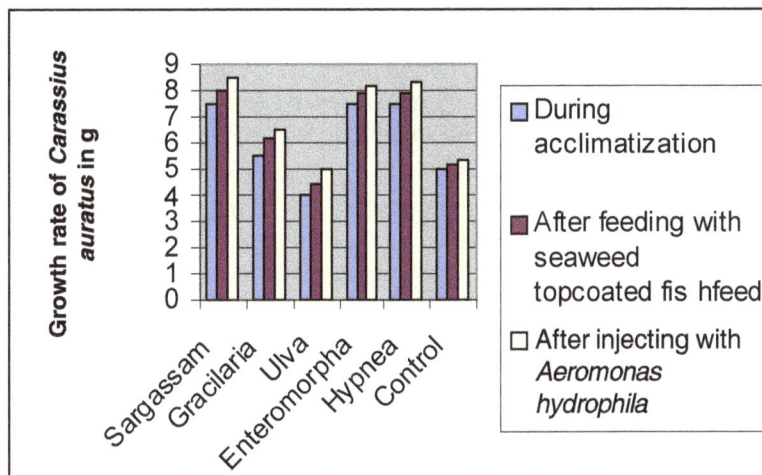

Figure 22.1: Absolute Growth Rate of *Carassius auratus*

the minimal survival rate(44 per cent). The absolute growth rate of the gold fishes fed with the extract of *S. wightii* was found to be 0.017g/day.

In case of *Enteromorpha* extract top-coated fish feed consumed fishes, the lowest concentration (1mg/100g of fish feed) of macroalgal extract, when incorporated into the fish feed for 30 days. Out of 10, seven were alive and after injection with the test pathogen *A. hydrophila* only 2 out of 7 remained alive. Whereas in case of 10mg/100g of macroalgal extract (*Enteromorpha*) incorporated fish feed fed fishes they showed a similar survival rate *i.e.* out of 10, 7 were alive. In this set of fishes, after injection with the test pathogen, 3 remained healthy without any disease symptoms.

In 20mg/100g of macroalgae top-coated fish feed fed fishes, the survival has been gradually decreased to 5 out of 10. In such cases, after injection with the test pathogen nothing remained alive. Among the 3 concentrations of *Enteromorpha* extract incorporated fish feed consumed fishes, the highest survival rate (43 per cent) was noted with 10mg concentration of macroalgae. 20mg concentration showed highest mortality rate (100 per cent). The absolute growth rate of the fishes fed with the extract of *Enteromorpha* was found to be 0.01g/day.

In case of *H. musciformis* extract incorporated fish feed consumed fishes, the lowest concentration 1mg/100g of fish feed of macroalgal extract, when incorporated into the fish feed for 30 days, Out of 10, six were alive and after injection with the test pathogen *A.hydrophila* only one out of 6 remained alive. Whereas in case of 10mg/100g of macroalgal extract topcoated fish feed fed fishes they showed a better survival rate *i.e.*, out of 10, 8 were alive. In this set of fishes, after injection with the test pathogen, 2 remained healthy without any disease symptoms.

In (20mg/100g) of macroalgae incorporated fish feed fed fishes, the survival has been gradually decreased to 5 out of 10. In such cases, after injection with the test pathogen nothing remained alive.

Among the 3 concentrations of *H. musciformis* extract incorporated fish feed consumed fishes, the highest survival percentage(25 per cent) was noted with 10mg concentration of macroalgae and 20mg showed highest mortality rate(100 per cent).The absolute growth rate of the fishes fed with the extract of *Hypnea* was found to be 0.013g/day. The absolute growth rate of the control fishes were found to be 0.006g/day which is relatively lesser than the absolute growth rate of gold fishes fed with the extract of the selected 5 macroalgae.

Discussion

Bioactive compounds from macroalgae have a potential role in treating various types of diseases of human beings. Marine biomass acts as a potential reservoir for wonderful therapeutic compounds. Macroalgae have potential antioxidant and antimicrobial activity which is used to cure even dreadful diseases, physiological disorders and malignancies.

Both invitro and invivo studies on the therapeutic value of macroalgae against the fish pathogen have been carried out using the selected macroalgae such as *U. fasciata, G. corticata, S. wightii, H. musciformis* and *E. compressa*. These macroalgae have been produced commercially for food and feed purpose by many people. At the same time, these macroalgae were also studied for their therapeutic activity

In the current study, the extract of the selected macroalgae were tested against the fish pathogen *A. hydrophila* under *in vivo* conditions. Feeding the fishes with macroalgae top-coated fish feed has showed considerable effect in extending the survival rate of the ornamental gold fish *C. auratus*, when compared with the control fishes which had been fed with the normally available commercial fish feed Trio.

Among the 3 concentrations of macroalgae selected for studying its therapeutic value against the fish pathogen *A. hydrophila*, 10mg concentration of macroalgal extract per 100g of fish feed Trio exerted the prolonged survival of the goldfishes when compared to 1mg and 20mg concentration.

The 20mg concentration of macroalgal extract topcoated fishfeed fed fishes showed 100 per cent mortality in case of *G. corticata* and *U. fasciata*. This indicates that 10mg/100g of macroalgal extract topcoated fishfeed is the suitable concentration for feeding the fishes to get rid of the infection from *A. hydrophila*. The high concentration of macroalgal (20mg/100g) extract was found to be the lethal dose for these gold fishes.

The gram negative bacteria *A. hydrophila* develop furunculosis and septicaemia in the cultivated fish stocks of both freshwater and marine fishes. The pathogenecity of *A. hydrophila* due to its toxin production and virulence factors has been explained by Ian *et al.* (1985). Ramasamy *et al.* (2010) reported that *A. hydrophila* can induce ulcerative dermatitis on intramuscular injection in gold fishes. In the current study due to *A. hydrophila* infection, *C. auratus* has developed dropsy, fungal infection, ulcerative disease, White spots etc., in the control set of fishes. But in macroalgal extract top-coated fish feed fed fishes, no such symptoms have been developed and they were found to be free from infection.

In addition to this, the macroalgal extract top-coated fish feed consumed fishes showed remarkable improvement in their body weight when compared to the control set of fishes. All the fishes were fed with 5 per cent of the body weight of the fishes. The enhanced growth rate of this test fishes indicates that some of the growth promoting compounds, *i.e.*, growth factor may be present in the macroalgae.

Acknowledgements

First author is grateful to the authorities of her college for the facilities provided.

References

Chesters, C.G. and Stott, J.A., 1956. Production of antibiotic substances by seaweeds. In: *Seaweed Symp.*, 2: 49–54.

Chengkui, Z. and Junfu, Z., 1984. Chinese seaweeds in herbal medicine. *Hydrobiologia*, 116/117: 152–154.

Ian, M. Watson, Jennifer O. Robinson, Valerie Burke. and Michael Gracey, 1985. Invasiveness of *Aeromonas* sps. in relation to biotype, virulence factors and clinical features. *J. Clinical Microbiol.,* 22: 48–51.

Ramasamy Harykrishnan, Chellam Balasundaram, Young-Gun Moon, Man Chul Kim, Subramanian Dharaneedaran and Moon-Soo Heo, 2010. Phytotherapy of *Ulcerative dermatitis* by *Aeromonas hydrophila* infection in gold fish *Carassius auratus. Acta Veterinaria Hungarica,* 10: 29–37.

Ravi Kumar, S., Anburajan, L., Ramanathan, G. and Kaliaperumal, N., 2002. Screening of seaweed extracts against antibiotic resistant post operative infectious pathogens. *Seaweed Res. Utiln.,* 24: 95–99.

Chapter 23

Inverse Relationship of Energy Reserves and Body Water Content in *Anabas testudineus* (Bloch) Under Starvation Stress

☆ *Padmavathi Godavarthy and Y. Sunila Kumari*

ABSTRACT

Anabas is a migratory fish known to tolerate prolonged periods of starvation. In the present starvation study on *Anabas*, we have observed a increase in body water content, which may be due to selective mobilization and utilization of proteins and lipids, as their oxidation yields water and CO_2. A close relationship seems to exist between body water content and utilization of energy reserves such as proteins and lipids and therefore measurement of body water content acts as is a good indicator of nutritional status of fish. Body water content was estimated by measuring the difference in weight before and after oven drying. Brief fasting of a fortnight and prolonged fasting for two months lead to an increase in the body water content which however was not statistically significant. The fish adapted well for the fasting stress and survived by selectively mobilizing and utilizing carbohydrates, lipids and proteins as indicated by the elevation in body water content. Perhaps if the starvation period was prolonged further there may be a significant increase observed in the body water content with greater utilization of energy reserves.

Keywords: Body water content, Anabas, Protein-water relationship, Lipid-water relationship, Cyclic temperature.

Introduction

Fish forms an important constituent of human diet. Surviving successful, prolonged periods of starvation is often seen in most of the species of fish. Many species of teleosts demonstrate the ability to withstand prolonged periods of starvation (Larsson and Lewander, 1973, Loughna and Goldspink, 1984; Machado *et al.*, 1988). At low temperatures some fish are known to survive for survive months or even years (Abraham, *et al.*, 1984, Love, 1970). Mobilization of energy reserves may also be related to nutritional status (Foster, *et al.*, 1993) diet composition, Machado *et al.*, 1988) endogenous rhythms such as reproductive cycle (Love, 1970) or to any other environmental factors such as temperature (Pastoureand, A., 1991) and salinity (Woo and Murat, 1981, Jurrs *et al.*, 1986).Irrespective of particular environment of individual species, mobilization of endogenous energy sources would be expected to proceed such that energy requirements are met in the most efficient manner. Proximate composition of body constituents which includes analysis of water is a good indicator of physiology which is needed for routine analysis of fisheries (Cui and Wootton, 1988). Starvation studies, by authors such as Sargent *et al.* (1989); Salem and Davies, (1994); Sinclair and Duncan, (1977); have observed lipid or protein mobilization with a corresponding increase in the body water contents so as to provide energy for the starving animal. In another study, by Naeem *et al.* (2011) on *Notopterus notopterus*, where a proximate body composition analysis revealed a loss of protein and increase in body water content, suggesting a correlation between energy reserves and water content.

Based upon these studies it is evident that there exists a correlation between utilization of protein/ lipids with body water content in a starving animal and so the present investigation is an attempt to study the fat-water relationship and protein water-relationship, when *Anabas* was subjected to brief and prolonged fasting.

Materials and Methods

Experimental Procedure

Fish weighing of 20-25g were obtained from Kolleru Lake of Eluru. Care was taken to ensure quick transport to the laboratory. Overcrowding was avoided during packing to minimize the mortality rate.

They were carefully transferred into Durex storage tank of capacity 500 lits. made of a material which was corrosive resistant, polypropylene. The closed plastic lid of the tank was replaced by a grill lid made of iron. This helped in proper ventilation and aeration of the tank. Fish which were injured or dead were removed from the tank from time to time. Disinfectant (KMnO$_4$) was used to avoid infection. They were given boiled egg, rice bran meal and commercial fish feed *ad. libitum*. Any leftover feed and fecal matter were removed daily. Water in the tank was changed every day. Fish were brought to the laboratory and sufficient time was allowed for acclimatization. Experimentation was done thereafter.

Fish measuring about 3'-4" in length and in the same range of weight were selected carefully. They were grouped together and kept in circular tubs made of plastic. The mouth of these tubs was covered with fine mesh and appropriately placed such that they were properly ventilated and well aerated. Two types of experimental set up were designed.

In the first set up, fishes were allowed to starve for 15 days and parallel control was also maintained. The control animals were fed regularly both in the morning and evening. On the 16th day both experimental and control animals were used for experimentation. The II experimental set up consisted of fishes which were allowed to starve for 2 months (long-term). Experimental group and a corresponding control group were maintained. Control group was fed regularly as in the case of short term. On the 61st day, the animals were used for experimentation.

Measurement of Body Water Content

Fishes were removed carefully from the experimental and control tubs, gently blotted using filter paper. Initial weights of both control and experimental animals were noted. The animals were placed in suitable containers and placed in an oven at a temperature of 800–1000°C and left for incubation over night.

The final weights of both control and experimental animals were noted the following day. Body water content can be calculated from the difference in the two weights and is expressed as gain in water content in grams.

Results

Body water content showed an increase during starvation period. Starvations for a fortnight lead to an increase of +12.24 per cent in the water content. This increase was not statistically significant. There was a corresponding decrease in the body weight of the animal.

Body water content showed an increase during long-term fasting. The increase in the water content was found to be 4.65 per cent (NS). The loss in body wt. of the animal coincided with increase in water content of the fish.

Table 23.1: Body Water Content of *A. testudineus* during Short Term Starvation

Sl.No.	Control Group (6)			Experiment Group (6)		
	Wt. Before Drying (W1)	Wt. after Drying (W2)	Gain in Water Content W1-W2	Wt. Before Drying (W1)	Wt. after Drying (W2)	Gain in Water Content W1-W2
1.	20.333	6	14.333	25.833%	9.5 Variation =	16.333NS +12.24

Table 23.2: Body Water Content of *A. testudineus* during Long Term Starvation

Sl.No.	Control Group (6)			Experiment Group (6)		
	Wt. Before Drying (W1)	Wt. after Drying (W2)	Gain in Water Content W1-W2	Wt. Before Drying (W1)	Wt. after Drying (W2)	Gain in Water Content W1-W2
1.	20.333	6	14.333SE = 1.115	25	10	15[NS]SE = 1.2111

% Variation = +4.65

Values expressed as gain in water content in g.

Each value is mean of SE ± 6 individual observations.

NS = Not significant

Discussion

The present study showed an insignificnat elevation of total body water content during both the starvations regimes,suggesting both lipid and protein mobilization and utilization by the starvaing *Anabas*, so as to over come the stressful period of staravtion. Lipid mobilization and fatty acid oxidation yields water and CO_2 and therefore explains the increase in water content during fasting, as suggested by Sargent *et al.* (1989), and this rise of water levels may be used as an indicator to measure the nutritional condition of fioh (Lovo, 1970).

Our observations coincide with that of the Naeem *et al.* (2011), who have also observed an increase in the body water content with a concomitent loss of muscle protein in *Notopterus notopterus* from Indus river.A similar kind of observations have also been reported by Collins and Anderson (1995) and Ince and Thorpe (1976) who have all observed a non significant increase in the water content of the carcass or muscle of fish - golden perch, which was starved for 210 days. Starvation of European eel(*Anguilla anguilla*) for 96 days also resulted in an increase in body water content of liver and muscle, with a corresponding decline in triglyceride content.(Dave *et al.,* 1975) These authors opined that their results are in accordance with the rule of fat-water relationship observed in fatty fish as suggested by Love, (1970)

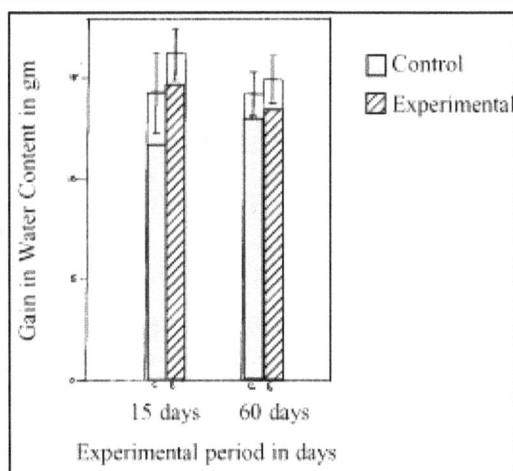

Figure 23.1: Body Water Content

A starvation experiment by Jobling (1980) has also reported a decrease in the carcass lipid content coupled with an increase in the percentage of moisture and ash content of the tissue of *Pleuronectes platessa*. Similarly in *Gadus morhua* there was an increase in the water content and sodium content during starvation with a corresponding sequential mobilization of liver lipid and muscle lipid (Love, 1970). Stirling (1976) has recorded an inverse relationship, both with that of lipid and proteins to the moisture content of the whole fish in *Dicentracus labrax.* Tanaka (1969a) has demonstrated a clear relationship between, the rise in muscle water with starvation and a fall in muscle glycogen, supporting the concept that protein is mobilized only when, the more readily available energy reserves have been used. In accordance with these studies, our own observations on starving *Anabas,* revealed a significant decline in the total tissue lipids (Godavarthy and Premakumari, 2006), suggesting a lipid utilization, and so the present observation of rise in body water content may be attributed to this lipid depletion. In another study by Mehner *et al.* (2011) it was observed that there was a correlation existing between, biochemical body composition and exposure of fish to cyclic temperatures. These authors opined that fish exposed to cyclic temperatures had a lower lipid and higher water content than that of the fish which were maintained at high temperatures. Apart from this, it has also been suggested by Love, (1960) that the tissue hydration in fish is known to be affected by size factor and also by the seasonal variation. These factors may also seem to contribute to the body water content of *Anabas* which needs to be investigated. Therefore, it may be concluded that the present rise in body water content with a corresponding decline in certain energy reserves suggests a close relationship between energy reserves and body hydration. The rise was found to be insignificant; perhaps if *Anabas* was starved further, we may observe a significant rise in the body water content.

Conclusion

In conclusion, it may be said that the general rise in the body water content observed in starving *Anabas*, during brief and prolonged fasting, may be due to selective mobilization of different energy resources such as glycogen, lipid and or proteins there by suggesting a lipid- water/protein water relationship. Factors such as temperature and seasonal variations may also contribute to this change.

References

Abraham, S., Heinz, J.M. Hansen and Hansen, Finn N., 1984. The effect of prolonged fasting on total lipid synthesis and enzyme activities in the liver of the European eel (*Anguilla anguilla*). *Comparative Biochemistry and Physiology*, 79B(2): 285–289.

Cui, Y. and Wootton, R.J., 1988. Effects of ration, temperature and body size on the body composition, energy content and condition of Minnow (*Phoxinus phoxinus* L.). *J. Fish Biol.*, 32: 749–764.

Collins, A.L. and Anderson, J.A., 1995. The regulation of endogenous energy stores during starvation and refeeding in the somatic tissues of golden perch. *J. of Fish Biology*, 47: 1004–1015.

Dave Goran, Maj-Lis Johansson–Sjobeck, Ake Larsson, Kerstin Lewander and Ulf Lidman, 1975. Metabolic and haematological effects of starvation in the European Eel, *Anguilla anguilla* L.–I. Carbohydrate, lipid protein and inorganic ion metabolism. *Comp. Biochem. Physiology*, 52A: 423–430.

Foster, A.R., Houlihan, D.F. and Hall, S.J., 1993. Effects of nutritional regime on correlates of growth rate in juvenile atlantic cod (*Gadus morhua*): Comparison of morphological and biochemical measurements. *Canadian Journal of Fisheries and Aquatic Science*, 50: 502–512.

Godavarthy, P. and Premakumari, R., 2006. Impact of starvation on total tissue Lipid content in Climbing Perch, *Anabas testudineus* (Bloch). *J. Aquatic Biology*, 21(1): 147–150.

Ince, B. and Thrope, A., 1976. The effects of starvation and force feeding on the metabolism of northern pike, *Esox lucius* L. *Journal of Fish Biology*, 8: 79–88.

Jurss, K., Bittorf, T. and Volker, T., 1986. Influence of salinity and food deprivation on growth, RNA/DNA ratio and certain enzyme activities in the rainbow trout, *Salmo gairdneri* (Richardson). *Comparative Biochemistry and Physiology*, 83B: 425–433.

Jobling, M., 1980 Effects of starvation on proximate chemical composition and energy utilization of plaice, *Pleuronectes platessa* L. *J. Fish Biology*, 17: 325–334.

Larsson, Ake and Lewander, Kerstin, 1973. Metabolic effects of starvation in the eel, *Anguilla anguilla* L. *Comp. Biochem. Physiol.*, 44A: 367–374.

Loughna, P.T. and Goldspink, G., 1984. The effects of starvation upon protein turnover in red and white myotomal muscle of the rainbow trout *Salmo gairdneri* Richardson. *Journal of Fish Biology*, 25: 223–230.

Love, R.M., 1960. Water content of Cod (*Gadus Callarias* L.) Muscle. *Nature*, 185: 692.

Love, R.M., 1970. *The Chemical Biology of Fishes*. Academic Press, London and New York.

Machado, C.R., Garofalo, M.A.R., Roselino, J.E.S., Kettelhut, I.C. and Migliorh R.H., 1988. Effects of starvation, refeeding and insulin on energy linked metabolic processes in cat fish (*Rhamdia hilarii*) adapted to a carbohydrate rich diet. *General and Comparative Endocrinology*, 71: 429–437.

Mehner Thomas, Suzzane Schiller, George Staaks and Jan Ohlberger, 2011. Cyclic temperatures influence growth efficiency and biochemical body composition of vertically migrating fish. *Freshwater Biology*–doi: 10.1111/j.1365–2427.2011.02594.x

Naeem, Muhammad, Azhar Rasal, Abdus Salam, Shahidiqbal, Abir Istiaq, Muhammed Ather, 2011. Proximate analysis of female population of wild feather back fish (*Notopterus notopterus*) in relation to body size and condition factor. *African Journal of Biotechnology*, 10(19): 3867–3871.

Pastoureand, A., 1991. Influence of starvation at low temperatures on utilization of energy reserves, appetite recovery and growth character in sea bass, *Dicentrarchus labrax*. *Aquaculture,* 99: 167–178.

Sargent, J., Henderson, R.J. and Tocher, D.R., 1989. The lipids. In: *Fish Nutrition,* 2nd Edn. (Ed.) J. Halver. Academic Press, London, pp. 153–216.

Stirling, H.P., 1976. Effects of experimental feeding and starvation on the proximate composition of European bass. *Dicentraehus labrax. Mar. Biol.,* 34: 85–91, 93, 134, 182, 184.

Tanaka, T., 1969a. Biochemical studies on wintering of culture carp. III. Changes in moisture content and water holding capacity of carp muscle. *Bull. Tokai Reg. Fish. Res. Lab.,* 59: 29–47, 184, 185, 361.

Chapter 24

Physico-chemical Parameters and Zooplankton Population in a Tamdalge Tank, Maharashtra

☆ *S.A. Manjare and D.V. Muley*

ABSTRACT

Study on hydrobiological status of Tamdalge tank water was made to assess the potability of water from January 2009 to December 2009. Some physico-chemical parameters were considered such as temperature, turbidity, pH, dissolved oxygen, carbon-dioxide, hardness, alkalinity, phosphate and nitrate. Among the zooplankton population four group of rotifera, cladocera, copepoda and ostracoda were studied.

Keywords: Perennial tank, Physico-chemical parameters, Monthly variation, Zooplankton.

Introduction

The interrelationship between the physico-chemical parameters and plankton production of tank water and its relation with fluctuation of zooloplankton are of great importance and basically essential in fish culture. Fishes are dependent on physico-chemical parameters. Any change of these parameters may affect the growth, development and maturity of fish (Jhingran, 1985). Different causal influences, which determine the quality of water, show a characteristic change from season to season (Zafar, 1964; Munwar, 1970; Islam *et al.*, 1974; Nasar, *et al.*, 1991).

Zooplankton constitutes important food item of many fishes. The larvae of carps feed mostly on zooplankton (Dewan *et al.*, 1977). Zooplankton also plays an important role in the food chain as they are second in tropic level as primary consumers and also as contributers to next tropic level (Qasim, 1977).

Sunkand and Patil (2004) and Islam (2007) have worked on the physico-chemical condition and seasonal variation of zooplankton.

The present study was made on water quality and occurrence of some zooplankton with respect to physicochemical parameters in Tamdalge tank of Kolhapur district (Maharashtra).

Materials and Methods

The study was conducted on Tamdalge tank situated 45km away from Kolhapur city which lies between 16°49′ 29.01″ N latitude and 74°27′ 49.94″E longitude, from January 2009 to December 2009. The area of the tank is 100 ha. Rainfall is the only source of water for tank. The water of the tank is clear and few aquatic plants are present.

Water samples were collected monthly in the morning at 8 am to 10 am from surface layer of the tank. Water temperature was recorded by a centigrade thermometer and pH was recorded by pocket digital pH meter. Turbidity was recorded by Nephloturbidity meter. Dissolved oxygen was determined by Winkler's method, free carbon-dioxide was determined by titrimetric method. Hardness was estimated by EDTA titrimetric method. Alkalinity was estimated by titrimetric method. Phosphate was estimated by ammonium molybdate method and Nitrate was estimated by Brucine method. Results were expressed in mg/l. (APHA, 1985 and 1991and Trivedy and Goel 1986). The numerical estimation of zooplankton was made by Sedgewick Rafter cell and expressed in number per litre. All data were statistically analyzed (standard deviation).

Results and Discussion

The data on physico-chemical analysis of Tamdalge tank water has been given in Table 24.1.

Temperature

During the study period water temperature varied from 22.42 ± 1.18°C to 26.24 ± 21°C. The maximum value was recorded in the month of December (winter) and minimum in May (summer). Jayabhaye *et al.* (2006), Salve and Hiware (2008), observed that during summer, water temperature was high due to low water level and clear atmosphere. Similar, results were obtained in the present study.

Turbidity

The turbidity of water fluctuates from 0.96 ± 0.97 NTU to 13.09 ± 0.37 NTU. The maximum value was recorded in the month of February (summer) and minimum value in the month of November (winter). It may be due to human activities.

pH

The pH values ranges from 7.38 ± 0.15 to 8.49 ± 0.93. The maximum pH value (8.49 ± 0.93) was recorded in the month of March (summer) and minimum (7.38 ± 0.15) in the month of October (winter).

Dissolved Oxygen

The values of dissolved oxygen fluctuate from 6.36 ± 0.49 mg/l to 14.91 ± 1.03 mg/l. The maximum values (14.91 ± 1.03 mg/l) was recorded in the month of May (summer) and minimum values (6.36 ± 0.49 mg/l) in the month of November (winter). The high dissolved oxygen in summer is due to increase in temperature and duration of bright sunlight has influence on the per cent of soluble gases (O_2 and CO_2). The long days and intense sunlight during summer seem to accelerate photosynthesis by phytoplankton, utilizing CO_2 and giving off oxygen. This possibly accounts for the greater qualities of

Table 24.1: Monthly Variations in Physico-Chemical Parameters (mg/l) of Water Samples of Tamdalge Tank (Jan 2009-Dec 2009)

Stations/Months	Temp °C	pH	Dissolved Oxygen	Free CO_2	Hardness	Alkalinity	Phosphate	Nitrate
Jan-2009	23.25 ± 0.20	7.79 ± 1.04	9.07 ± 0.44	0.00 ± 0.00	83.41 ± 0.71	121.83 ± 0.99	1.92 ± 0.19	8.03 ± 0.71
Feb	24.74 ± 1.07	8.10 ± 0.61	9.03 ± 0.55	0.00 ± 0.00	80.58 ± 1.08	121.08 ± 0.55	2.93 ± 0.27	11.90 ± 0.86
Mar	25.33 ± 0.85	8.49 ± 0.93	11.88 ± 0.77	4.42 ± 0.00	103.83 ± 0.52	140.83 ± 0.99	3.37 ± 0.77	13.85 ± 0.77
Apr	24.23 ± 1.20	7.99 ± 0.28	14.50 ± 1.02	4.26 ± 0.41	174.57 ± 0.71	150.08 ± 0.79	3.88 ± 0.48	26.04 ± 1.07
May	26.24 ± 0.21	8.01 ± 0.17	14.91 ± 1.03	4.05 ± 1.08	159.74 ± 0.95	194.16 ± 0.72	7.99 ± 0.59	36.48 ± 0.79
Jun	23.82 ± 0.44	8.09 ± 0.32	11.18 ± 0.67	6.64 ± 1.00	148.83 ± 0.77	109.32 ± 0.84	1.68 ± 1.04	14.02 ± 0.81
Jul	23.80 ± 0.73	8.24 ± 0.27	9.25 ± 1.07	7.34 ± 1.07	62.66 ± 0.72	161.24 ± 0.94	0.73 ± 0.75	39.83 ± 0.82
Aug	22.64 ± 0.61	8.38 ± 0.20	8.69 ± 0.48	11.69 ± 0.70	93.83 ± 0.70	187.08 ± 0.83	0.29 ± 0.34	11.42 ± 0.72
Sept	24.77 ± 1.07	8.30 ± 0.08	8.83 ± 0.38	21.19 ± 1.08	102.83 ± 0.65	203.49 ± 0.80	4.25 ± 0.59	4.32 ± 0.48
Oct	25.19 ± 0.18	7.38 ± 0.15	8.64 ± 0.67	12.52 ± 1.15	67.49 ± 0.86	220.41 ± 0.78	0.13 ± 0.04	5.39 ± 0.63
Nov	23.46 ± 1.20	7.49 ± 0.11	6.36 ± 0.49	16.15 ± 0.51	126.16 ± 0.32	168.74 ± 0.97	0.23 ± 0.13	4.24 ± 0.46
Dec	22.42 ± 1.18	8.18 ± 0.12	8.53 ± 1.04	26.19 ± 0.55	83.91 ± 0.68	249.16 ± 1.13	5.02 ± 0.58	5.25 ± 0.53

O_2 recorded during summer. The quantity is slightly lesser during winter were reported by Masood Ahmed and Krishnamurthy (1990).

Carbon-dioxide

The values of free CO_2 range from 0.0 mg/l to 26.19 ± 0.55 mg/l. The maximum value (26.19 ± 0.55 mg/l) was recorded in the month of April (summer) and absent in the month of January and February (winter). High carban dioxide it might be due to increase in the decomposition of organic matter, low temperature and photosynthetic activities of phytoplankton.

Hardness

The values of hardness fluctuate from 62.66 ± 0.72 mg/l to 174 ± 0.71 mg/l. The maximum value (174 ± 0.71 mg/l) was recorded in the month of April (summer) and minimum value (62.66 ± 0.72 mg/l) in the month of July (monsoon).

Alkalinity

Total alkalinity ranges from 109.32 ± 0.84 mg/l to 220.41 ± 0.78 mg/l. the maximum value (220.41 ± 0.78 mg/l) was recorded in the month of October (winter) and minimum value (109.32 ± 0.84 mg/l) in the month of June (summer).

Phosphates

The value of phosphate fluctuates from 0.13 ± 0.04mg/l to 7.99 ± 0.59 mg/l. the maximum value (7.99 ± 0.59 mg/l) was recorded in the month of May (summer) and minimum (0.13 ± 0.04mg/l) value in the month of October (winter). The high values of phosphate in September (monsoon) months are mainly due to rain, surface runoff, agriculture runoff; washer man activity could have also contributed to the inorganic phosphate content. Similar results reported by Arvind Kumar (1995).

Nitrates

The value of nitrate ranges from 4.24 ± 0.48 mg/l to 39.83 ± 0.82 mg/l. The maximum value (39.83 ± 0.82 mg/l) was observed in the month of July (monsoon) and minimum (4.24 ± 0.48 mg/l) in the month of November (winter). Swaranatha and Narsingrao (1998) reported nitrates are in low concentration in summer and high during monsoon which might be due to surface runoff, due to rain.

In present investigation, zooplankton consists of Rotifera, Cladocera, Ostracoda and Copepoda in the Tamdalge tanks (Table 24.2). Sixteen species from 3 orders and 2 families were identified. Out of them 5 species of Rotifera, 4 species of Cladocera, 5 species of Copepoda 2 species of Ostracoda was observed. Copepoda is dominant group in Tamdalge tank. In the Tamdalge tank total zooplanktons ranged from 62 to 717 organism/l.

Table 24.2: List of Zooplankton Species from Tamdalge Tank (Jan 2009-Dec 2009)

Rotifera	Cladocera	Copepoda	Ostracoda
Branchionus angularis	Moina macrocopa	Mesocyclops hyalinus	Spirocypris
Branchionus caudatus	Moina rectirostris	Neodiaptomus Strigilipes	Hyocypris gibba
Branchionus falcatus	Daphnia pulex	Paracyclops fimbriatus	
Branchionus calyciflorus	Euryalona oriantalis	Rhinediaptomus indicus	
Branchionus vulgaris		Diaptomus copepod	

In Tamdalge tank, Rotifera was recorded with their percent compositions ranges between 3.47 per cent in August and 41.08 per cent in September. Species of Rotifera recorded are *Brachionus angularis*, *B. caudatus*, *B. fulcatus*, *B. calyciflorus*, *B. valgaris* in the tanks. Among them *B. falcatus* and *B. calyciflorus* were most dominant and abundant from Tamdalge tank.

Copepoda was also dominant and abundant but next to Rotifera and its per cent composition ranged between 9.35 per cent in October and 52.15 per cent in May at Tamdalge tank. It was represented by *Mesocyclop hyalinus, Neodiaptomus, Strigilipes, Paracyclops fimbriatus, Rhinediaptomus indicus, Diaptomus copepods*, were recorded from Tamdalge tank.

Cladocera was third dominant species and its percent composition varied from 1.89 per cent in June and 61.47 per cent in September, 15.38 per cent in June was recorded from Tamdalge tanks. It was represented by *Monia macrocopa, M rectirostris, Daphnia pulex, Euryalona oriantalis* from Tamdalge

Ostracods was fourth species and its percent composition varied from 13.94 per cent in April and 24.68 per cent in June was recorded. It was represented by *Spirocypris, Hyocypris gibba* from this tank.

In Tamdalge tank – Copepoda (37.15 per cent) > Rotifera (26.15 per cent) > Cladocera (18.50 per cent) > Ostracoda (17.70 per cent).

Acknowledgements

The authors are thankful to University Grants Commission, New Delhi for providing FIP, and Prof. G. P. Bhawane, Head, Department of Zoology, Shivaji University, Kolhapur, (M.S.), INDIA, for providing necessary research facilities

References

APHA, 1985. *Standard Methods for Examination of Water and Wastewaters*, 16th Edn. American Public Health Association, Washington D.C.

APHA, 1991. *Standard Methods for Examination of Water and Wastewater*, 20th Edn. American Public Health Association, Washington, D.C.

Dewan, S., Ali, M. and Islam, M.A., 1977. Study on the size and patterns of feeding of fries and fingerlings of three major carps, *e.g. Labeo rohita* (Ham), *Catla catla* (Ham) and *Cirrhina mrigala* (Ham). *Bangaladesh J. Agril.*, 2(2): 223–228.

Islam, A.K.M.N., Haroon, A.K.Y. and Zaman, K.M., 1974. Limnological studies of the river Buriganga. *Dhaka Univ. Stud. Pt.* B., 22(2): 99–111.

Islam, S.N., 2007. Physico-chemical condition and occurrence of some zooplankton in a pond of Rajshahi University. *Res. J. Fish. and Hydrobiol.*, 2(2): 21–25.

Jayabhaye, U.M., Pentewar, M.S. and Hiware, C.J., 2006. A study on physico-chemical parameters of a minor reservoir, Sawana, Hingoli district, Maharashtra.

Jhingran, V.G., 1985. *Fish and Fisheries of India*. Hindustan Publishing Corporation, Delhi, India.

Kaushik, S. and Sharma, N., 1994. Physico-chemical characteristics and zooplankton population of a perennial tank, Matsya Sarowar, Gwalior. *J. Eniviron. Ecology*, 1: 429–434.

Masood, Ahmed and Krishnamurthy, R., 1990. Hydrobiological studies of Wohar reservoir Aurangabad (Maharashtra State) India. *J. Environ. Biol.*, 11(3): 335–343.

Munwar, M., 1970. Limnological studies on freshwater ponds in Hyderabad, India. I: The biotype. *Hydrobiol.*, 35: 127–162.

Nasar, M.N., Safi, M., Shana, M.S. and Barua, G., 1991. On the productivity of catfish, *Clarias batrachus* (L.) fry rearing ponds at Mymengsingh, Bangaladesh. *Bangaladesh J. Zool.*, 9(2): 229–235.

Patil, H.S. and Karikal, S.M., 2001. Zooplankton diversity of Bhutnal water reservoir at Bijapur, Karnataka state. In: *Water Quality Assessment and Zooplankton Diversity,* (Ed.) B.K. Sharma, pp. 236–249.

Qasim, S.Z., 1977. Contribution of zooplankton in the water environments. *Proc. Symp. Water Zool. P. Gao. India,* p. 700–708.

Singh, D.N., 2000. Seasonal variation of zooplankton in a tropical lake. *Geobios,* 27: 92–100.

Sunkad, B.N. and Patil, H.S., 2004. Water quality assessment of fort lake of Belgaum (Karnataka) with special reference to zooplankton. *J. Env. Biol.*, 25(1): 99–102.

Trivedy, R.K. and Goel, P.K., 1986. *Chemical and Biological Methods for Water Pollution Studies.* Environmental Publication, Karad, Maharashtra.

Chapter 25

Physico-chemical Condition and Occurrence of Some Zooplankton in a Laxmiwadi Tank, Maharashtra

☆ *S.A. Manjare, S.A. Vhanalakar and D.V. Muley*

ABSTRACT

Study on hydrobiological status of Laxmiwadi tank water was made to assess the potability of water from January 2009 to December 2009. Some physico-chemical parameters were considered such as temperature, turbidity, pH, dissolved oxygen, carbon-dioxide, hardness, alkalinity, phosphate and nitrate. Among the zooplankton population four group of rotifera, cladocera, copepod and ostracoda were studied.

Keywords: Perennial tank, Physico-chemical parameters, Monthly variation, Zooplankton.

Introduction

The interrelationship between the physicochemical parameters and plankton production of tank water and its relation with monthly fluctuation of zooloplankton are of great importance and basically essential in fish culture. Fishes are dependent on physico-chemical parameters any change of these parameters may affect the growth, development and maturity of fish (Jhingran, 1985). Different causal influences, which determine the quality of water, show a characteristic change from season to season (Zafar, 1964., Munwar, 1970., Islam, *et al.,* 1974, Nasar, *et al.,* 1991).

Zooplankton constitutes important food item of many fishes. The larvae of carps feed mostly on zooplankton (Dewan *et al.,* 1977). Zooplankton also plays an important role in the food chain as they are second in tropic level as primary consumers and also as contributors to next tropic level (Qasim, 1977).

Sunkand and Patil (2004) and Islam (2007) have worked on the physico-chemical condition and seasonal variation of zooplankton.

The present study was made on water quality and occurrence of some zooplankton with respect to physico-chemical parameters in Laxmiwadi tank of Kolhapur district (Maharashtra).

Materials and Methods

The study was conducted on Laxmiwadi tank situated 35 km away from Kolhapur city which lies between 16°47′ 49.04"N latitude and 74°22′ 57.47" E longitude, from January 2009 to December 2009. The area of the tank is 100 ha. Rainfall is the only source of water for tank. The water of the tank is clear and few aquatic plants are present.

Water samples were collected monthly in the morning at 8 am to 10 am from surface layer of the tank. Water temperature was recorded by a centigrade thermometer and pH was recorded by pocket digital pH meter. Turbidity was recorded by Nephloturbidity meter, dissolved oxygen was determined by Winkler's method, Free carbon-dioxide was determined by titrimetric method, Hardness was estimated by EDTA titrimetric method, Alkalinity was estimated by titrimetric method Phosphate was estimated by ammonium molybdate method and Nitrate was estimated by Brucine method. Results were expressed in mg/l. (APHA, 1985, 1991; Trivedy and Goel, 1986).

The numerical estimation of zooplankton was made by Sedgewick Rafter cell and expressed in number per liter. All data were statistically analyzed (standard deviation).

Results and Discussion

The data on physico-chemical analysis of Vadgaon tank water has been given in Table 25.1.

Temperature

During the study period water temperature varied from 22.47 ± 0.50°C to 26.33 ± 1.02°C. The maximum water temperature was recorded in May (summer) and minimum in December (winter). Jayabhaye *et al.* (2006), Salve and Hiware (2008), observed that during summer, water temperature was high due to low water level and clear atmosphere. Similar results were obtained in the present study.

Turbidity

The turbidity of water fluctuated from 0.36 ± 0.07 NTU to 10.93 ± 1.06 NTU. The maximum value was recorded in the month of June (monsoon) and minimum value in the month of March (summer). It may be due to human activities like washing, bathing etc.

pH

The pH values ranges from 7.59 ± 0.08 to 9.12 ± 0.12. The maximum pH value was recorded in the month of September (monsoon) and minimum in the month of April (summer). pH remain alkaline throughout study period.

Dissolved Oxygen

The values of dissolved oxygen fluctuated from 6.72 ± 0.54 mg/l to 13.08 ± 0.57 mg/l. The maximum values was recorded in the month of May (summer) and minimum values in the month of February (winter). The high DO in summer is attributed to increase in temperature and duration of bright sunlight. The long days and intense sunlight during summer seems to accelerate photosynthesis

Table 25.1: Monthly Variations in Physico-chemical Parameters (mg/l) of Water Samples of Laxmiwadi Tank (Jan 2009–Dec 2009)

Stations/Months	Temp °C	pH	Dissolved Oxygen	Free CO_2	Hardness	Alkalinity	Phosphate	Nitrate
Jan-2009	23.66 ± 0.57	8.36 ± 0.07	7.67 ± 0.46	4.13 ± 0.46	102.49 ± 1.04	136.66 ± 0.57	0.49 ± 0.03	1.65 ± 0.25
Feb	25.00 ± 1.00	8.34 ± 0.16	6.72 ± 0.54	4.46 ± 0.11	102.66 ± 0.77	149.99 ± 0.63	0.50 ± 0.07	1.55 ± 0.21
Mar	25.66 ± 0.57	8.42 ± 0.26	9.55 ± 0.55	4.46 ± 0.22	143.33 ± 0.55	146.66 ± 0.77	0.77 ± 0.30	2.29 ± 0.55
Apr	23.33 ± 0.57	7.59 ± 1.08	9.10 ± 0.26	7.00 ± 1.03	82.33 ± 1.05	108.33 ± 0.63	0.34 ± 0.32	4.64 ± 0.38
May	26.33 ± 1.02	8.21 ± 0.14	13.08 ± 0.51	4.46 ± 0.75	136.66 ± 0.77	148.33 ± 0.88	0.17 ± 0.06	5.03 ± 0.55
Jun	23.5 ± 0.50	8.44 ± 0.34	10.58 ± 0.48	5.83 ± 1.02	104.16 ± 1.14	149.33 ± 1.15	0.79 ± 0.07	7.67 ± 1.05
Jul	23.82 ± 0.76	8.41 ± 0.19	8.52 ± 1.06	2.93 ± 1.08	75.00 ± 0.50	156.66 ± 1.07	0.22 ± 0.16	5.52 ± 1.02
Aug	25.00 ± 1.00	8.62 ± 0.19	10.71 ± 0.57	0.00 ± 0.00	76.66 ± 1.05	163.33 ± 1.20	1.65 ± 0.93	4.63 ± 0.61
Sept	25.11 ± 1.03	9.12 ± 0.12	10.88 ± 0.28	0.00 ± 0.00	43.33 ± 0.77	216.66 ± 0.77	0.06 ± 0.01	4.76 ± 1.00
Oct	24.66 ± 0.57	8.24 ± 0.21	11.61 ± 0.70	0.00 ± 0.00	61.33 ± 0.57	223.33 ± 1.16	0.97 ± 0.49	7.50 ± 1.06
Nov	24.49 ± 1.05	8.23 ± 0.11	9.86 ± 1.04	3.66 ± 1.05	93.33 ± 0.85	146.66 ± 0.77	0.20 ± 0.03	4.51 ± 0.59
Dec	22.47 ± 0.50	8.36 ± 0.05	10.39 ± 0.80	5.11 ± 0.66	118.33 ± 0.68	161.66 ± 1.04	7.61 ± 0.72	5.24 ± 1.08

by phytoplankton, utilizing CO_2 and giving off oxygen. This accounts for the greater quantity of dissolved oxygen recorded during summer. The quantity is slightly less during winter as reported by Masood Ahmed and Krishnamurthy (1990).

Carbon Dioxide

The values of free CO_2 range from 0.0 mg/l to 7.00 ± 1.06 mg/l. The maximum value was recorded in the month of April (summer) and absent in the month of August, September and October (monsoon). High carban dioxide is due to increase in the decomposition of organic matter, low water level and density of phytoplanktons. Absence of free carbon dioxide is due to its utilization by algae during photosynthesis or carbonates present.

Hardness

The values of hardness fluctuate from 43.33 ± 0.77 mg/l to 143.33 ± 0.55 mg/l. The maximum value was recorded in the month of March (summer) and minimum value in the month of September (monsoon). High values in summer is due to decrease in water volume and increase in rate of evaporation at high temperature.

Alkalinity

Total alkalinity ranges from 108.33 ± 0.63 mg/l to 223.33 ± 1.16mg/l. The maximum value was recorded in the month of October (winter) and minimum value in the month of April (summer). Maximum alkalinity in winter it might be due to presence of carbonates and bicarbonates.

Phosphates

The value of phosphate fluctuates from 0.06 ± 0.01mg/l to 7.61 ± 0.72 mg/l. the maximum value was recorded in the month of December (winter) and minimum value in the month of September (monsoon). The high values of phosphate in September (monsoon) months are mainly due to rain, surface runoff, agriculture runoff; washing activities that contributed to the inorganic phosphate content. Similar results reported by Arvind Kumar (1995).

Nitrates

The value of nitrate ranges from 1.55 ± 0.21 mg/l to 7.67 ± 1.05 mg/l. The maximum value was observed in the month of June (monsoon) and minimum in the month of February (winter). Swaranlatha and Narsingrao (1998) reported that nitrates are in low concentration in summer and high during monsoon which might be due to surface run off, due to rain. Similar results obtain in the present study.

The zooplanktons were represented by four groups viz rotifera, cladocera, copepoda and ostracoda (Table 25.2). From these groups 20 species were identified and recorded. Among copepoda 9 species belongs to order Eucopepoda. These are *Mesocyclops hyalinus, Paracyclops fimbriatus, Neodiaptomus strigilipes, Rhinedioptomus indicus, Diaptomus copepod, Calanoid copepods Eucyclopoid species, Cyclopoid copepod*, whereas Cladocerans were represented by *Monia macrocopa, M. rectirostrix, M. brachiatatris, Daphnia pulex, Euryalona oriantalis, Diaphanosoma sarsi, D. excisum, Macrothrix laticornis*. The Ostracodas are represented by *Spirocypris, Hyocypris gibba, Hemicypris fossulate, Stenocypris* from the plankton samples of the tanks.

The rotifera species like *Brachionus angularis, B. cadatus, B. falcatus, B. calyciflorus, B. vulgaris* was observed as common forms throughout the investigation period while *B. rubens* and *Keratella tropica* were recorded infrequently in the plankton samples of the tanks.

Table 25.2: List of Zooplankton Species from Laxmiwadi (January 2009 to December 2009)

Rotifera	Cladocera	Copepoda	Ostracoda
Branchionus angularis	*Moina macrocopa*	*Mesocyclops hyalinus*	*Spirocypris*
Branchionus caudatus	*Moina rectirostris*	*Neodiaptomus Strigilipes*	*Hyocypris gibba*
Branchionus falcatus	*Daphnia pulex*	*Paracyclops fimbriatus*	*Stenocypris*
Branchionuscalyciflorus	*Euryalona oriantalis*	*Rhinediaptomus indicus*	*Hemicypris fossulate*
Branchionus vulgaris		*Heliodiaptomus viduus*	
		Calanoid copepod	
		Eucyclops spp	
		Mesocyclops leuckarti	

In Laxmiwadi tank, the rotifera group represented distinct peak in the month of July (61.08 per cent). With its maximum density in June (43.64 per cent). Among these rotifrea *B. angularis, B. caudatus, B. falcatus, B. calyciflorus and B. vulgaris* were most commonly observed in winter and summer season. Comparatively *B. angularis, B. falcatus, B. calyciflorus* were dominant to other species such as *B. caudatus* and *B. vulgaris* during the year 2008 and 2009.

The high percentage of copepod was observed in the month of April (42.02 per cent) with their high density in March (41.53 per cent) whereas it's low percentage in August (38.91 per cent).

Cladocera was recorded maximum percentage in July (39.60 per cent) and minimum in the month of June (15.38 per cent), with their absence was also noticed in the month of August at Laxmiwadi tank,

The maximum percentage of ostracoda was recorded in the month of August (71.79 per cent) with their minimum appearance in the month of February (8.71 per cent), where as Ostracoda were absent in the month of June and July.

The order of dominance of various groups of zooplanktons were represented as Copepoda (38.45 per cent) > Rotifera (29.15 per cent) > Cladocera (19.65 per cent) > Ostracoda (12.74 per cent).

Acknowledgements

The authors are thankful to University Grants Commission, New Delhi for providing FIP, and Head, Department of Zoology, Shivaji University, Kolhapur, (M.S.), INDIA, for providing necessary research facilities.

References

APHA, 1985. *Standard Methods for Examination of Water and Wastewaters,* 16th Edn. American Public Health Association, Washington, D.C.

APHA, 1991. *Standard Methods for Examination of Water and Wastewater,* 20th Edn., American Public Health Association, Washington, D.C.

Dewan, S., Ali, M. and Islam, M.A., 1977. Study on the size and patterns of feeding of fries and fingerlings of three major carps, *e.g. Labeo rohita* (Ham), *Catla catla* (Ham) and *Cirrhina mrigala* (Ham). *Bangaladesh. J. Agril.,* 2(2): 223–228.

Islam, A.K.M.N., Haroon, A.K.Y. and Zaman, K.M., 1974. Limnological studies of the river Buriganga. *Dhaka Univ. Stud. Pt. B.,* 22(2): 99–111.

Islam, S.N., 2007. Physico-chemical condition and occurrence of some zooplankton in a pond of Rajshahi University. *Res. J. Fish. and Hydrobiol.*, 2(2): 21–25.

Jayabhaye, U.M., Pentewar, M.S. and Hiware, C.J., 2006. A study on physico-chemical parameters of a minor reservoir, Sawana, Hingoli district, Maharashtra.

Jhingran, V.G., 1985. *Fish and Fisheries of India*. Hindustan publishing Corporation, Delhi, India.

Kaushik, S. and Sharma, N., 1994. Physico-chemical characteristics and zooplankton population of a perennial tank, Matsya Sarowar, Gwalior. *J. Eniviron. Ecology*, 1: 429–434.

Kumar, Arvind, 1995. Some limnological aspects of the freshwater tropical wetland of Santhal Pargana (Bihar) India. *J. Eniv. and Poll.*, 2(3): 137–141.

Masood, Ahmed and Krishnamurthy, R., 1990. Hydrobiological studies of Wohar reservoir Aurangabad(Maharashtra state) India. *J. Environ. Biol.*, 11(3): 335–343.

Munwar, M., 1970. Limnological studies on freshwater ponds oin Hyderabad, India. I. The biotype. *Hydrobiol.*, 35: 127–162.

Nasar, M.N., Safi, M., Shana, M.S. and Barua, G., 1991. On the productivity of catfish, *Clarias batrachus* (L.) fry rearing ponds at Mymengsingh, Bangaladesh. *Bangaladesh J. Zool.*, 9(2): 229–235.

Patil, H.S. and Karikal, S.M., 2001. Zooplankton diversity of Bhutnal water reservoir at Bijapur, Karnataka state. In: *Water Quality Assessment and Zooplankton Diversity*, (Ed.) B.K. Sharma, pp. 236–249.

Qasim, S.Z., 1977. Contribution of zooplankton in the water environments. *Proc. Symp. Water Zool. P. Gao. India*, p. 700–708.

Salve, B.S. and Hiware, C.J., 2006. Studies on water quality of wanparakalpa reservoir Nagpur, near Parli Vaijnath Dist. Beed, Marathwada region. *J. Aqua Biol.*, 21(2): 113–117.

Singh, D.N., 2000. Seasonal variation of zooplankton in a tropical lake. *Geobios*, 27: 92–100.

Sunkad, B.N. and Patil, H.S., 2004. Water quality assessment of fort lake of Belgaum (Karnataka) with special reference to zooplankton. *J. Env. Biol.*, 25(1): 99–102.

Swarnalatha, P. and Rao, A.N., 1998. Ecological studies of Banjara lake with reference to water pollution Hyderabad. *J. Env. Biol.*, 19(2): 179–186.

Trivedy, R.K. and Goel, P.K., 1986. *Chemical and Biological Methods for Water Pollution Studies*. Environmental Publication, Karad, Maharashtra.

Zafar, A.R., 1967. On the ecology of algae in certain fish ponds of Hyderabad, India. III. The periodicity. *Hydrobiologia*, 30: 96–112.

Chapter 26

Ichthyaofauna of Osmanabad District, Maharashtra

☆ *J.S. Mohite, V.B. Sakhare and S.G. Rawate*

ABSTRACT

Altogether 31 species of fishes belonging to 11 families were recorded from Osmanabad district of Maharashtra.Out of 31 species 23 were endemic and 8 were introduced species. Among the species *Catla catla,Cirrhinus mrigala, Labeo rohita, Puntius sarana, P.ticto, Chela phulo, Mystus seenghala, Wallago attu, Clarias batrachus, Glossogobius giuris, Channa marulius, C.striatus,and C.gachua* were recorded from all the reservoirs. Next to that, *Cyprinus carpio, Notopterus notopterus, Notopterus kapirat, Puntius sophore, Rohtee cotio, Mystus vittatus, Mystus cavasius, Ompak bimaculatus* and *Ambassis nama* were recorded from three reservoirs. Among the other species, *Heteropneustes fossilis, Anabus testudineus* and *Mastacembelus pancalus* were recorded from only two reservoirs. While *Labeo boggut, Ctenopharyngodon idella, Hypophthalmichthys molitrix, Oreochromis mossambicus, Mystus tengra* and *Mastacembelus armatus* were recorded from only a single reservoir.

Keywords: *Ichthyofauna, Osmanabad district, Maharashtra.*

Introduction

Fish constitutes about half of the total number of vertebrates in the world. They live in almost all conceivable aquatic habitats; 21,723 living species of fish have been recorded out of 39,900 species of vertebrates. Out of these 8,411 are freshwater species and 11,650 are marine India is one of the mega biodiversity countries in the world and occupies the ninth position in terms of freshwater mega biodiversity. In India there are 2,500 species of fishes of which 930 live in freshwater and 1,570 are marine.

In the field of freshwater ichthyodiversity of Maharashtra there is valuable contribution by many workers (Sakhare, 2001, Sakhare and Joshi 2002, 2005).

The climate of Osmanabd district is mainly dry extent in June when humidity rises. The dryness is little lesser till the month of November. On an average the district receives 767.4mm of rainfall. The important rivers in district are Wan, Manjra, Bori, Tavrja, and Sina.Due to increased demands for reliable supplies of drinking water, and irrigation the number of small and medium sized reservoirs were constructed across the rivers.

As far as Osmanabad district is concerned, there are many minor reservoirs, which are contributing significantly to the total inland fish production, and scarcely any attention seems to have been devoted to any systematic investigation on either of the variety abundance catch of the fishes from reservoirs. So it is felt that there is an urgent need to provide information about various kinds of fishes available in reservoirs. Since there is no information on the ichthyofauna of the reservoirs of Osmanabad district, it has been attempted to prepare checklist of the fishes recorded from reservoirs of the Osmanabad district.

Materials and Methods

Fishes were collected regularly on monthly basis for one year (year 2010 to 2011) from different wetlands of Osmanabad district of Maharashta. The fishes for the study were collected with the help of local fishermen besides purchasing them from local fish market. After systematic identification the specimens were preserved in 4 per cent formalin and deposited in the laboratory of Department of Zoology and Fisheries of Yeshwantrao Chavan Mahavidyalaya,Tuljapur.The identification of fishes were carried out with the help of standard literature (Day 1878, Jayaram 1981 and Talwar and Jhingran, 1991).

Results and Discussion

During the present investigation 31 species of fishes belonging to 11 families were recorded (Table 26.1). Among the species *Catla catla,Cirrhinus mrigala,Labeo rohita,Puntius sarana, P.ticto, Chela phulo, Mystus seenghala, Wallago attu, Clarias batrachus, Glossogobius giuris, Channa marulius, C.striatus,and C.gachua* were recorded from all the reservoirs. Next to that, *Cyprinus carpio, Notopterus notopterus, Notopterus kapirat, Puntius sophore, Rohtee cotio, Mystus vittatus, Mystus cavasius, Ompak bimaculatus* and *Ambassis nama* were recorded from three reservoirs. Among the other species, *Heteropneustes fossilis, Anabus testudineus* and *Mastacembelus pancalus* were recorded from only two reservoirs. While *Labeo boggut, Ctenopharyngodon idella, Hypophthalmichthys molitrix, Oreochromis mossambicus, Mystus tengra* and *Mastacembelus armatus* were recorded from only a single reservoir.

In Achler tank, *Oreochromis mossambicus* has adversely affected the indigenous fishes and has nearly contributed up to 40 per cent of the total fish catch. Such type of impact of *O.mossambicus* on indigenous fishes has been reported by many workers (Sreenivasan 1967,Sugunan 1997 and Moyle 1976). Hence, it is strongly recommended not to stock *O.mossambicus* in reservoirs of district as this fish is unsuitable for culture along with Indian Major Carps because of adverse effect it causes on the growth and production of carps and its depredation on carp fry.

Out of total 31 fish species, *Catla catla, Cirrhinus mrigala, Labeo* sp, *Wallago attu, Channa* spp, *Mastacembelus* sp, *Clarias batrachus* and *Heteropneustes fossilis* have a very high demands in fish markets of Osmanabad, Tuljapur and Naldurg. The super larger sized fishes fetch very fancy prices, but unfortunately the catch of such large fish is very meager. In fish markets of Osmanabad, Tuljapur and Naldurg the price of Indian Major Carps remain almost constant at Rs. 70 to 100/kg; while the fishes

Table 26.1: List of Fishes Recorded from Reservoirs of Osmanabad District, Maharashtra

Sl.No.	Introduced Fishes	Terna	Bennetura	Achler	Makni
	Family: Cyprinidae				
	Subfamily: Cyprininae				
1.	Catla catla	✓	✓	✓	✓
2.	Cirrhinus mrigala	✓	✓	✓	✓
3.	Labeo rohita	✓	✓	✓	✓
4.	Labeo boggut	✓	✗	✗	✗
5.	Ctenopharyngodon idella	✓	✗	✗	✗
6.	Cyprinius carpio	✓	✓	✗	✓
7.	Hypophthalmichthys molitrix	✗	✗	✗	✓
	Family: Cichlidae				
8.	Oreochromis mossambicus	✗	✗	✓	✗
	Endemic fishes				
	Family: Notpteridae				
9.	Notopterus notopterus	✓	✗	✓	✓
10.	Notopterus kapirat	✓	✓	✗	✓
	Family: Cyprinidae				
	Subfamily: Cyprininae				
11.	Puntius sophore	✗	✓	✓	✓
12.	Puntius sarana	✓	✓	✓	✓
13.	Puntius ticto	✓	✓	✓	✓
14.	Rohtee cotio	✓	✓	✗	✓
15.	Chela phulo	✓	✓	✓	✓
	Family: Bagridae				
16.	Mystus seenghala	✓	✓	✓	✓
17.	Mystus vittatus	✗	✓	✓	✓
18.	Mystus cavasius	✓	✗	✓	✓
19.	Mystus tengra	✓	✗	✗	✗
20.	Wallago attu	✓	✓	✓	✓
21.	Ompak bimaculatus	✓	✓	✓	✗
	Family: Claridae				
22.	Clarias batrachus	✓	✓	✓	✓
23.	Heteropneustes fossilis	✓	✗	✗	✓
	Family: Ambassidae				
24.	Ambassis nama	✗	✓	✓	✓
	Family: Gobiidae				
25.	Glossogobius giuris	✓	✓	✓	✓
	Family: Anabantidae				
26.	Anabus testudineus	✗	✓	✗	✓
	Family: Channidae				
27.	Channa marulius	✓	✓	✓	✓
28.	Channa striata	✓	✓	✓	✓
29.	Channa gachua	✓	✓	✓	✓
	Family: Mastacembelidae				
30.	Mastacembelus pancalus	✓	✓	✗	✗
31.	Mastacembelus armatus	✓	✗	✗	✗

like *Wallago attu, Clarias batrachus, Heteropneustes fossilis* and *Mastacembelus* sp were sold in all the three fish markets constantly at higher prices, as compared to Indian Major Carps.

References

Datta Munshi, J.S. and Srivastava, M.P., 1988. *Natural History of Fishes and Systematics of Freshwater Fishes of India*. Narendra Publishing House, Delhi.

Day, F.S., 1878. *The Fishes of India*. William Dowson and Sons Ltd., London.

Jayaram, K.C., 1981. *The Freshwater Fishes of India*. Zoological Survey of India, Calcutta.

Jhingran, A.G., 1991. *Performance of Tilapia and its Possible Impact on the Ichthyofauna*. FAO Fisheries Report No. 458, Suppl. Rome, FAO, 1992, pp. 218.

Moyle, P.B., 1976. *Fish Introduction and Fisheries of Rihand Reservoir*. Dept. of Fisheries, Uttar Pradesh, Lucknow, India.

Sakhare, V.B., 2001. Reservoir fisheries in Solapur district of Maharashtra. *Fishing Chimes*, 21(5): 29–30.

Sakhare, V.B. and Joshi, P.K., 2002. Ecology and Ichthyofauna of Bori reservoir in Maharashtra. *Fishing Chimes*, 22(4): 40–41.

Sakahre, V.B. and Joshi, P.K., 2005. Ichthyofauna of reservoirs in Osmanabad district, Maharshtra. *Fishing Chimes*, 25(9): 43–44.

Sreenivasan, A., 1967. *Tilapia mossambica*: Its ecology and status in Madras state, India. *Madras J. Fish.*, 3: 33–39.

Sugunan, V.V., 1997. Fisheries management of small water bodies in seven countries in Africa, Asia and Latin America. FAO Fisheries Circular No. 933 FIRI/C933.

Talwar, P.K. and Jhingran, A.G., 1991. *Inland Fishes of India and Adjacent Countries*, Vols. 1 and 2. Oxford and IBH Publishing Co. Pvt. Ltd., New Delhi.

Chapter 27

Plankton Diversity in Hangarga Reservoir in Osmanabad District, Maharashtra

☆ *V.B. Sakhare, J.S. Mohite and S.G. Rawate*

ABSTRACT

During the present investigation plankton diversity of Hangarga reservoir near Tuljapur town in Osmanabad district was studied. Among planktonic groups cyanophyceae dominated the reservoir at 37.04 per cent, followed by chlorophyceae (27.19 per cent), copepoda (11.14 per cent), eugelnophyceae (8.03 per cent), bacillariophyceae (5.14 per cent), cladocera (4.40 per cent), rotifera (2.85 per cent), ostracoda (2.20 per cent) and protozoans (1.71 per cent) of the total planktonic population during year 2007-09.

Keywords: Phytoplankton diversity, Hangarga reservoir, Maharashtra.

Introduction

The term 'Plankton' refers to those minute aquatic forms which are non motile or insufficiently motile to overcome the transport by currents and living suspended in the open or pelagic water. Those of plant origin are called phytoplankton, the producers belonging to first trophic level while those of animal origin are the zooplankton which are the primary consumers belonging to second trophic level.

Plankton productivity of water differs considerably from very high to scanty plankton production. The quantitative measure of the production can be expressed by the standing crop, annual crop and average individual weight of each kind of plankton in different environmental conditions.

The larval forms of nearly all fish species are plankton feeder and growth of a particular fish can be successfully carried out in ponds with good growth of planktons. The presence of a specific type of plankton on which a particular fish lives is absolutely necessary. Thus, the knowledge of specific plankton is essential for the development of the fishes as it forms the essential diet of particular fish which feeds on the water surface.

The density of plankton in a water body determines the stocking rate of fishes because they are the chief source of food of many economically important fishes. Plankton, due to its key role in the ecosystem of the environment, is directly related to the fish catch potential of a reservoir. An insight into the distribution, composition and succession of plankton gives valuable clue for determining the fishing grounds, selection of suitable species for stocking and determining the level of utilization of the available food by the existing fish stock.

The ability of water to allow production of plankton depends on many factors, but the most important is usually the availability of inorganic nutrients for phytoplankton growth. Essential elements for phytoplankton growth include C,O,P,N,S,K,Na,Ca,Mg,Fe,Mn,Cu,Zn,B,Co,Cl and possibly few others. Phosphorus is the most important element that regulates the phytoplankton growth.

To monitor the aquatic ecosystems and integrity of water,plankton has been used recently as bioindicator (Beaugrand *et al.*, 2000).Bioindicators and biotic indexes are being used by Europeans to assess water quality of water bodies (Sousa *et al.*, 2008) for last 100 years.

Phytoplankton forms the very basis of aquatic food chain. The water quality especially the nutrients influence its population. Phytoplankton survey thus indicates the trophic status and the presence of organic population in the ecosystem. Nutrients enrichment in water bodies is known as eutrophication, which is a common phenomenon with algal blooms. Phytoplanktons are ecologically significant as they trap radiant energy of sunlight and convert into chemical energy.

Zooplankton support the economically important fish population. They are the major mode of energy transfer between phytoplankton compositions, abundance and seasonal variations is helpful in planning and successful fishery management (Jhingran, 1974).

Potentiality of zooplankton as bioindicator is very high because their growth and distribution are dependent on some abiotic (e.g., temperature, salinity, stratification, pollutants) and biotic parameters (e.g., food limitation, predation, competition).Pandey and Verma (2004) revealed that abiotic parameters (e.g., pH, transparency, temperature, dissolved oxygen and some micro-nutrients) in relation to seasonal fluctuation influence zooplankton abundance.

Zooplankton are the central trophic link between primary producers and higher trophic levels. The freshwater zooplankton comprise of Protozoa, Rotifers, Cladocerans, Copepods and Ostracods. Most of them depend to a large extent,on various bacterioplankton and phytoplankton for food. Many of the larger forms feed on smaller zooplankton, forming secondary consumers. Some of them are detritivore feeders, browsing and feeding on the substrate attached organic matter, phytoplankton or concentrating on the freely suspended organic matter particles or those lying on the bottom sediment. Many of these organisms are also fish food organisms and are consumed by the other aquatic macrofauna.The freshwater zooplankton is mainly constituted of five groups:

Zooplankton constitutes an important link in food chain as grazers (primary and secondary consumers) and serves as food for fishes directly or indirectly. Therefore any adverse effect to them will be indicated in the wealth of the fish populations. Thus, monitoring them as biological indicators of pollution could act as a forewarning for the fisheries particularly when the pollution affects the food chain (Mahajan, 1981). Thus, the use of zooplankton for ecological biomonitoring of the water bodies helps in the analysis of water quality trends, development of cause-effect relationships between water quality and environmental data and judgement of the adequacy of water quality for various uses.

Zooplankton communities of freshwaters constitute an extremely diverse assemblage of organisms represented by nearly all the phyla of invertebrates. The most significant feature of zooplankton is its immense diversity over space and time. In an ecosystem 90 per cent of zooplankton species are herbivorous, remaining 10 per cent being carnivore. Studies on the aquatic systems reveal that the structure of the zooplankton community is quite characteristic and is the product of the environmental conditions prevailing there. Several zooplankton species have been classified as indicative of polluted conditions like *Keratella, Brachionus, Cephalopoda, Monostyla, Lecane, Bosminia, Adona, Alonella, Cyclops* and several members of the group protozoa (Rawson 1956 and Pennak 1978).

The zooplankton study has been a fascinating subject for a long time. Enough literature exists on the plankton of various water bodies (Angadi *et al.,* 2005,Ganapathi 1941,Saha and Pandit, 1990,singh and Ahmed 1990,Sugunan 1980, Singh 1960, More and Nandan 2000,Bahura 2001 and Shukla *et al.,* 1989).Such studies from state of Maharashtra have been very recently initiated and the only contributions on this aspect are those of Deshmukh (2001),Meshram and Dhande (2000), Pulle and Khan (2003), Sakhare (2007) and Sakhare and Joshi (2002, 2006).

The aim of the present investigation is to know the plankton diversity and seasonal variations in the Hangarga reservoir near Tulajpur town in Osmanabad district of Maharashtra.The reservoir is very close to Tuljapur-Latur state highway.The investigation was made for the period of two years *i.e.,* February 2007 to January 2009.

Materials and Methods

Standard methodology after Welch (1948) with suitable modifications to suit local availability was used. Preliminary identification was made by using Pennak (1978), Battish (1992), Michael (1973) and Dhanapathi (2000) as basic references.

The rotifers were ordinarily distinguished at species level by trophic characteristics in whole animal. Although in some difficult cases animals were crushed with the help of 2 micro-needles to force animal out.

Generally, the cladocerans were identified by carapace characteristics, although in a number of cases a minor dissection of post abdominal segment with the help of 2 micro-needles was carried out. After dissection, these abdominal segments were observed under high power in a glycerine mount.

Copepods were generally identified with the help of body appendages and for species identification a minor dissection on the 5th leg was carried out in a glycerine mount. Animal or dissected part was observed under high power.

For identification of ostracods, body shell was removed to study detailed structure. These animals also required a minor dissection of furca.After dissection, furca was observed under high power in glycerine.

For quantitative estimation of plankton, a Sedgwick-Rafter counting cell of mm^2 and micropipette of glass van grade I of 1, 2 and 5 ml capacity were used.

Results

Phytoplankton Diversity

The various plankton groups occurred in the reservoir during year 2007-2009 is listed in Table 27.1 and different species of phytoplankton in Table 27.2. 12 Species of chlorophyceae 8 of cyanophyceae, 11 of bacillariophyceae and 2 of euglenophyceae were recorded from the reservoir. The phytoplankton were represented by chlorophyceae, cyanophyceae, bacillariophyceae and euglenophyceae at 35 per cent, 47.65, 7 per cent, and 10.34 per cent of the total phytoplankton population respectively (Figure 27.1).

Table 27.1: Plankton Groups and their Composition in Hangarga Reservoir during year 2007–09

Groups	Range (Units l⁻¹)	Average (per cent)
Chlorophyceae	450 to 1250	1050 (27.19%)
Cyanophyceae	950 to 1840	1430 (37.04%)
Bacillariophyceae	150 to 270	210 (5.44%)
Euglenophyceae	Nil to 750	310 (8.03%)
PHYTOPLANKTON		**3000 (77.70%)**
Protozoans	31 to 103	66 (1.71%)
Rotifera	70 to 175	110 (2.85%)
Cladocera	25 to 225	170 (4.40%)
Copepoda	50 to 860	430 (11.14%)
Ostracoda	70 to 90	85 (2.20%)
ZOOPLANKTON		**861 (22.30%)**
TOTAL PLANKTON		**3861**

During year 2007-09, from chlorophyceae, *Cosmarium* sp, *Pediastrum* sp, and *Ulothrix* sp dominated the reservoir. The maximum population of chlorophyceae was recorded in February (1250 l⁻¹).

8 species of cyanophyceae were identified.*Microcystis sp* and *Lyngbya sp* dominated the reservoir. Out of 8 species recorded, 6 species exhibited their presence during September to November and 5 in February.

11 species from bacillariophyceae were identified. *Navicula* sp, *Synedra* sp and *Cyclotella* sp dominated the reservoir. Out of 11 species recorded, 8 species exhibited their presence during October to December 2007.Thus,these three months showed a maximum population of members of bacillariophyceae.The maximum population of bacillariophyceae during year 2007-08 was recorded in November (220 l⁻¹).

Table 27.2: Phytoplankton Diversity in Hangarga Reservoir

Chlorophyceae: *Ankistrodesumus* sp., *Coelastrum* sp., *Closterium* sp., *Pediastrum* sp., *Scenedesmus* sp., *Staurastrum* sp., *Cosmarium* sp., *Chlorella* sp., *Spriogyra* sp., *Ulothrix* sp., *Volvox* sp., *Oedogonium* sp.

Cyanophyceae: *Anabaena* sp., *Chroococcus* sp., *Spirulina* sp., *Microcystis* sp., *Lyngbya* sp., *Nostoc* sp., *Merismopedia* sp., *Oscillatoria* sp.

Bacillariophyceae: *Cyclotella* sp., *Gyrosigma* sp., *Diatomas* sp., *Cymbella* sp., *Melosira* sp., *Fragillaria* sp., *Tabellaria* sp., *Navicula* sp., *Nitzschia* sp., *Pinnularia* sp., *Synedra* sp.

Euglenophyceae:*Eugelna* sp., *Phacus* sp.

Euglenophyceae was represented by 2 species with dominance of *Euglena* sp. This group attained a highest peak in month of November.

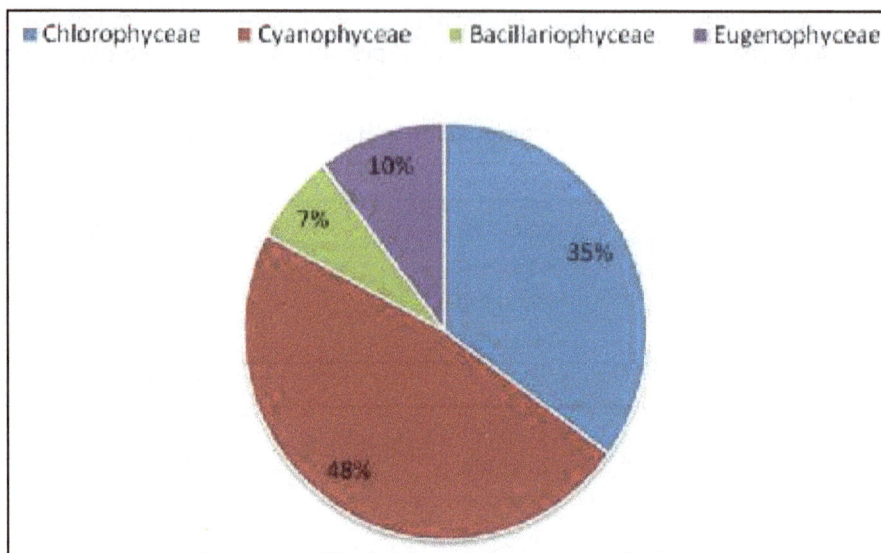

Figure 27.1: Phytoplankton Composition during Investigation Period

During year 2008-09, 11 species of chlorophyceae, 8 of cyanophyceae, 11 species of bacillariophyceae and 2 of euglenophyceae were recorded from Hangarga reservoir.

Chlorophyceae was abundant in April with density of 1150 l^{-1}.It was mainly represented by *Cosmarium* sp, *Chlorella* sp, *Pediastrum* sp and *Ulotrix* sp.

Cyanophyceae was mainly represented by *Microcystis* sp,*Merismopedia* sp, and *Lyngbya* sp. Their population was maximum (1840 l^{-1}) in January 2009.

Bacillariophyceae was richly represented in months of August (270 l^{-1}), followed by September (195 l^{-1}).The group was dominated by *Synedra* sp, *Navicula* sp, *Cymbella* sp. and *Melosira* sp.

Euglenophyceae was represented by 8 species with dominance of both, *Euglena* and *Phacuss* sp. This group attained a highest peak in monsoon season, with maxima in August 2008 at 750 $l^{-1.}$

Among planktonic groups cyanophyceae dominated the reservoir at 37.04 per cent, followed by chlorophyceae (27.19 per cent), copepoda (11.14 per cent), eugelnophyceae (8.03 per cent), bacillariophyceae (5.14 per cent), cladocera (4.40 per cent), rotifera (2.85 per cent), ostracoda (2.20 per cent) and protozoans (1.71 per cent) of the total planktonic population during year 2007-09.

Zooplankton Diversity

Zooplanktons were represented by protozoa,rotifera, cladocera, copepoda and ostracoda.

Protozoans accounted for 7.66 per cent of the total zooplankton,with seven species showing maxima in monsoon.*Colpidium* spp were totally absent in first year of investigation.

Rotifers were accounted for 12.78 per cent of the total zooplankton, with five species showing maxima in summer. *Brachinous* sp was dominant among rotifers during both the years.

During present investigation cladocera was represented by *Ceriodaphnia* sp, *Daphnia* sp, *Moina* sp, *Alona* sp, *Macrothrix* sp and *Indialona* sp. The maximum population of cladocera was recorded in November 2008 at 225 l^{-1}.

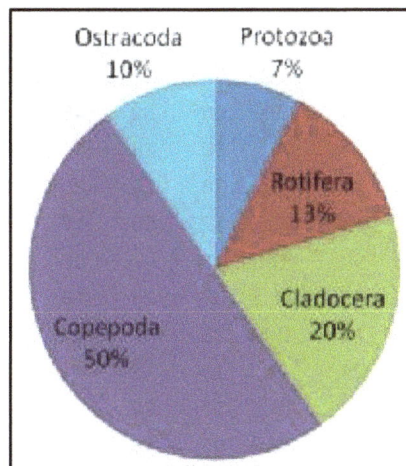

Figure 27.2: Zooplankton Composition during Investigation Period

Table 27.3: List of Zooplankton in Hangarga Reservoir, Maharashtra

Protozoans: *Diffugia* spp, *Arcella* spp, *Vorticella* spp, *Opercularia* spp, *Paramecium caudatum*, *Heleopera* spp, *Colpidium* sp.

Rotifera: *Branchionus* sp, *Filina longiseta*, *Keratella* sp, *Lecane* sp, *Trichocera* sp.

Cladocera: *Ceriodaphnia* sp, *Daphnia* sp, *Moina* sp, *Alona* sp, *Macrothrix* sp, *Indialona* sp.

Copepoda: *Cyclops* sp, *Mesocyclops* sp, *Diaptomus* sp, *Phylladiaptomus* sp, *Nauplius larva*

Ostracoda: *Cypris* sp, *Stenocypris* sp, *Cyclocypris* sp, *Candocypria osborni*

Table 27.4: Composition of Orotozoans (Density : Organisms/litre) during Year 2007-08

	Feb	Mar	Apr	May	Jun	Jul	Aug	Sept	Oct	Nov	Dec	Jan
Difflugia sps.	—	—	08	04	02	20	38	26	24	05	11	05
Arcella sps.	—	—	—	—	12	17	20	16	15	10	10	10
Voritcella sps.	—	—	03	—	06	15	10	09	08	02	—	04
Opercularia sps.	04	09	12	19	—	—	—	—	06	—	03	15
Paramecium, caudatum	24	34	20	19	14	20	10	05	10	05	05	22
Heleopera sps.	03	09	04	08	10	03	02	05	06	02	05	05

During present study, copepods were represented by five species and among them *Cyclops* sp, and *Diaptomus* sp, were present throughout the investigation period. They exhibited a major peak in summer in both the years and minor peak in winter and rainy season during first and second year respectively. This is in confirmation with the findings of Deshmukh (2001) and Bini *et al.* (1997).

Ostracods were accounted for 9.88 per cent of the total zooplankton population. This group was represented by four species. In the present study higher population of ostracods was recorded during monsoon.

Table 27.5: Composition of Rotifera (Density : Organisms/litre) during Year 2007-08

	Feb	Mar	Apr	May	Jun	Jul	Aug	Sept	Oct	Nov	Dec	Jan
Brachionus falcatus	58	50	63	72	65	52	45	40	25	28	35	45
Brachionus calyciflorus	16	22	14	28	32	28	15	10	10	—	—	25
Brachionus diversicornis	42	50	54	42	18	26	20	15	15	42	18	27
Euchlanis dilatata	23	08	12	23	17	09	15	22	25	23	11	15
Filinia longiseta	21	15	18	22	17	—	05	—	—	20	15	10
Keratella tropica	15	32	15	18	33	29	19	27	15	24	14	08
Keratella quadrata	08	10	15	18	23	10	—	—	12	10	05	10
Lecane bulla	30	32	28	—	19	15	—	—	—	22	20	18
Trichocera porcelus	15	21	22	18	30	22	—	—	05	—	15	23
Trichocera longiseta	12	23	27	30	05	18	21	10	—	—	17	18

Table 27.6: Composition of Cladocera (Density : Organisms/litre) during Year 2007-08

	Feb	Mar	Apr	May	Jun	Jul	Aug	Sept	Oct	Nov	Dec	Jan
Indialona ganapati	11	13	18	05	20	07	10	08	15	10	12	15
Moina micrura	18	22	14	25	07	16	15	10	15	—	—	14
Diaphanosoma sarsi	18	28	20	16	11	21	20	08	20	12	18	09
Alona rectangular	26	06	23	19	18	—	—	—	—	05	10	15
Biapertura karna	16	21	25	18	23	20	15	05	—	—	05	11

Table 27.7: Composition of Copepoda (Density : Organisms/litre) during Year 2007-08

	Feb	Mar	Apr	May	Jun	Jul	Aug	Sept	Oct	Nov	Dec	Jan
Diaptomus marshianus	15	28	35	22	15	—	—	—	15	18	09	15
Phyllodiaptomus annae	15	18	22	28	15	18	20	—	—	10	07	10
Neodiaptomus lindbergi	29	27	44	26	20	15	10	16	23	14	10	15
Nauplius larva	18	27	24	29	14	10	18	05	—	09	—	16
Mesocyclops leukarti	30	44	35	22	16	28	16	18	15	26	20	23
Mesocyclops hyalinus	48	53	30	20	20	09	15	23	15	10	18	33
Cyclops viridis	13	20	05	29	15	10	09	07	—	10	—	05

Table 27.8: Composition of Ostracoda (Density : Organisms/litre) during Year 2007-08

	Feb	Mar	Apr	May	Jun	Jul	Aug	Sept	Oct	Nov	Dec	Jan
Stenocypris sps.	23	19	29	24	13	18	22	10	—	—	—	12
Cypris obensa	23	18	32	18	11	14	19	05	10	04	05	09
Cyclocypris globosa	24	27	21	29	28	20	16	10	15	09	05	20
Candocypria osborni	19	23	30	18	10	15	—	05	—	05	10	15

Table 27.9: Composition of Protozoans (Density : Organisms/litre) during Year 2008-09

	Feb	Mar	Apr	May	Jun	Jul	Aug	Sept	Oct	Nov	Dec	Jan
Difflugia sps.	—	—	—	08	20	27	30	22	29	14	10	05
Arcella sps.	05	—	—	—	15	18	20	34	24	20	24	12
Voritcella sps.	—	—	—	—	10	14	30	22	18	15	09	05
Opercularia sps.	17	09	29	22	18	—	—	—	08	02	11	05
Paramecium, caudatum	29	40	34	24	15	15	—	—	18	20	07	20
Heleopera sps.	08	12	05	17	12	11	05	02	06	08	05	05
Colpidium sps.	12	09	05	11	—	—	—	—	—	—	—	10

Table 27.10: Composition of Rotifera (Density : Organisms/litre) during Year 2008-09

	Feb	Mar	Apr	May	Jun	Jul	Aug	Sept	Oct	Nov	Dec	Jan
Brachionus falcatus	35	45	58	35	18	22	28	30	30	27	15	38
Brachionus calyciflorus	22	15	28	32	35	10	20	15	—	—	20	19
Brachionus diversicornis	30	40	32	19	33	16	19	22	18	08	15	20
Euchlanis dilatata	17	28	35	18	12	19	14	22	08	15	21	22
Filinia longiseta	08	22	17	09	18	12	—	—	16	23	—	15
Keratella tropica	28	20	37	26	24	19	24	18	18	29	20	30
Keratella quadrata	15	24	20	29	10	—	15	—	21	16	08	09
Lecane bulla	20	09	18	27	16	10	—	05	15	08	05	14
Trichocera porcelus	18	14	21	05	09	15	—	—	18	09	12	20
Trichocera longiseta	10	08	12	05	19	23	32	—	—	14	18	25

Table 27.11: Composition of Cladocera (Density : Organisms/litre) during Year 2008-09

	Feb	Mar	Apr	May	Jun	Jul	Aug	Sept	Oct	Nov	Dec	Jan
Indialona ganapati	15	20	27	10	09	10	15	12	10	05	10	19
Moina micrura	22	32	24	19	15	10	17	19	16	14	10	20
Diaphanosoma sarsi	15	20	15	10	05	08	04	05	09	05	05	12
Alona rectangular	28	22	20	15	—	—	—	—	12	10	10	19
Biapertura karna	25	30	15	05	12	18	10	05	—	05	05	22
Ceriodaphnia cornuta	18	10	05	04	02	05	05	09	04	08	05	10
Diaphanosoma excisum	15	12	04	04	12	08	06	04	10	05	09	17

Discussion

In the present study the increased phytoplankton density in post monsoon period is due to rich nutrients received through rainwater, as reported by Sreenivasan (1974) for Bhavanisagar reservoir. A total of 31 phytoplankton species were recorded during first and second year of the study. Mishra and Tripathi (2000) reported algae belonging to different groups such as chlorophyceae, bacillariophyceae, cyanophyceae and euglenophyceae at different sampling stations of river Ganga at Varanasi. The winter months showed higher phytoplankton density followed by summer and rainy months. The lower densities during rainy months may be due to high turbidity, low light intensity, cloudy weather and more water coverage with rains.

Table 27.12: Composition of Copepoda (Density : Organisms/litre) during Year 2008-09

	Feb	Mar	Apr	May	Jun	Jul	Aug	Sept	Oct	Nov	Dec	Jan
Diaptomus marshianus	41	33	23	17	30	—	—	—	—	25	32	18
Phyllodiaptomus annae	20	07	16	23	36	15	19	20	—	15	—	22
Neodiaptomus lindbergi	05	22	38	31	30	24	20	16	27	12	19	18
Nauplius larva	18	14	18	12	20	05	23	—	08	10	12	05
Mesocyclops leukarti	37	44	78	32	30	19	24	—	18	15	29	22
Mesocyclops hyalinus	42	12	68	27	15	—	—	—	14	27	08	23
Cyclops viridis	20	15	24	30	20	14	10	—	05	05	09	18

Table 27.13: Composition of Ostracoda (Density : Organisms/litre) during Year 2008-09

	Feb	Mar	Apr	May	Jun	Jul	Aug	Sept	Oct	Nov	Dec	Jan
Stenocypris sps.	15	15	33	28	20	12	—	—	—	—	08	06
Cypris obensa	18	16	27	20	24	—	—	20	09	14	08	15
Cyclocypris globosa	19	16	22	16	05	05	12	20	15	14	11	15
Candocypria osborni	16	10	19	33	22	—	10	—	18	04	09	18

Govind (1963) observed a spurt in phytoplankton immediately after the south-west monsoon (July) in Tungabhadra reservoir. He found that the zooplankton were comparatively higher in density in summer months than the phytoplankton which was higher in winter. This observation is in confirmation with the present findings.

Kaushik *et al.* (1991) reported 63 species of planktonic algal forms from Viveknagar pond at Gwalior. Out of which 26 species belonged to chlorophyceae, 19 species to euglenophyceae, and 9 species each to cyanophyceae and bacillariophyceae. Chlorophyceae was most abundant group in this pond as far as number of species was concerned, while euglenophyceae was numerically abundant.

Devi (1997) reported that the communities of cyanophyceae, chlorophyceae, bacillariophyceae and euglenophyceae constituting the phytoplankton bulk in Shathamraj and Ibrahimbagh reservoirs of Hyderabad. Among the phytoplankton community, the cyanophyceae was found to be rich and dominated in both the reservoirs.

Yousuf and Parveen (1990) recorded 84 taxa of phytoplankton from Dal lake. Of these 50 belonged to chlorophyceae, 16 each to cyanophyceae and bacillariophyceae and two to euglenophyceae. Chlorophyceae was the most dominant group among the phytoplankton population qualitatively as well as quantitatively. The group contributed on an average about 48.08 per cent of the total population, recording a range of 10.25 per cent to 93.43 per cent. Bacillariophyceae contributed 4.09 per cent to 87.09 per cent with a mean of 26.33 per cent. Euglenophyceae was next in order contributing on an average 15.22 per cent of the total population. Cyanophyceae was the most scarce group with a mean contribution of 10.38 per cent.

Sugunan (1980) reported myxophyceae as the most dominant group followed by diatoms and green algae in Nagarjunasagar reservoir. He reported two peaks (October and April) of myxophyceae, which increased steadily till April and declined from May reaching minimum in August. Bacillariophyceae peaks were largely constituted by *Melosira. Navicula* dominated the peaks in February and July. During the blooming of myxophyceae in summer, the diatoms were in very few numbers and they picked up soon, after the bloom diminished (June-July).

The phytoplankton constituted 84.03 per cent of the total plankton, while the zooplankton was recorded at 15.97 per cent of the total plankton. Sugunan and Yadava (1991) reported 96.41 per cent of phytoplankton and 3.59 of zooplankton in the Nongmahir reservoir of Meghalaya.

Temperature range of 15 °C to 25°C was best suited for diatoms and a further rise in temperature affected the population adversely (Jha, 1990). Centric diatoms like *Melosira, Cyclotella* and *Staphanodisus* preferred to grow below 20°C, and therefore these diatoms may be termed as 'coldwater stenothermic forms'. However, requirement of temperature varies from species to species. Diatoms *like Cymbella turgid, Gammatophora serpentine, Pinnularia viridis* and *Navicula cuspidate* were found thriving well at high temperature and thus may be called as warm water forms. In Gobindsagar reservoir, the centric diatoms which otherwise use to be prevalent towards the upper strata of the water column, were found growing in the deeper strata in summer months. This phenomenon is suggestive of the fact that the centric diatoms in general cannot withstand high temperature (Jha, 1990).

Myxophyceae dominate in many Indian reservoirs, the necessity of introducing fishes to utilize them as food was stressed by Natarajan (1975). The report of impressive growth of *Hypopthalmichthys molitrix* in Getalsud reservoir (Anon, 1977 b) indicate that the fish can be successfully introduced to other myxophyceae dominated lakes. During present investigation, the myxophyceae dominated the reservoir. Hence *Hypothalmichthys molitrix* can be introduced in reservoir as discussed above. The dominance of myxophyceae was also recorded by Devi (1985) in Osmansagar and Himayathsagar, and found maximum domination in Mir-alam lake. Rao and Choubey (1990) reported that myxophyceae was third dominant group among phytoplankton in Gandhsagar reservoir.

According to Sugunan (1995) the ubiquitous blooms of *Microcystis* in reservoirs in peninsular India are an example of a lacustrine biocoenose giving way to fluviatile ones in an impoundment. On the reservoir formation and the consequent transformation of lotic environment into the lentic system, saprophobes disappear from the scene giving room for the rapid multiplication of saproxenes. Microcystis, finding a favorable note with the new environment, bursts into blooms, outnumbering all other forms into insignificance. In many reservoirs, orientation of lacustrine and fluviatile plankton can be clearly discerned from the composition of plankton in lotic, lentic and the cove sectors. The fluviatile lotic sector, although recording a lower plankton density, often shows better diversity and evenness indices, compared to the lentic and bay sectors, the still waters of which are characterized by higher concentration of dominance and low evenness (Sugunan, 1991).

Zooplankton comprising of rotifers, cladocerans and copepods are considered to be most important in terms of population density biomass production, grazing and nutrient regeneration in any aquatic ecosystem. Their diversity and density is mainly controlled by availability of food and favourable water quality (Chandrasekhar, 1996).

Maruthanayagam *et al.* (2003) studied the season specific zooplankton diversity in Thirukkulam pond, Mayiladuthurai, Tamil Nadu. This study showed that community size of zooplankton was the highest in rainy season while the lowest density of zooplankton was in summer due to the higher temperature. Among the all zooplankton copepods were the dominant group followed by cladocera,rotifer and ostracoda.Five species of rotifer,four species of cladocera and three species each of ostracoda and copepods were recorded.

Charjan *et al*. (2001) recorded 23 species of rotifers belonging to 7 genera from reservoirs of Washim district of Maharashtra.Deshmukh (2001) reported 28 species of rotifer with maxima in summer, which corroborate with the present investigation.

Arora and Mehra (2003b) while analyzing seasonal dynamics of rotifers in relation to physico-chemical conditions of lotic waterbody observed increased densities of zooplanktons in summers and reduced densities in winters. In summer season, the absence of inflow of the water brings stability to the waterbody.The availability of food is more due to production of organic matter and decomposition.The above factors contribute for the high species density in that season.

Moitra and Bhowmik (1968) observed members of three main zooplanktonic groups *i.e.*, rotifer, cladocera and copepod, which dominate in freshwater fish pond in Kalyani (West Bengal). Agrawal (1978) reported five genera amongst zooplankton population of Janatal at Gwalior.However, Pathak and Mudgal (2004) observed five genera of rotifer, three genera of cladocerans and ostracodans and two genera each in respect of protozoans and copepodans in Virla reservoir (Madhya Pradesh).

The numerical variation in peak periods of different group of zooplankton might be due to different biological parameters. In Hangarga reservoir rotifers were found dominant over the other groups. This findings is in agreement with Kohli *et al*. (1982) and Balkhi *et al*. (1987). The numerical variations in rotifers may apparaently be influenced by the water quality.

Hutchinson (1967) observed that *Brachionus* species are very common in temperate and tropical waters, which indicates alkaline nature of the water bodies.

Investigation of CIFRI (1997) on Bhatghar reservoir of Maharashtra accounted rotifer average at 5.95 per cent of the total plankton. A peak in their production was observed in March followed by October and July. This group was absent during August –September. The dominant forms were *Keratella* sp., *Brachionus* sp., *Asplanchna* sp. and *Cephalodella* sp.

Sharma and Diwan (1993) reported species of rotifers from Yeshwant sagar reservoir, which showed dominance during summer months. This observation corroborate with present findings on rotifers in the Hangarga reservoir.*Brachionus* sp. was dominant among rotifers during both the years. The present findings are in agreement with Sakhare (2007).

Cladocera constitute an important group. The greater significance of cladocera in the aquatic food chain as food for both young and adult fish was emphasized much earlier (Pennak, 1978).

During present investigation cladocera was represented by *Ceriodaphnia* sp, *Daphnia* sp, *Moina* sp, *Alona* sp, *Macrothrix* sp and *Indialona* sp. The maximum population of cladocera was recorded in November 2008 at 225 l$^{-1.}$ Devi (1997) reported the maximum cladocerans population during postmonsoon and premonsoon in Ibrahimbagh and premonsoon and post monsoon in Shathamraj reservoir of Hyderabad.

Sakhare (2007) accounted plankton diversity in Yeldari reservoir in which cladocera was represented by seven species. The similar species were recorded during the present investigation.

Free living copepods are an essential link in food chain occupying the intermediate trophic level between bacteria, algae and protozoa on one hand and small and large plankton predators on the other. Though, they are not as important as cladocerans in the diet of fish.

During present study, copepods were represented by five species and among them *Cyclops* sp, and *Diaptomus* sp, were present throughout the investigation period. They exhibited a major peak in summer in both the years and minor peak in winter and rainy season during first and second year respectively. This is in confirmation with the findings of Deshmukh (2001) and Bini *et al.* (1997).

Sigh *et al.* (1993) reported three species of copepod including *Mesocyclops leukarti, Diaptomus* sp and *Cyclops* from Nanak sagar reservoir of Nainital which were also recorded from the Hangarga reservoir.

Ostracods are bivalve crustaceans found in both freshwater and marinewater.There are over 1700 species of known ostracods of which about one-third are freshwater forms.They inhibit a wide variety of freshwaters likes lakes, pools, swamps, streams and heavily polluted areas (Edmondson, 1959). Patil and Goudar (1989) reported occurrence of seven species of ostracods in Dharwad district.In the present study, only four species of ostracods were found.

In the present study higher population of ostracods was recorded during monsoon. This is in confirmation with the findings of Chandrasekhar (1996).

Sharma and Sahai (1990) described the observations on plankton population of Jari reservoir of Uttar Pradesh and their significance to fisheries. Among zooplankton *Keratella* sp, *Asplanchna* sp, *Notholca* sp, *Mesocyclops* sp and *Naupli* were the dominant zooplankter.

Pandey *et al.* (2004) studied the seasonal fluctuation of zooplankton community in relation to physicochemical parameters in river Ramjan of Kishanganj, Bihar. The collections were dominated by rotifera, followed by cladocera and copepoda. Rotifera showed a negative correlation with pH, dissolved oxygen and transparency and copepods showed negative correlation with water temperature, nitrate and phosphate. The cladocerans also revealed negative correlation with pH, transparency and phosphate. This indicates several abiotic factors exert a considerable influence on the zooplankton abundance.

The physico-chemical and zooplankton analysis of the Shendurni river, Kerala was studied by Sahib (2004). The dissolved oxygen levels were observed to be highly saturated and a direct correlation between dissolved oxygen level and zooplankton populations were observed.Chakrapani (1996) compared the zooplankton diversity and physico-chemical analysis of both urban and non-urban lakes. 19 urban and 24 non-urban lakes were selected for the study. The zooplankton diversity of some of the urban and non-urban lakes was compared with the earlier study. The changes in the populations have indicated the influences of pollution on these lakes. Biological analysis indicated that lakes such Anekepalya, Bellandur, Chilkkahulimam, Harohalli, Kengeri, Kalkere, Nagavara, Nelamangala, Puttenahalli, Rachenahalli, Rampura, Tavarakere, Ulsoor, Varthur, Vengaiah, Yellahehalli, and Yellamallappuchetty were threatened ecologically and unsuitable for human usage.

Studies on the zooplankton diversity in the evaluation of the pollution status of water bodies were carried out by Khan and Rao (1981). The potential effects of thermal pollution, nutrient enrichment in eutrophication, interaction of ions,toxic substances like heavy metals, halogens, solids, reducing agents, and radioactive wastes on aquatic protozoan communities had been studied. Rotifers were found to be more sensitive to pollution than other groups of zooplankton.

Among crustaceans, cyclopoids and cladocerans were found to be associated with increasing productivity. The ratio of calanoids to cyclopoids plus cladocerans was found to be good indication of trophic condition and valuable index of pollution.

Mahajan (1981) made preliminary studies of the identification of species among the zooplankton community, which could serve as indicators of different types of pollution. Species of zooplankton which could serve as indicators of thermal pollution and stress pollution, Eutrophication, Heavy metal pollution, Pesticidal pollution and miscellaneous pollution activities were studied. Toxicity tests conducted for the selected species indicate, different groups of zooplankton were found to be sensitive to different types of pollutants.

Chakrabarty *et al*. (1959) studied the plankton and the physicochemical conditions of river Jumna at Allahabad. The zooplankton community was represented by rotifera, protozoa, copepoda, cladocerans and ostracoda. The rotifers were found to be dominant group followed by protozoa and crustaceans.The collections were made during the early morning, midday and in the night and vertical migration was noticed by the presence or absence of certain forms during the morning, midday, and night collections of the same day. The study shows that water quality, turbidity and temperature have an influence on plankton populations.

Chakrapani (1996) studied the plankton diversity of Sixty one lakes of Bangalore and reported zooplankton falling under five major groups - protozoa, rotifera, cladocera, copepoda and ostracoda. Fourteen forms of protozoa, twenty nine forms of rotifera, six forms of cladocera, four forms of copepoda and five forms of ostracoda were recorded. On the overall sixty two forms of zooplankton were observed including five unidentified forms. The appearance of intermediary stages of *Rabditis* species in five lakes indicates the presence of potential human parasites.

Sampaio *et al*. (2001) studied the species composition and abundance of zooplankton community of 7 reservoirs of Paranapanema river, Brazil. Plankton samples were collected from the limnetic region of all reservoirs in each season. Diversity was evaluated using Shannon-Weiner index and Sorensen index. From 27 analysed samples in 7 reservoirs, a total of 76 species of rotifers, 2species of calanoid copepods, 5 species of cyclopoid copepod and 26 species of cladocerans were recorded. The values for the Shannon-Weiner index varied from 1.5 to 3.0 among the reservoirs. The Sorensen index showed that the similarity among the reservoirs was low. A positive relationship between the trophic state of the reservoir and the diversity of zooplankton community was recorded. Among the reservoirs they surveyed, the oldest reservoirs had the highest species diversity and hence indicated a stable environment.

Das (2002) studied the dynamics of net primary production and zooplankton diversity in brackish water shrimp culture pond in northern part of Ganjam district, Orissa. Significant negative correlation was noticed between net primary production and zooplankton population. Copepods and rotifers were found to be the dominant groups among zooplankton. The zooplankton population varied with different seasons of the year with rainy and summer seasons showing the minimal density in zooplankton population. Patil and Shrigur (2004) studied the morphology and identification characteristics of four copepods species namely *Thermocyclops crassus, Mesocyclops leukarti, Apocyclops royi* (order cyclopidae) and *Eudiaptomus gracilis* (order Calanoida and family Diaptomidae). The study was conducted in government fish farm, Goregaon, Mumbai. *Eudiaptomus gracilis* and *Apocyclops royi* were the first record from Mumbai.

Zooplanktonic diversity of 6 ponds of Durg-Bhillai city, Chhatisgarh state was studied by Anil kumar *et al*. (2004). Rotifers and copepods were found to be predominant group. The rotifers were represented by 5 species of *Brachionus* and 1 species of *Tesdinella*, indicating eutrophicated status. Copepods were mainly dominated by *Mesocyclops* species and cladocerans, the least abundant group comprise *Moina* species and *Ceriodaphnia* species. Ostracoda were also observed in their collection with *Cypris* species being the dominant organism. Ostracods were also found to show diurnal variation between day (206 organisms/m^3) and night samples (555 organisms/m^3). The predominance of rotifers and copepods indicate the nutrient availability in these ponds.

References

Agarwal, S.S., 1978. Hydrobiological survey of Janaktal tank, Gwalior (Madhya Pradesh), India. *Proceedings of All India Seminar on Ichthyology*, p. 20–26.

Angadi, S.B., Shidda Mallaiah, N. and Patil, P.C., 2005. Limnological studies of Papnash pond, Bidar, Karnataka. *J. Environ. Biol.*, 26: 213–216.

Anon., 1977b. *CICFRI Newsletter*, 1(3).

Arora, J. and Mehra, N., 2003b. Seasonal dynamics of the rotifers in relation to physical and chemical conditions of the river Yamuna (Delhi), India. *Hydrobiol.*, 491: 101–109.

Balkhi, M.H., Yousut, A.R. and Quadri, M.R., 1987. Hydrobiology of Ancher Lake, Kashmir. *Comparative Physiology and Ecology*, 12: 131–139.

Battish, S.K., 1992. *Freshwater Zooplankton of India*. Oxford and IBH Publishing Co. Pvt. Ltd., New Delhi.

Beaugrand, G., Ibanez, F. and Reid, P.C., 2000. Spatial, seasonal and long-term fluctuations of plankton in relation to hydroclimatic features in the English channel, Celtic Sea and Bay of Biscay. *Marine Ecology Progress Series*, 200: 93–102.

Bahura, C.K., 2001. Phytoplanktonic community of a highly eutrophicated temple tank, Bikaner, Rajasthan. *J. Aqua. Biol.*, 16(1&2): 1–4.

Bini, L.M., Tundisi, J.G. and Yundisi, T.M., 1997. Spatial variatiation of zooplankton groups in a tropical reservoir (Broa Reservoir: Sao Paulo State–Brazil). *Hydrobiologia*, 357(1–3): 89–98.

Chandrasekhar, S.V.A., 1996. Ecological studies on Saroornagar Lake, Hyderabad. *Ph.D. Thesis*, Osmania University.

Chankarabarty, R.D., Roy, P. and Singh, S.B., 1959. A quantitative study of the plankton and the physico-chemical conditions of the river Jamuna at Allahabad in 1954–55. *Indian J. Fisheries*, 6(1): 186–203.

Chakrapani, B.K., Krishna, M.B. and Srinivasa, T.S., 1996. A report on the water quality, plankton and bird populations of the lakes in and around Bangalore and Maddur, Karnataka, India. Department of Ecology and Environment, Karnataka State Government.

Charjan, A.P., Dabhade, D.S. and Malu, R.A., 2001. Population trends of rotifers in Washim District, Maharshtra. In: *Souvenir of Nat. Conf. on Fish and Fisheries: Challenges in the Millennium*, P.G. College of Science, Osmania University, Hyderabad, 14–15 March, Abstract No. 66.

CICFRI, 1997. *Ecology and Fisheries of Bhatghar Reservoir*. Golden Jublee spl. Bulletin No. 73.

Das, S.K., 2002. Primary production and zooplankton biodiversity in brackish water shrimp culture pond. *J. Ecobiol.*, 14(4): 267–271.

Dhanapathi, M.V.S.S.S., 2000. Taxonomic notes on the rotifers from India (from 1889 – 2000). *Indian Association of Aquatic Biologists*, Publ. No. 10.

Deshmukh, U.S., 2001. Ecological studies of Chhatri lake, Amravati with special reference to planktons and productivity. *Ph.D. Thesis*, Amravathi University, Amravathi.

Devi, M.J. 1985. Ecological studies of the limnoplankton of three freshwater bodies of Hyderabad. *Ph.D. Thesis*, Osmania University.

Devi, Sarla, B., 1997. Present status, potentialities, management and economics of fisheries of two minor reservoirs of Hyderabad. *Ph.D. Thesis*, Osmania University, Hyderabad.

Edmondson, W.T., 1959. *Freshwater Biology*, 2nd Edn, John Wiley and Sons, New York, USA.

Ganapathi, S.V., 1941. Studies on the chemistry and biology of ponds in the Madras city: Seasonal changes in the physical and chemical conditions of a garden pond containing aquatic vegetation. *J. Madras Univ.*, 13: 55–59.

Govind, B.V., 1963. Preliminary studies on plankton of the Thungabhadra Reservoir. *Indian J. Fish.*, 10: 148–158.

Hutchinson, G.E., 1967. *A Treatise in Limnology, Vol. II: Introduction to Lake Biology and Limnoplankton.* John Wiley and Sons, N.Y., 1115 pp.

Jha, B.C., 1990. The periphyton of Gobindsagar reservoir and allied waters, Himachal Pradesh. I. Abundance and periodicity of diatms. In: *Contributions to Fisheries of Inland Open Water Systems in India,* (Ed.) A.G. Jhingran, V.K. Unnithan and Amitabha Ghosh. Inland Fisheries Society of India, pp. 122–129.

Jhingran, V.G., 1974. *Fish and Fisheries of India*. Hindustan Publishing Corpoaration, New Delhi.

Kaushik, S., Agarkar, M.S. and Saksena, D.N., 1991. On the planktonic algae of sewage fed vivek Nagar pond at Gwalior. *Poll. Res.*, 10(1): 25–32.

Khan, M.A. and Rao, I.S., 1981. Zooplankton in the evaluation of pollution. In: Paper presented at *WHO Workshop on Biological Indicators and Indices of Environmental Pollution*. Cent.Bd. Prev. Cont. Poll., Osm. Univ, Hyderabad, India.

Kohli, M.P.S., Thakur, N. Kumar and Munnet, S.K., 1982. Seasonal changes in plankton population of some freshwater ponds at Patna. *Journal of Inland Fisheries Society of India*, 14: 69–76.

Kumar, Anil, Tripathi, Seema and Ghosh, P., 2004. Status of freshwater in 21st century: A review. In: *Water Pollution: Assessment and Management,* (Eds.) Arvind Kumar and G. Tripathi. Daya Publishing House, Delhi.

Mahajan, C.L., 1981. Zooplankton as indicators for assessment of water pollution. In: Paper presented at *WHO Workshop on Biological Indicators and Indices of Environmental Pollution*. Cent.Bd. Prev. Cont. Poll., Osm. Univ, Hyderabad, India.

Maruthanayagam, C., Sasikumar, M. and Senthilkumar, C., 2003. Studies on zooplankton population in Thirukkulam pond during summer and rainy seasons. *Nature Environ. Pollut. Technol.*, 2: 13–19.

Meshram, C.B. and Dhande, 2000. Algal diversity with respect to pollution status of Wadali lake, Amravati, Maharashtra, India. *J. Aqua. Biol.*, 15 (1&2): 1–5.

Michael, R.G., 1973. A guide to the study of freshwater organisms. *Madurai Univ. J. Suppl.*, (Ed.), 1: 186 pp.

Mishra, B.P. and Tripathi, B.D., 2000. Changes in algal community structure and primary productivity of river Ganga as influence by sewage discharge. *Ecol. Env. and Cons.*, 6(3): 279–287.

Moitra, S.K. and Bhowmik, M.L., 1968. Seasonal cycle of rotifers in freshwater fish pond in Kalyani, West Bengal. *Proceedings of Symposium on Recent Advances in Tropical Ecology*, p. 359–367.

More, Y.S. and Nandan, S.N., 2000. Hydrobiological study of algae of Panzara river (Maharashtra). *Ecol. Env. and Cons.*, 6(1): 99–103.

Pandey, B.N., Hussain, S., Jha, A.K., and Shyamanand, 2004. Seasonal fluctuation of zooplankton community in relation to certain parameters of river Ramjan of Kishanganj, Bihar. *Nature, Environment and Pollution Technology*, 3(3): 325–330.

Pandey, J. and Verma, A., 2004. The influence of catchment on chemical and biological characteristics of two freshwater tropical lakes of Southern Rajasthan. *J. Environ. Biol.*, 25: 81–87.

Pathak, S.K. and Mudgal, L.K., 2004. Biodiversity of zooplankton of Virla reservoir Khargone (M.P.), India. In: *Biodiversity and Environment*, (Ed.) Arvind Kumar. APH Publishing Corporation, New Delhi, pp. 317–321.

Patil, S.D. and Shirgur, G.A., 2004. Morphology and identification characteristics of copepod species occurring in the government fish farm, Goregaon, Mumbai. *J. Ecobiol.*, 16(1): 45–52.

Pennak, R.W., 1978. *Freshwater Invertebrates of the United States*, 2nd Edn. John Wiley Sons, New York, 803 pp.

Pulle, J.S. and Khan, A.M., 2003. Phytoplanktonic study of Isapur dam water. *Ecol. Env. Conser.*, 9: 403–406.

Rawson, D.S., 1956. Algal indicators of trophic lake types. *Limnol. Oceanogr.*, 1: 18–25.

Saha, L.C. and Pandit, B., 1990. Dynamics of primary productivity between lentic and lotic systems in relation to abiotic factors. *J. Indian Bot. Soc.*, 69: 213–217.

Sahib, S.S., 2004. Physico-chemical parameters and zooplankton of the Shendurni River, Kerala. *Journal of Ecobiology*, 16(2): 159–160.

Sakhare, V.B., 2007. *Reservoir Fisheries and Limnology*. Narendra Publishing House, Delhi, pp. 187.

Sakhare, V.B. and Joshi, P.K., 2002. Ecology of Palas–Nilegaon reservoir in Osmanabad district, Maharashtra. *J. Aqua. Biol.*, 18(2): 17–22.

Sakhare, V.B. and Joshi, P.K., 2006. Plankton diversity in Yeldari reservoir, Maharashtra. *Fishing Chimes*, 25(12): 23–25.

Sampaio, E.V., Rocha, O., Matsumura-Tundisi, T. and Tundisi, J.G., 2001. Composition and abundance of Zooplankton in the limnetic zone of seven reservoirs of the Paranapanema river, Brazil. *Braz. J. Biol.*, 62(3): 525–545.

Sharma, Neelima and Sahai, Y.N., 1990. Some observations on the plankton population of Jari reservoir near Allahabad (U.P.) and their significance to fisheries. In: *Reservoir Fisheries*, (Ed.) Arun G. Jhingran and V.K. Unnithan, 3–4 January. Spl. Publ. 3, Asian Fisheries Society, Indian Branch, Mangalore, India, p. 131–138.

Sharma, Rekha and Diwan, A.P., 1993. Limnological studies of Yeshwantsagar reservoir. I. Plankton population dynamics. In: *Recent Advances in Freshwater Biology*, (Ed.) K.S. Rao. 1: 199–211.

Shukla, Suresh, C., Kant, Rajani and Tripathi, B.D., 1989. Ecological investigation on physico-chemical characteristics and phytoplankton productivity of river Ganga at Varanasi. *Geobios*, 16: 20–27.

Singh, A.K. and Ahmed, S.H., 1990. A comparative study of the phytoplankton of the river Ganga and pond of Patna (Bihar), India. *J. Indian Bot. Soc.*, 69: 153–158.

Singh, C.S., Sharma, A.P. and Deorari, B.P., 1993. Plankton population in relation to fisheries in Nanak sagar reservoir, Nainital. In: *Recent Advances in Freshwater Biology*, Vol. 1, (Ed.) K.S. Rao, pp. 66–79.

Singh, V.P., 1960. Phytoplankton ecology of the inland waters of Uttar Pradesh. In: *Proc. Symposium on Algology*, ICAR, New Delhi.

Sousa, W., Attayde, J., Rocha, E. and Eskwazi-Santanna, E., 2008. The response of zooplankton assemblages to variations in the water quality of four man-made lakes in semi-arid northeastern Brazil. *J. Plankton Res.*, 30: 699–708.

Sreenivasan, A., 1974. Limnological features of a tropical impoundment, Bhavanisagar reservoir, (Tamil Nadu) India. *Int. Revueges Hydrobiol.*, 59(3): 327–342.

Sugunan, V.V., 1980. Seasonal fluctuations of plankton of Nagarjunasagar reservoir, A.P., India. *J. Inland Fish. Soc., India*, 12(1): 79–91.

Sugunan, V.V. and Yadava, Y.S., 1991b. Feasibility studies for fisheries development of Nongamahir reservoir. CICFRI, Barrackpore, pp. 30.

Welch, P.S., 1948. *Limnological Methods*. McGraw-Hill Book Co., New York.

Yousuf, A.R. and Parveen, M., 1990. Phytoplankton dynamics in Dal Lake, Kashmir. In: *Contributions to Fisheries of Inland Open Water Systems in India*, (Eds.) A.G. Jhingran, V.K. Unnithan and Amitabha Ghosh. Inland Fisheries Society of India, pp. 58–63.

Chapter 28

Studies on Water Quality Parameters of Siddeshwar Dam, Hingoli District, Maharashtra

☆ *D.C. Deshmukh and S.O. Bondhare*

ABSTRACT

The present investion was carried out to study the water quality and physico-chemical characteristics of Siddeshwar Dam. Attempts were made to study and analyze the physico-chemical characteristics of the water like tempreture pH, toal alkalinity, dissolved oxygen, carbon dioxide, total hardness, calcium, chloride, etc. The significant range variation in various physico-chemical parameters of the dam were recorded which depicted the suitability of the dam for fish culture.

Keywords: Siddheshwar dam, Physico-chemical parameters, Water quality.

Introduction

Water is vital for the substance of life. Water is crucial for energy production wheather it is hydro, nuclear or thermal. Water continues to be an important means of transportation. It is used for recreation, for cleaning for maintaing ecological habitats, and numerous other economic, environmental, and social uses. There is no substitute for this essential resource. Much of the current concern with regards to environmental quality is focused on water because of its importance in maintaining the human health and health of the ecosystem. Freshwater is finite resource, essential, without freshwater of adequate quantity and quality, sustainable developments will not possible.It has been estimated that only 0.00192 per cent of the total water on the earth is avaible for human consumption. It is necessary to know the physico-chemical properties of water to study the rearing practices of the fish in water bodies. Assessment of water resource quality of any region is an important aspect for development activities of the region because rivers, lakes, man made reservoirs and ponds are used for water supply to domestic, industrial, agricultural and fish cultural purpose (Khatri, 1985).Present investigation was undertaken to ascertain the physico-chemical characteristics of Siddeshwar dam in Hingoli district of Maharashtra.

Materials and Methods

Siddheshwar dam is situated near village Siddheshwar in Hingoli district.The dam receives rain water from Godavari river. In the present investigation water samples was collected for physico-chemical analysis from three stations *i.e.* S_1, S_2 and S_3 in early morning from second week of each month of October 2009 to September 2010. Water samples were collected in presterlized plastic bottles. The atmospheric and water temperature were measured at the sampling site by using centigrade mercury thermometer and pH was measured with standard pH meter. For estimation of dissolved oxygen the sample was fixed on the sampling site by Winkler's solutions, while other parameters were estimated in the laboratory by standards methods (APHA, 1992; Kodarkar *et al.*, 1998; Pant *et al.*, 2006; APHA, 1998; and Trivedi and Goel, 1986).

Results and Discussion

Temperature

The analytical report of water quality characteristics are presented in Table 28.1. Air tempreture ranged from 25.37°C to 32.70°C throughout the study period. Similarly water temperature varied from 23.02°C to 30.52°C. The highest water temperature was recorded during summer season and lowest was recorded during winter season.

Table 28.1: Mean and Standard Deviation of Physico-chemical Characteristics during Winter Season

Parameters	Station I		Station II		Station III	
	Mean	±SD	Mean	S.D.	Mean	S.D.
Air Temp. (°C)	24.37	±1.40	25.60	±1.27	2561	±1.32
Water Temp. (°C)	22.02	±0.93	23.18	±1.03	23.4	±1.06
pH	6.2	±0.046	7.3	±0.13	7.2	±0.12
Total Alkalinity	360	±10.42	332.3	±6.75	349.65	±6.90
Dissolved Oxygen	6.35	±0.3	5.66	±0.36	5.68	±0.71
Free carbon dioxide	Absent	-	Absent	-	Absent	-
Total Hardness	66.26	±1.66	70.99	±1.96	72.25	±2.84
Calcium	16.09	±0.72	16.15	±0.28	16.40	±0.85
Chloride	33.50	±14.00	32.28	±1.81	32.18	±1.18

pH

It is considered as an important ecological factor and is the result of the interaction of varios substances *i.e.* solution in the water and also numerous biological phenomenon's. The acidic and alkaline nature of water is indicated by pH. Most of the biological processes and biochemical reactions are pH dependent. The avarege of pH varies from 7.1 to 7.3, being minimum in summer and maximum in Winter. Similar results were also reported by Khare (2002) and Jayabhaye *et al.* (2008).

Table 28.2: Mean and Standard Deviation of Physico-chemical Characteristics during Summer Season

Parameters	Station I		Station II		Station III	
	Mean	±SD	Mean	S.D.	Mean	S.D.
Air Temp. (°C)	31.37	±1.64	32.60	±1.73	32.73	±1.73
Water Temp. (°C)	30.37	±0.93	28.18	±1.03	28.52	±1.50
pH	6.8	±0.046	6.9	±0.13	6.2	±0.3
Total Alkalinity	289	±13.42	300	±6.75	24.65	±1.35
Dissolved Oxygen	4.35	±1.50	4.66	±1.36	4.82	±0.026
Free carbon dioxide	Absent	-	Absent	-	Absent	-
Total Hardness	66.11	±6.66	63.99	±5.26	63.25	±4.84
Calcium	16.11	±10.72	16.12	±0.60	16.45	±1.28
Chloride	43.50	±14.25	44.28	±3.81	44.18	±1.18

Parameters	Mean	BIS Standard	
		P	E
Air Temp.(°C)	28.42	-	-
Water Temp. (°C)	26.08	-	-
pH	7.2	6.6	9.0
Total Alkalinity	316.76	200	600
Dissolved Oxygen	4.9	4	10
Free carbon dioxide	23.03	-	-
Total Hardness	70.05	300	600
Calcium	15.21	75	200
Chloride	36.67	250	999

P: Permissible limit; E: Excessive limit.

Total Alkalinity

The alkalinity in water is usually interpreted as the quality and kinds of compounds such as bicarbonates, carbonates and hydroxides present which collectively shift the pH to the alkaline side of neutrality. The total alkalinity ranged from 289.4mg/1 to 349.65 mg/l. The minimum value was recroded during summer season. Alkalinity of water increases with an increased pH and calcium. The results were compared with Deshmukh (2001).

Dissolved Oxygen

Oxygen is one of several dissolved gases important to aquatic life. It is a primary and comprehensive indicator of water quality in surface in surface water. In the present study the dissolved oxygen value ranges between 3.89 mg/1 to 6.38 mg/1 indicate that the water body is polluted. So less dissolved oxygen was observed in the study period. It shows that the minimum dissolved oxygen was recorded during winter season. The variation in dissolved oxygen may be due to exposure of water to atmosphere in different seasons. The same results are in concurrent with Chavan *et al*. (2005).

Carbon Dioxide

Free CO_2 in water is the source of carbon that can be assimilated and incorporated into the living matter of all the aquatic autrophs (Hutchinson,1957) During the present study free CO_2 was rarely observed. The absence of the free CO_2 was recorded during the winter season. It may be due to complete utilization in photosynthetic activitiy or its inhabition by the presence appreciable amount of carbonates in water. In the present investigation the free CO_2 values ranges between 19.64mg/1 and 29.87 mg./l. The minimum value was recorded during rainy season and maximum value was recorded during rainy season. And maximum value was recorded during the summer season. In the present investigation high values of carbonate hardness were recorded through out of the year and hence the absence of free CO_2 for most of the months may be due to utilization of carbon in carbonate formation.

Total Hardness

The total Hardness often employed as indicator of waste, water quality depends on the concentration of carbon ate and bicarbonate salts of calcium and magnesium or sulphate, chloride or other anions of maineral acids. It is a measures of the capacity to precipitate soap. It is the sum of the polyvalent cations present in water. The total hardness ranged from 63.80 mg/l to 72.30mg./l. The relatively minimum values were recorded in Rainy season. The values are within the permissible limits prescribed by WHO. Similar observations are also made by Jayabhaye (2008).

Calcium

Calcium is essential for metabolic process in all living organisms. The present study showed a range of 15.34mg/l to 16.52mg/l. Maximum calcium value were recorded in summer and minimum rainy season.

Chloride

Chlorides are important in detecting the concentration of ground water by waste by wastewater. Chlorides occur rally in all types of waters. High concentration of chlorides is to be the indicatior of pollution due to high organic wastes of animal or industrial origin. The present study found a range of 32.18 mg/l to 45.97 mg/l. The lower value of chloride in winter was recorded in summer seasons. In natural water 4 to 10 mg/l of chloride indicates its purit.The present observation are in confirmation with that of George (1961) and Khan *et al.* (1970).

References

APHA, 1992. *Standard Method for the Examination of Water and Wastewater*, 18th Edn. APHA, AWWA, WPET, Washington DC, U.S.A.

APHA, 1998. *Standard Method for the Examination of Water and Wastewater*, 20th Edn. American Public Health Association, Washigton D.C.

Bureau of Indian Standards (BIS), 1991. *Indian Standard Specification for Drinking Water*. IS 10500, pp. 2–4.

Chavan, R.J., Mohekar, A.D., Sawant, R.J. and Tat, M.B., 2005. Seasonal variations of abiotic factors of Manjara Project Water Reservoir in Dist. Beed, Maharashtra, India. *Poll. Res.*, 24(3): 705–708.

Deshmukh and Ambore, 2006. Seasonal variations in physical aspects of pollution in Godavari river at Nanded, Maharashtra India. *J. Aqua. Biol.*, 21(2): 93–96.

Deshmukh, Ujjwala S., 2001. Ecological studies of Chhatri lake, Amravati, with special reference to planktons and productivity. *Ph.D. Thesis,* Amaravati University, Amravati.

Goerge, M.G., 1961. Hydrobiological studies on the lower lake of Bhopal with special reference to the productive of economic fish. *Ph.D. Thesis,* Bhopal University, Bhopal.

Hutchinson, 1957. *A Treatise on Limnology, Vol. 1.* John Wiley and Sons, New York, London.

Jayabhaye, U.M., Salve, B.S. and Pentewar, M.S., 2008. Some physico-chemical aspects of Kayadhu River, District Hingoli Maharashtra. *J. Aqua Biol.,* 23(1) : 64–68.

Khan, K.A., Siddiqui A.G. and Nazir, M., 1970. Diumalariation in a shallow tropical freshwater fish pond in Shajanpur, U.P., India. *Ibid,* 35: 279–304.

Katri, T.C., 1985. Physico-chemical features of ladki reservoir, Kerala, during premonsoon period. *Env. and Ecol.,* 3: 134–137.

Khare, P.K., 2002. Assessement of organic pollution and water quality of Satri tank Chhatrapur. *I.J. Environ. Ecol.,* 6(1): 39–44.

Kodarkar, M.S., Diwan, A.D., Murungan, N., Kulkarni, K.M. and Anuradha, R., 1998. *Methodology for Water Analysis.* Indian Association of Aquatic Biologists, Hyderabad.

Chapter 29

Fishing Crafts and Gears in River Godavari at Nanded, Maharashtra

☆ *R.P. Mali and S.O. Bondhare*

ABSTRACT

River Godavari is also called as 'Dakshin Ganga'. It flows from western to southern India. In river Godavari, fishing is take place. Every water body has its unique pattern for fishing crafts and gears. The present paper deals with fishing crafts and gears used in river Godavari at Nanded.

Keywords: Crafts, Gears, Godavari river, Nanded, Maharashtra.

Introduction

Godavari river originates in Deolai (Bhramgiri) near Nashik. Its length is about 1450 km. Vishnupuri dam constructed on it is one of the largest life irrigation project in Asia. The Idea of this project was put forward and pursued by former Chief Minister of Maharashtra Late Shri Shankarrao Chavan. Hence, it his remembrance the government of Maharashtra has named the water reservoir 'Shankar Sagar Jalashaya' (Vishnupuri Reservoir).In Godavari river fishing take place. There is a valuable contribution by several workers (Sakhare, 2007; Niture and Chavan, 2009) on types of crafts and gears used in freshwater fisheries of Maharashtra.

Materials and Methods

Information was collected by personal interview method of fishermen who perform the fisheries activities in this area. A questionnaire was also prepared in vernacular language to know the cost of fishing devices and their operations.

Results and Discussion

Crafts

Masula Boat

In Godavari river fishermen use this type of boat. It is made from iron metal sheets and wooden planks. The length of craft is 10 to 15 feet and width at the centre is 4 to 5 feet. It carries 4 to 5 persons, weight is about 60 to 90 kg. craft is operated by using 1 to 2 flipper by fishermen. Its cost is Rs. 8000/- to 10000/- in market.

Thermocoal Craft

These boat is made from thermocoal sheet of size 7 x 2.5 feet long and 6 to 8 inch in thick. It is called 'Nav'. Generally for its construction fishermen used continued sheet of thermocoal during 2005 to 2007 study duration almost all crafts used were of this type. These crafts are not suitable to carry large sized or heavy nets. It is navigated by single fishermen with the help of a hand flipper, made from single bamboo stick of 4 to 5 feet and to which plastic flippers are fitted at both the end of bamboo. It is locally called as 'Chatu'. Good balance technique is necessary to drive the thermocoal craft. This craft carry 30 to 35 kg stones to tie with the nets as sinkers. Its cost is about Rs. 400/- to 500/-.

Coracle

It is a simple circular basket with a wide mount about 4 meter in diameter. The frame basket is made up of bamboo and is made water tight by covering leather. It is used for fishing by single fishermen. Its cost is Rs. 1000/- to 1500/-.

Rubber Tube Platform

In Godavari river some young fishermen use rubber tube platform. To this tube they tie pieces of stones as the sinkers. It is used as craft for fishing purpose.

Banana Stem Craft

Some fishermen made their own craft by scooping banana stem of 6 to 10 meter long. It is provided with planks and rubber tight. It is traditional method.

Gears

Cast Net (Ghagaria Jal)

In Vishnupuri reservoir it is commonly used. It is a circular net having the shape of a large umbrella about 5 to 15 meter wide circumference. A strong cord is attached to the apex of umbrella and number of iron or lead weight are fixed all along the margin. The fishermen through the net fully spread over the water keeping the long rope in his left hand. This has to be done very skillfully so that, the net falls on the surface of water fully expanded. The net sinks to the bottom and the circumference closes due to the weight attached to it.

All kinds of small fisher are entangled in this net which then pulled out by means of the cord. In single operation fishermen catches about 30 to 35 kg fishes. Size of mesh is 2.5 cm. Its cost is Rs. 1800/- to 2000/- in market.

Drag Net

It is used in Godavari river. It is also called 'Zorli' or 'Wadap' and 'Pandya' or 'Purai'. The upper margin of the net is supported by a head rope and is provided with floats. Along lower margin is the foot rope, to which a number of weights are tied, to keep the net in position. One end of the net remains on the bank of river while a boat carries the rest of the net to spread it in water in semicircular way. When fishes and prawns are trapped in central bag then the net is slowly dragged by two parties of fishermen.

The length of bag varies from 10 to 50 meter. Zorli net made up from mosquito net and Purai from nylon thread.5to10 fishermen are required to operate drag net.

Hook and Line

It contain a long nylon wire or rope, about 100 to 150 ft. length have knots at 1 to 1.5 ft. distance all along the rope. In between 2 knots a nylon wire of 2-3 ft. length is loosely fixed. So as to move this wire in between 2 knots. At free end of the wires rounded, metallic hooks are fixed. To each hook fresh weed fishes are attached by fisherman to attract the predatory fishes. Thermocol floats are fixed to main long ropeat some intervals the weight in the form of stone is also fixed to the main line. For baiting fisherman use only weed fishes. One end of main line is fixed at the shore of reservoir to a large stone and other with baited hooks is carried in a bamboo basket. Fisherman releases the hooks in reservoir. Predatory fishes attracted towards the baited fish and engulf the bait along with hook and get trapped.

Gill Net

These are wall like nets, with floats attached to the head rope made from thermocol and sinkers of metal are fixed to the foot rope. The net is set in transvers direction of flow so that when the fishes tries to swim through a net wall,the meshes form a noose round its head and the fish is caught. As the fish tries to escape it gets stuck up behind the opercle. Mesh size varies from 1.5 to15 cm, made from nylon threads. 'Ekbotijal', 'Do botijal', 'Teen botijal' etc. are other names to different nets on basis of mesh size.It is generally fixed in reservoir at morning and hauled up in evening. 5 to 10 kg fishes are trapped in one attempt in it. Its cost is Rs. 400/- to 500/-

Hand Net

In Godavari river at shallow costal region it is used by man and women also to catch the crabs. It have triangular frame of bamboo of height 5 to 6ft. A small piece of bamboo of height 5 to 6ft. A small piece of bamboo is fixed in anterior region to frame of the net. At starting of bamboo frame, a net piece of 2 to 4 cm mesh size is fixed. At the center a conical mosquito net bag of 8 to 12 feet length is fixed. The catch is trapped in the central bag, because a rope is fixed at 2 ends of the bottom of bamboo. By lifting whole frame fisherman remove the central bag. Its cost is Rs. 400/- to 500/-.

After use of all crafts and gears they are dried in sunlight and store for next use. All these crafts and gears indicated the creativity of fisherman. Generally all fisherman use thermocol crafts. Diversity was observed in gears to exploit all elements of the fish community despite the dominance of gill net.

References

Desai, S.S., 1980. Fisheries of Nath Sagar reservoir. *India: Today and Tomorrow,* 8(4): 181 and 161.

Niture, S.D. and Chavan, S.P., 2009. Crafts and gears used in Yeldari reservoir, Maharashtra. *Ecology and Fisheries,* 2(1): 113–120.

Sakhare, V.B. and Joshi, P.K., 2002. Ecology of Palas–Nilegaon reservoir in Osmanabad district, Maharashtra. *J. Aqua. Biol.,* 18(2): 17–22.

Sakhare, V.B., 2007. *Reservoir Fisheries and Limnology.* Narendra Publishing House, Delhi.

Valsangkar, S.V., 1980. Economic rehabilitation of fishermen in Yeldari reservoir. *India: Today and Tommorow,* 8(4): 162–163.

Chapter 30

Studies on Groundwater Quality in Villages of Kaij Tehsil in Beed District, Maharashtra

☆ *V.B. Sakhare and J.S. Mohite*

ABSTRACT

Groundwater quality from different five places in Kaij tehsil of Beed district in Maharashtra was evaluated from June 2011 to May 2012.The water quality parameters like pH, Total hardness, Total dissolved solids, Chlorides, Carbonates, Bicarbonates and Sulphates were determined. Content of the above parameters in majority of the samples were within the standard limits. The values of total hardness and Total dissolved solids at some places are above the standard limits indicating sources of sanitary disposal near the studied area causing the health problems.

Keywords: Groundwater, Physico-chemical analsysis, Ambajogai city.

Introduction

Water is the most precious and important natural resource on the earth since without it life cannot exist. Water is used in day to day activities for drinking, bathing, washing, irrigation, recreation and industrial purposes.

Groundwater is an integral part of the environment and is the major source of drinking water in both urban and rural India. Besides, it is an important source of water for the agricultural and the industrial sector. Pollution of groundwater resources has become a major problem today. The pollution of air, water, and land has an effect on the pollution and contamination of groundwater.

According to some estimates, groundwater accounts for nearly 80 per cent of the rural domestic water needs, and 50 per cent of the urban water needs in India. Ground water in more than a third of Indian districts is not fit for drinking. The iron levels in ground water are higher than those prescribed in 254 districts while fluoride levels have breached the safe level in 224 districts Times of India (2010).

The present chapter deals with the physico-chemical analysis of groundwater in five villages of Kaij tehsil in Beed district in Maharashtra. Most of the land of the tehsil is full of rocks and thin layers of soil.In some parts there are few strips of black rich land. Majority of the people in Kaij tehsil use ground water for daily use.

Materials and Methods

Water samples for quality assessment were collected from hand pumps, wells and bore wells. The investigation was carried out for a period of one year (June 2011 to May 2012). Water samples like Total hardness, Chlorides, Carbonates, Bicarbonates and Sulphates were analyzed as per the procedure given in APHA (1985), Kodarkar *et al.* (1998) and Trivedy and Goel (1986). pH and total dissolved solids were determined by using the Hanna made digital pH meters.

Results and Discussion

The transparency of water samples is in general clear and does not show colour on visual observation. They have no colour or taste. The results obtained are depicted in Table 30.1.

Table 30.1: Water Quality of Samples Collected from different Villages in Kaij Tehsil, Maharshtra

Sl.No.	Water Parameters	Pathra	Awasgaon	Jawalban	Anandgaon	Soni Jawla
1.	pH	7.9 to 8.0	7.96to 7.8	7.7 to 8.0	7.7 to 8.0	7.6 to 7.9
2.	Total hardness	190 to 220	199 to 581	217 to 477	358 to 400	225 to 367
3.	Total Dissolved Solids	325 to 396	318 to 403	475 to 559	320 to 409	345 to 420
4.	Chlorides	47 to 54	39 to 55	69 to 72	47 to 59	52 to 60
5.	Carbonates	Nil to 08	Nil to 08	Nil to 09	Nil to 07	Nil to 05
6.	Bicarbonates	125 to 211	128 to 161	289 to 314	205 to 267	128 to 161
7.	Sulphates	17 to 19	11to 17	11 to 20	12 to 20	13 to 19

* All parameters except pH expressed in mg/l.

The pH of the water samples are to be the mild alkaline range of 7.6 to 8.0 (with a permissible limit) indicating the presence of very weak basic salts. The pH range of all the selected places was in the range of suitable drinking water standards as described by ISI (6.0 to 9.0).

Hardness of water is mainly imparted by alkaline earth metallic cations, mainly calcium and magnesium present in it. Hardness also restricts water use; hard water is unsuitable for cooking, washing and bathing due to boiling point in the first, while poor lather forming capacity in the latter two uses (Kodarkar *et al.*, 1998). Total hardness of the present investigation varies between 199 to 477 mg/l. The ISI (1983) limit for it is 300 mg/l. The values of total hardness at all the stations except Pathra village are higher as compared to ISI standard value. High concentration of hardness may cause the problem of heart disease and kidney stones (Jain, 1996). Sinha *et al.* (1990) also reported the higher values of hardness in tube wells and well waters which crossed the range of maximum permissible standard of ISI.

Total Dissolved Solids (TDS) during the present investigation ranged between 318 to 603 mg/l. The limit recommended by ISI (1983) for TDS is 500 mg/l. The values are higher at Jawalban village (475 to 559 mg/l). High level of dissolved solids may be aesthetically unsatisfactory for bathing and washing (Jameel, 2002). He also reported that all the ground waters in Trichirapalli except that of Cauvery are having high amounts of suspended, dissolved and total solids. Higher content of TDS can be attributed to the contribution of salts from the thick mantle of soil and the weathered media of the rock and further due to higher residence time of groundwater in contact with the aquifer body Maheswari and Sankar, 2011).

High Chloride level indicates pollution from domestic sewage and industrial effluents. Though chlorides level as high as 250 mg/l is safe for human consumption, a level above this imparts a salty taste to the potable water (Kodarkar *et al.*, 1998). The chlorides concentration in present investigation varied between 39 to 72 mg/l. The ISI (1983) permissible limit for chlorides concentration is 250 mg/l. Compared to this standard the values of chlorides are well below permissible limit.

The concentration of bicarbonates alkalinity was highest in deep tube wells. The values of these are well below the limits except the water samples collected at Jawalban.

Natural waters contain higher level of sulphates contributed from weathering of the rocks. In addition to this, domestic sewage and industrial waste also contribute sulphates to an aquatic ecosystem and hence high levels of sulphates are an indicator of pollution from organic matter (Kodarkar *et al.*, 1998). ISI (1983) limit for sulphate is 250 mg/l. The values of sulphates in all the five villages were well below the permissible limit.

Thus, the present study revealed that majority of the water parameters of drinking water of different places in Kaij tehsil of Beed district are well below the ISI limits. But the total hardness at all the places except groundwater in Pathra area was more above the standard permissible limits suggested by ISI. Total Dissolved Solids were also high at Jawalban.

References

APHA, 1985. *Standard Methods for the Examination of Water, Sewage and Industrial Wastes*, 16th Edn. American Public Health Association, Washington, USA.

ISI, 1983. *Indian Standard Specification for Drinking Water*, ISI 10500.

Jain, P.K., 1986. Hydrochemistry and groundwater quality of Singhari river basin Dist. Chhattarpur (M.P.). *Poll. Res.*, 15(4): 407–409.

Jameel, A.A., 2002. Evaluation of drinking water quality in Trichirapalli, Tamil Nadu. *Indian J. Environ. Hlth.*, 44(2): 108–112.

Kodarkar, M.S., Diwan, A.D., Murugan, N., Kulkarni, K.M. and Ramesh, Anuradha, 1998. *Methodology for Water Analysis: Physico-chemical, Biological and Microbiological*. IAAB, Hyderabad, Publication No. 2.

Maheswari, J. and Sankar, K., 2011. Groundwater quality assessment in Vaippar river basin Tamil Nadu (India). *International Journal of Current Research*, 3(12): 149–152, Available online at http://www.journalcra.com.

Sinha, D.K., Roy, S.P. and Datta Munshi, J.S., 1990. Assessment of drinking water quality of Santhal Paraganas, Bihar. *Environ. Ecol.*, 8(3): 937–941.

The Times of India, 2010. Groundwater in 33 per cent of India undrinkable. News available on articles.timesofindia.indiatimes.com

Trivedy, R.K. and Goel, P.K., 1986. *Chemical and Biological Methods for Water Pollution Studies.* Environmental Publications, Karad, Maharashtra.

Chapter 31

Serum Acetylcholinesterase Levels of Freshwater Fishes

☆ *Sudhish Chandra*

ABSTRACT

Normal serum acetylcholinesterase levels were observed in 18 species of freshwater fishes belonging to 5 orders of class Teleostomi. The enzyme level ranged widely, showing maximal level in *Heteropneustes fossilis* and minimal in *Notopterus notopterus*, revealing a difference of 50.96 per cent between the two. Comparative observations clearly indicated that the enzyme level primarily correlated with their body size, growth and activity, besides their taxonomic relationship. Alteration in acetylcholinesterase levels may provide an early indication of degrading aquatic environment, which causes consequences and compensations to fishes.

Introduction

Acetylcholinesterase is an enzyme known for terminating the action of acetylcholine at cholinergic synapses (Ronald *et al.*, 1999) and as neurotransmitter maintaining physiological configuration of organisms, besides acting as a key transducer at cell membrane level. Environmental pertuberances in teleost fishes causes consequences and compensations (Jensen *et al.*, 1993). Acetylcholinesterase is a major target enzyme of variety of pollutants which disturb the natural physiological function of the aquatic organisms, acting as stress inducing agent which affects the functional state of different body organs (Praveen *et al.*, 2004; Shaikh and Negi, 2004). Blood being a pathophysiological reflector of whole body, its characteristics are among the important indices of the state of internal environment of fish. The changes in biochemical blood profile indicate alterations in metabolism and biochemical processes of the organism due to various ecophysiological changes and can be used as bioindicator and constitute a promising approach to the problem of detecting the impact of environmental hazards at the earliest (Luskova *et al.*, 2002). Such observations are useful in cultural propagation practices, besides detection and diagnosis of metabolic disturbances and investigating diseases in fishes (Edsall, 1999), provided complete sets of ranges of blood parameters for all important species of fish are available (Mulcahy, 1970). Present paper embodies observations on normal serum acetylcholinesterase levels of 18 species of economically important common freshwater teleost fishes of this area.

Materials and Methods

Live fish were collected from river Gomti and suburbs with the help of fishermen during different parts of the year. Fish were transported in water filled plastic containers, transferred to large glass aquaria and acclimatized to laboratory conditions as described earlier (Tandon and Chandra, 1976). Fish were taken out of aquaria with least stress, blotted dry using a clean turkish towel and their weight and size noted. Blood was collected by amputating its caudal end in a clean dry tube and allowed to clot at room temperature for 20 minutes, then centrifuged at 2000rpm, clear serum decanted in a vial and kept frozen until analyzed. The body cavity, visceral organs and fresh blood film were also examined for any possible infection. Method of Hestrin (1949) was followed for the biochemical estimation of serum acetylcholinesterase level using Bausch and Lomb spectronic-20 spectrophotometer at 540µm against blank.

Results

Data of normal serum acetylcholinesterase (AchE) levels observed in 18 species of freshwater fishes belonging to 14 genera and 10 families of 5 orders of class Teleostomi have been presented in Table 31.1. The enzyme level in general ranged widely amongst the fishes examined. Highest mean serum acetylcholinesterase level (29.98+3.80µ moles acetylcholine hydrolysed/ml/hr) was observed in stinging catfish *Heteropneustes fossilis,* while lowest level (14.70+1.26µ moles acetylcholine hydrolysed/ml/hr) was noticed in featherback *Notopterus notopterus,* revealing a difference of 50.96 per cent between the maximal and minimal value. Amongst cypriniformes, the enzyme level in carps *Labeo rohita, L. calbasu, Catla catla* and *Cirrhina mrigala* revealed narrow variations (5.97 per cent) as compared to catfishes *Wallago attu, Rita rita, Mystus seenghala, Clarias batrachus, Heteropneustes fossilis* and *Bagarius bagarius* (17.07 per cent). Murrels belonging to order ophiocephaliformes also indicated least range amongst their serum enzyme levels. Low serum acetylcholinesterase levels were observed in *Anabas testudineus* and *Nandus nandus* of order perciformes. Interestingly spiny eel *Mastocembelus armatus* and giant featherback *Notopterus chitala* showed almost identical AchE levels, although belonging to different orders. A comparative observation clearly indicated that serum acetylcholinesterase levels in these fishes are primarily correlated with their body size, growth and activity besides their taxonomic relationship.

Discussion

Fishes normally experience natural environmental variations and may have evolved some biochemical adjustment which allowed them to maintain normal functioning of body (Gordon, 1972). Variety of pollutants reaching aquatic ecosystem also brings about changes in physiological characteristics of the receiving water (Tripathi *et al.,* 2001; Bhatnagar, 2003). Any unnatural harmful variation in fish environment trigger abnormal metabolic and physiological changes, behavioral stress and immunosupression (Gopal and Pathak, 1993). Fishes invariably inhabiting these environment having omnivorous feeding habits and using gills for respiration which are in intimate contact, becomes sensitive showing visible effects on their metabolism, leading to physiological, pathological and biochemical disorders (Arasta *et al.,* 1999; Karuppasamy, 2000; David *et al.,* 2003). Enzyme like esterases and transferases involving every tissue become primary target to toxic pollutants and interaction of these enzymes make a complex phenomenon between several metabolic pathways. Pesticides have been found to be toxic and lethal to fish even at relatively low concentration in long period. Due to such changing aquatic condition the fishes are prone to exhibit effective variations in tissues and its blood profile (Shafi, 2000). Acetylcholinesterase distribution was reported to be higher than normal levels in brain, liver and gill tissue following Rogor toxication in *Clarias batrachus* (Chandra, 2010), *Heteropneustes fossilis* (Chandra, 2008), *Cyprinus carpio* (Jeyarthi and Jebanessan, 2001) and *Tilapia mossamtica* (Sudheer Kumar *et al.,* 2006). Significant inhibition in AchE activity in different tissues of fish following higher concentration of organophosphate pesticides has been reported by Gupta *et al.,* 1975; Koundiya and Ramamurthy, 1978; Natarajan 1984; Thannipon *et al.,* 1995).

Table 31.1: Serum Acetylcholinesterase (AchE) Levels of 18 Species of Freshwater Fishes

Sl.No.	Name of the Fish (No. of Observation)	Weight Range (Grams)	AchE Levels (μ moles acetylcholine hydrolysed/ml/hr) Mean + Std. Deviation
1.	Notopterus chitala (16)	600-700	27.60+ 2.40
2.	N. notopterus (22)	175-210	14.70+1.26
3.	Catla catla (34)	580-630	27.44+6.20
4.	Labeo calbasu (38)	530-600	26.90+6.58
5.	L. rohita (32)	520-600	26.22+3.28
6.	Cirrhina mrigala (31)	500-560	25.80+4.84
7.	Wallago attu (40)	370-430	24.90+3.42
8.	Rita rita (40)	260-320	26.80+5.26
9.	Mystus seenghala (29)	380-450	26.10+4.06
10.	Heteropneustes fossilis (44)	190-260	29.98+3.80
11.	Clarias batrachus (42)	200-280	28.70+4.64
12.	Bagarius bagarius (20)	300-400	25.40+4.16
13.	Channa marulius (20)	240-300	22.20+4.60
14.	C. punctatus (38)	160-200	18.06+3.16
15.	C. striatus (36)	220-250	21.15+2.82
16.	Nandus nandus (20)	50-80	15.10+1.26
17.	Anabas testudineus (26)	60-80	16.40+1.60
18.	Mastocembelus armatus (21)	480-600	27.80+5.24

Comparative data on acetylcholinesterase levels of teleost fishes are very scarce. The enzyme level also vary in fishes due to variation in size and stamina (Nikolsky, 1963; Pyle, 1965). Varied AchE levels observed in 18 species of freshwater teleost fishes may plaussibly be correlated with rate of growth, feeding behaviour, activity level and metabolism, since a correlation coexhists between them, besides their taxonomic relationship. Identical trends have also been observed in total serum proteins (Tandon and Chandra., 1985), serum cholesterol (Chandra, 1982) and blood urea (Chandra, 1980) levels of freshwater fishes belonging to similar orders. Haematological profile depends upon fish species, age, sexual maturity and health condition but analytical blood constituents have shown useful informations in detection and diagnosis of metabolic disturbances and diseases in fish (Hrubec *et al.,* 2000). Repressed acetylcholinesterase level when compared to normal ones have also been attributed to decreased neurosecretory and hepatosecretory activities (Praveen *et al.,* 2004) which indicated unfavourable and degrading environment. Under such changing situations fish also exhibits evident behavioural changes like sluggishness, surfacing, increased opercular beat, copious mucus secretion followed by restlessness, hyperactivity and erratic movements (Nath, 2003). Acute toxicity may cause skeletal deformities in fishes causing inefficient swimming, less capability of acquiring food and escaping (Kesherwani *et al.,* 2007). These situations may destroy aquaculture potential and production of quality fishes. Thus observing any alteration in behavioural activities, morphological symptoms and by monitoring certain ecophysiological and biochemical blood constituents may provide early indication of degrading aquatic environment, which alarm to take up preventive measures to save healthy population of cultured fish and declining biodiversity in their natural ecosystem.

Acknowledgement

The author is thankful to Prof. R. S. Tandon for his valuable suggestion and encouragement.

References

Arasta, T., Bais, V.S. and Thakur, P.B., 1999. Changes in selected biochemical parameters in the liver and muscles of fish *Mystus vittatus* exposed to aldrin. In: *Environmental Pollution Management,* (Ed.) V.B. Bais. Creative Pub., Sagar, pp. 109–112.

Bhatnagar, M.C., Rajlaxmi, V. and Sharma, I., 2003. Toxicity of valour and match to common carp *Cyprinus carpio. J. Ecophysiol. Occup. Hlth.,* 3: 85–92.

Chandra, S., 1980. Blood urea levels of 20 species of freshwater fishes. *J. Ichthyol.,* 1: 11–13.

Chandra, S., 1982. Serum cholesterol levels of 20 species of freshwater fishes. *Intl. J. Acad. Icthyol.,* 3: 13–16.

Chandra, S., 2008. Toxic effects of malathion on acetylcholinesterase activity of liver, brain and gills of freshwater catfish *Heteropneustes fossilis. Environ. Conser. J.,* 9: 47–52.

Chandra, S., 2010. Impact of rogor toxicity on aldolase and acetylcholinesterase activity in liver, brain and gills of freshwater catfish, *Clarias batrachus. Rec. Adv. Fish Ecol. Limnol. Ecocons.,* 8: 165–175.

David, M., Mushiger, S. P., Prashant, M.S. and Mathod, S.G., 2003. Hepatotoxicity of malathion on protein metabolism in *Catla catla Adv. Biol.,* 22: 115–120.

Edsall, C.C., 1999. A blood chemistry profile for lake trout. *J. Ag. Anim. Health,* 11: 81–86.

Gopal, K. and Pathak, S.P., 1993. Possible causes of outbreaks of fish diseases and mortality in polluted water. *J. Adv. Zool.,* 14: 53–60.

Gordon, M.S., 1972. *Animal Physiology.* The McMilan Co., New York.

Hestrin, S., 1949. The reaction of acetylcholine and other carboxylic acid derivatives with hydroxylamine and its analytical application. *J. Biol. Chem.,* 180: 249–261.

Gupta, P.K., Gupta, R.C. and Kohli, J.D., 1975. Recent pesticides: A review. *Pesticides,* 9: 15–210.

Hrubec, T.C., Cardinale, J.L. and Smith, S.A., 2002. Haematology and plasma chemistry reference intervals for cultured *Tilapia (Oreochromis* hybrid). *Vet. Pathol.,* 29: 7–12.

Jensen, F.B., Nikinmaa, M. and Weber, R.E., 1993. Environmental perturbations of oxygen transport in teleost fishes: Causes, consequences and compensations. In: *Fish Ecophysiology,* (Eds.) J.C. Rankin and F.B. Nensen. Chapman and Hall, London, pp. 161–179.

Jeyarthi Shanti, T. and Jebanesan, A., 2001. Effect of sublethal concentration of chloropyriphos on acetylcholinesterase activity in freshwater fish *Cyprinus carpio. Convergence,* 3: 31–35.

Karuppasmy, R., 2002. Short and long term effect of phenyl mercuric acetate on protein metabolism in *Channa punctatus. J. Natcon.,* 12: 83–93.

Kesherwani. D., Verma, R.S., Shukla, S. and Sharma, U.D., 2007. Cadmium induced skeletal deformities in freshwater catfish *Heteropneustes fossilis, Environ. Ecol.,* 25: 248–351.

Koundiya, P.R. and Ramamurthi, S.R., 1978. Effect of sumithion on some selected enzyme system in the fish *Tilapia mossambica. Ind. J. Expt. Biol.,* 16: 809–811.

Luskova, V., Svoboda, M. and Kolarova, J., 2002. The effect of diazinon on blood plasma biochemistry in carp, *Cyprinus carpio. Aquacult,* 7: 159–171.

Mulcahy, M.F., 1970. Blood values in the pike, *Esox, lucius. J. Fish. Biol.,* 2: 203–209.

Natrajan. G.M., 1984. Effect of lethal concentration of metasystox on some selected enzymes system in air breathing fish *Channa punctatus. Comp. Physiol Ecol.,* 9: 29–32.

Nath, A., 2003. Bioaccumulation of various pesticides in fish muscle. *J. Ecophysiol. Occup. Hlth.,* 1: 40–46.

Nikolsky, G.V., 1963. *The Ecology of Fishes.* Academic Press, New York.

Praveen. M., Sharma, R. and Kumar, S., 2004. Effect of Neemta 2100 toxicity on acetylcholinesterase and serum glutamate oxalacetate transaminase enzyme in serum of fish *Oreochromis mossambica J. Ind. Fish. Assoc.,* 31: 97–100.

Pyle, E.A., 1965. Maintenance of relative swimming performance among brook trout grouped by initial swimming performance. *Fish. Res. Bull.,* New York, 28: 55–59.

Ronald, B.P., Javier, S.V. and David, H.S., 1999. Inhibition of neurite outgrowth from chick sympathetic neurons by cholinesterase inhibitors not mediated by binding of cholinesterase. *Neurosci. Letter,* 266: 77–80.

Shaikh. N. and Negi, S.G., 2004. Effects of rogor on protein and glycogen in the muscle of freshwater fish, *Lepidocephalichthyes thermalis. Aquacult.,* 5: 119–121.

Shafi, S.M., 2002. *Modern Ichthyology.* Inter India Publication, New Delhi.

Sudheer Kumar, D.J., John Sushma, N., Sivaiah, V., Madhav Rao, S. and Jayantha Rao, K., 2006. Effect of chlorpyriphos and azadirachtin on acetyl cholinesterase of fish *Tilapia mossambica. Aquacult.,* 7: 87–91.

Tandon, R.S. and Chandra, S., 1976. Serum alkaline phosphatase levels of some freshwater teleosts. *Z. Tierphysiol. Tierernahrg. U. Futtermittelkde,* 37: 330–333.

Tandon, R.S. and Chandra, S., 1985. Total serum protein levels of 18 species of freshwater fishes. *Kan. Univ. Res. J. (Sci.),* 6: 1–4.

Thannipon, W., Thangnipon, W., Luappaiboon, P. and Chinabut, S., 1995. Effect of organophosphate insecticide monocrotophos on acetylcholinesterase activity in the Nile *Tilapia* fish brain. *Neurochem. Res.,* 20: 587–591.

Tripathi, G., Verma, P., Pandey, G.C. and Tripathi, Y.C., 2001. Biological effects of fenvalerate. In: *Current Trends in Environmental Science,* (Eds.) G. Tripathi and G.C. Pande. ABD Publications, Jaipur, India.

Chapter 32

Role of Natural Ultraviolet Radiation (UVR) on Freshwater Phytoplankton Growth on Urban Wetland (Temuco, Chile)

☆ *Jacqueline Acuna, Patricio De los Ríos-Escalante, Patricio Acevedo and Aracelly Ulloa*

ABSTRACT

Currently it was observed an increase of penetration of natural ultraviolet radiation due to ozone depletion in polar and subpolar latitudes, with respective effects on terrestrial and aquatic ecosystems. In aquatic ecosystems, the natural ultraviolet radiation can penetrate in to surface when the water column is transparent, and this condition generates effects on biotic components, that develop photoprotective strategies. The aim of the present study is determine the effects of natural ultraviolet radiation exposure on phytoplankton growth in an urban wetland. For this purpose it was developed four experiments in limnocorrals, in southern summer, with four different treatments, exposed to natural ultraviolet radiation and the three remaining with different screens against natural ultraviolet radiation. It was found that the growth rate and density was inversely associated to total ultraviolet radiation and ultraviolet B radiation. The exposed results agree with similar descriptions for freshwater and marine phytoplankton and zooplankton. Ecological and photobiological topics were discussed.

Keywords: Ultraviolet radiation, Doses, Phytoplankton, Growth rate, Density.

Introduction

Currently it was observed an increase of natural ultraviolet radiation penetration mainly in polar and subpolar latitudes due ozone depletion (Díaz *et al.*, 2006; Marinone *et al.*, 2006). The natural ultraviolet radiation affects the aquatic ecosystems, because this can penetrate into water column when this is transparent due low concentration of dissolved organic carbon that is a natural screen against ultraviolet radiation (Díaz *et al.*, 2006; Marinone *et al.*, 2006). Under this scenario the biotic components when are exposed to natural ultraviolet radiation develops a photoprotective strategies such as vertical migration in a negative phototactism for avoid ultraviolet radiation exposure or synthesis of photoprotective substances for avoid ultraviolet radiation damage (Morris *et al.*, 1995; Villafañe *et al.*, 2001). In Patagonia (38-54° S) the increase of ultraviolet radiation and its effects in inland water ecosystems were reported mainly for zooplankton (Zagarese *et al.*, 1997a,b, 1998a,b; De los Ríos and Soto 2005; De los Ríos, 2005, 2007), fishes (Battini *et al.*, 2000) and amphibians (Barriga *et al.*, 2006). Whereas the effects on phytoplankton were documented mainly for marine environments (Villafañe *et al.*, 2001; Helbling *et al.*, 2008). The aim of the present study is determine the effects of natural ultraviolet radiation on a shallow urban pond close to Temuco (38° S, northern Patagonia) and determine the lethal doses of total ultraviolet radiation and ultraviolet B radiation.

Materials and Methods

The study site is a artificial wetland in the surrounding of Temuco (Catholic University of Temuco, 38° 42′ 11″ S; 72° 32′ 50″ W; Cf. Rivera *et al.*, 2010), the experiment consisted in limnocorrals with four treatments with three replicated each one, the first treatment, was exposed to natural ultraviolet radiation, and the remaining three were covered with different screens against natural ultraviolet radiation (Table 32.1). The experiments were done three times in January 2008 (Table 32.2). The natural ultraviolet radiation was measured in according to the descriptions from literature, considering total ultraviolet radiation (UV) and ultraviolet B radiation (UVB) doses (De los Ríos *et al.*, 2007a,b). The phytoplankton activity was measured considering growth rate and total cells (Stein, 1973), the growth rate and total cells were measured by cellular counting using an Newbauer counting chamera. For statistical analysis was applied a lineal regression using the software Xlstat 7.0. For the equations obtained in regression analysis was obtained the instantaneous values of final density and growth rates in function of UV and UVB doses by application of derivates from the obtained equations.

Table 32.1: Transmittance for Total UVR and UVB of Different Plastics Screens Used during the Study

Plastic screen	Transmittance	
	UV total	UVB
Monolytic policarbonate	0.003	0.009
Alveolar policarbonate	0.008	0.014
Transparent plastic	0.096	0.701

Results and Discussion

The results revealed that under increase of total UVR exposure there is an inverse association that denotes a lineal regression (Table 32.2; Figure 32.1; $R^2 = 0.8008$; $P < 0.001$) between with total cells, and logarithmic regression with growth rate (Table 32.2; Figure 32.1; $R^2 = 0.8715$; $P < 0.001$). Similar situation was reported for UVR-B radiation, where was reported an inverse relation denoted as logarithmic regression with total cells (Table 32.2; Figure 32.1; $R^2 = 0.6700$; $P < 0.001$) and growth rate

(Table 32.2; Figure 32.1; R² = 0.8303; P < 0.001). These results revealed that the ultraviolet radiation has a significant effect in phytoplankton activity, that is denoted in decrease of growth rate and in consequence a decrease of phytoplankton due mortality of cell with low density in surface layers under conditions of natural ultraviolet radiation exposure (Table 32.2; Figure 32.1; Helbling *et al.,* 2001). The results of instantaneous variation of final density denoted that the variation was constant in function to UV doses, whereas the variation was negative at low UVB doses and was close to zero

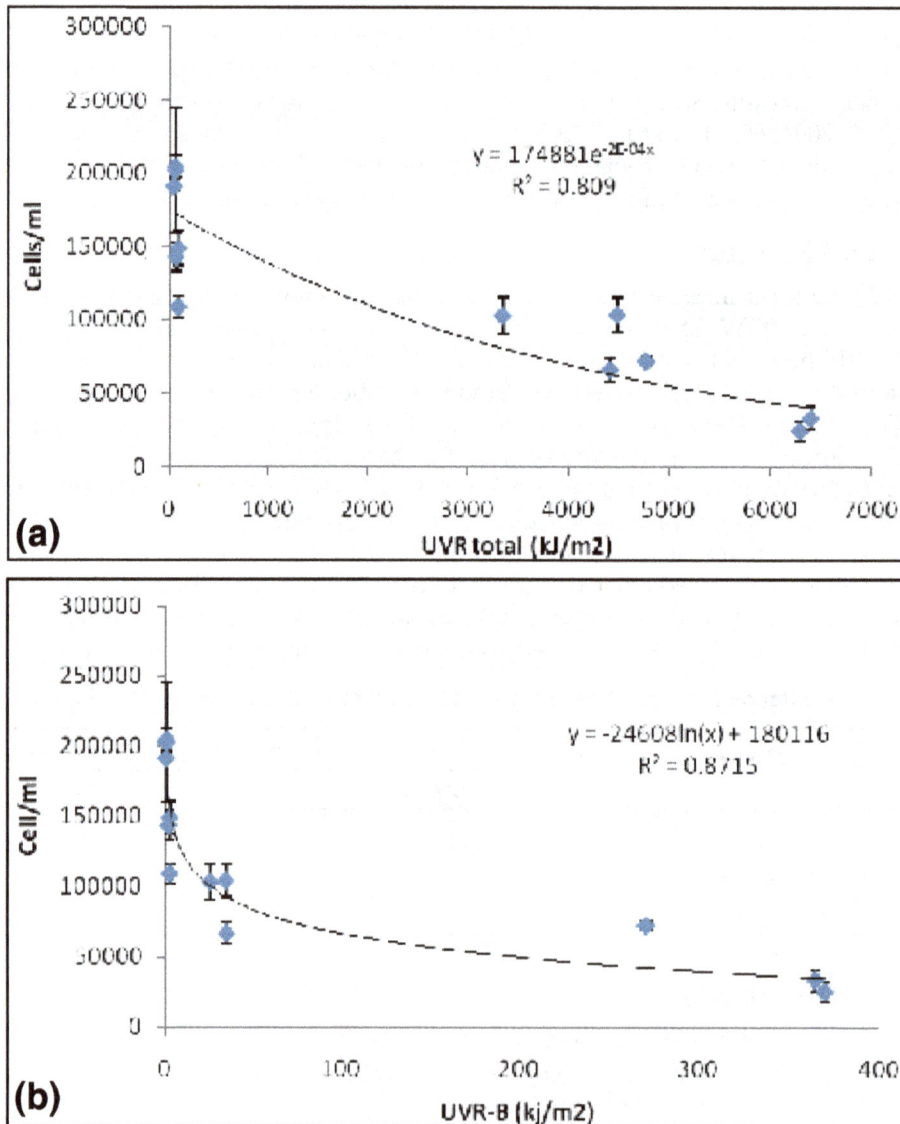

Figure 32.1: Results of Regression Analysis for Experiments done in the Present Study, Between (a), (b) Final density (cell*ml⁻¹) with UV and UVB Doses; (c) (d) Growth Rate (cell*day⁻¹) with UV and UVB Doses

Contd...

Figure 32.1–*Contd...*

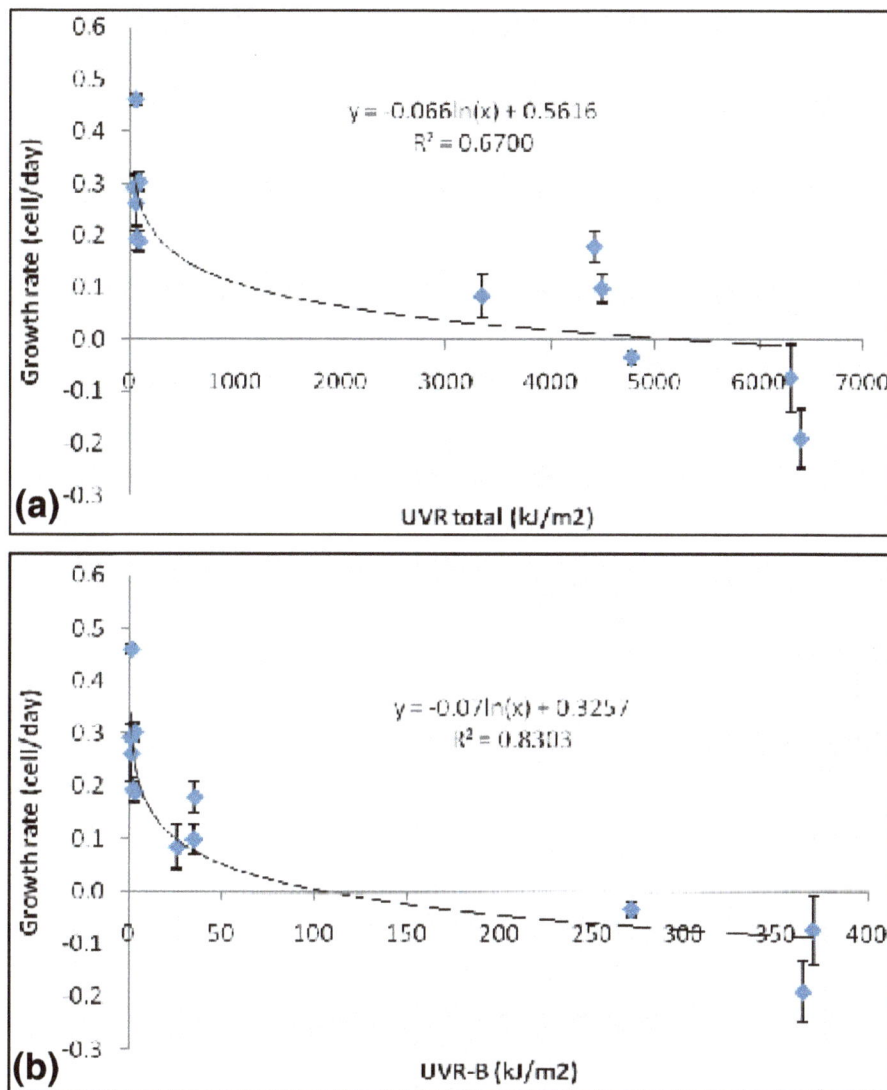

(a)

(b)

at high UVB-doses (Table 32.3). Finally the results of instantaneous variation of growth rate in function of UV and UVB doses was negative at low UVB doses and was close to zero at high UVB-doses (Table 32.3).

These results revealed that the ultraviolet radiation exposure can generate effects such as mortality and low photosynthetic activity (Helbling *et al.*, 2001, 2008; Villafañe *et al.*, 2001), that is due to cellular damage due exposure to UVR-B (Villafañe *et al.*, 2001; Helbling *et al.*, 2006). Although in this present study was considered the whole phytoplankton community, that was a similar procedure using for phytoplankton and nutrient limitation in Chilean lakes (Steinhart *et al.*, 2002), the literature agree with the obtained results about phytoplankton mortality, nevertheless the responses are different for each

Table 32.2: Average Final Density and Growth Rate (± Standard Deviation); Dose and Maximum UV for 280-400 nm and 280-320 nm, for Each Treatment Used during the Present Study

Treatment	Experiment	Date	Dose UV (280-400) ($kJ*m^{-2}$)	Dose UVB (280-320) ($kJ*m^{-2}$)	Initial Density ($cell*ml^{-1}$)	Final Density ($cell*ml^{-1}$)	Growth Rate ($cell*ml^{-1}$)
Monolytic policarbonate	1	14-18th January 2008	56.70	1.10	32500	205000.00 ± 7500.00	0.50 ± 0.01
Alveolar policarbonate	1	14-18th January 2008	88.20	2.96	32500	109166.67 ± 7637.62	0.30 ± 0.01
Transparent plastic	1	14-18th January 2008	4416.30	35.52	32500	66666.67 ± 7815.58	0.20 ± 0.03
Exposed	1	14-18th January 2008	6300.00	370.30	32500	24833.33 ± 6806.85	-0.10 ± 0.07
Monolytic policarbonate	2	18-22th January 2008	57.66	1.10	70000	202500.00 ± 42720.02	0.30 ± 0.05
Alveolar policarbonate	2	18-22th January 2008	89.70	2.92	70000	149166.67 ± 11814.54	0.20 ± 0.02
Transparent plastic	2	18-22th January 2008	4491.31	35.04	70000	104166.67 ± 11814.54	0.10 ± 0.02
Exposed	2	18-22th January 2008	6407.00	365.00	70000	33333.33 ± 8036.38	-0.20 ± 0.05
Monolytic policarbonate	3	22-25th January 2008	42.99	0.81	80000	191666.67 ± 3818.81	0.30 ± 0.01
Alveolar policarbonate	3	22-25th January 2008	66.88	2.17	80000	143333.33 ± 10103.63	0.20 ± 0.02
Transparent plastic	3	22-25th January 2008	3348.68	26.02	80000	103333.33 ± 12829.00	0.10 ± 0.04
Exposed	3	22-25th January 2008	4777.00	271.00	80000	72500.00 ± 2286.75	0.01 ± 0.01

phytoplankton group (Warnberg *et al.*, 2008). On a community ecology view point, the responses of phytoplankton assemblages are different under UVR exposure, because the phytoplankton with xantophyll pigments such as diatoms are more tolerant to UVR exposure in comparison to other phytoplankton groups without xantophyll pigments such as chlorophytes (Gao *et al.*, 2008). The exposed results would denote the presence of a photoinhibition effect due natural ultraviolet radiation although it is possible that phytoplankton generates photoprotective substances such as mycosporine like amino acids for avoid UVR damage (Laurion and Roy, 2009), these situation would not be observed because the high mortality under high ultraviolet radiation exposure (Figure 32.1; Table 32.2).

Table 32.3: Instantaneous Variation of Final Density (Cell*ml^{-1}) and Growth Rate (Cell*day^{-1}) in Function of UV Total and UVB Doses (in kJ*m^{-2}) Obtained from Derivates of Equations from Figure 32.1

UV total	(cell/ml)/ (kJ/m^2)$^{-1}$	(cell/day)/ (kJ/m^2)$^{-1}$	UVB	(cell/ml)/ (kJ*m^{-2})$^{-1}$	(cell/day)* (kJ*m^{-2})$^{-1}$
56.700	205000	-0.001182	1.100	-22370.90909	-0.06364
88.200	205000	-0.000760	2.960	-8313.51351	-0.02365
4416.300	205000	-0.000015	35.520	-692.79279	-0.00197
6300.000	205000	-0.000011	370.300	-66.45423	-0.00019
57.663	205000	-0.001162	1.095	-22473.05936	-0.06393
89.698	205000	-0.000747	2.920	-8427.39726	-0.02397
4491.307	205000	-0.000015	35.040	-702.28311	-0.00200
6407.000	205000	-0.000010	365.000	-67.41918	-0.00019
42.993	205000	-0.001558	0.813	-30268.14268	-0.08610
66.878	205000	-0.001002	2.168	-11350.55351	-0.03229
3348.677	205000	-0.000020	26.016	-945.87946	-0.00269
4777.000	205000	-0.000014	271.000	-90.80443	-0.00026

About the experiments these experiments were done in four days that is the optimal period for obtain results in phytoplankton assemblages (Soto, 2002; Jaramillo, 2004). The studies with limnocorrals as mesocosms revealed the effect of light availability, because in these kind of environments, the light penetrate into all experimental units, and this situation would generate the disadvantage of avoid the use of negative phototactism migration against natural ultraviolet radiation into depth zones (Jaramillo, 2004; Warnberg *et al.*, 2008). If we considered the UVR doses, in according to the literature (Llabres and Agusti, 2006), described that in doses between 151 to 659 kJ*m^{-2} occurs the 50 per cent mortality, and it was found a maximum mortality at 1280 659 kJ*m^{-2}, that agree with the results observed in the present study (Table 32.2, Figure 32.1). The exposed results revealed that the natural ultraviolet radiation generates, nevertheless it is necessary more studies for determine the role of different components at phytoplankton communities and the effects at level of primary productivity and photosynthetic activity.

Acknowledgements

The present study was founding by Research Direction of the Catholic University of Temuco (Project DGI-CDA-UCT-01) and projects DI-08-0040 of the Universidad de la Frontera.

References

Barriga, J.P., Battini, M.A., Macchi, P.J., Milano, D. and Cussac, V.E., 2002. Spatial and temporal distribution of landlocked *Galaxias maculatus* and *Galaxias platei* (Pisces: Galaxidae) in a lake in the South American Andes. *New Zealand J. Mar. Freshwat. Res.*, 36: 349–363.

Battini, M., Rocco, V., Lozada, M., Tartarotti, B. and Zagarese, H.E., 2000. Effects of ultraviolet radiation on the eggs of landlocked *Galaxias maculatus* in northwestern Patagonia. *Freshwat. Biol.*, 44: 547–552.

De los Ríos, P. and Soto, D., 2005. Survival of two species of crustacean zooplankton under to two chlorophyll concentrations and protection or exposure to natural ultraviolet radiation. *Crustaceana*, 78: 163–169.

De los Ríos, P., Hauenstein, E., Acevedo, P. and Jaque, X., 2007a. Littoral crustaceans in mountain lakes of Huerquehue National Park (38°S, Araucania Region, Chile). *Crustaceana*, 80: 401–410.

De los Ríos, P., Acevedo, P. and Verdugo, K., 2007b. Survival of *Ceriodaphnia dubia* (Crustacea, Cladocera) exposed to different screens against natural ultraviolet radiation. *Pol. J. Env. Stud.*, 16: 481–485.

De los Ríos, P., 2005. Survival of pigmented freshwater zooplankton exposed to artificial ultraviolet radiation and two levels of dissolved organic carbon. *Pol. J. Ecol.*, 53: 113–116.

De los Ríos, P., 2007. Short term effects of exposition to natural ultraviolet radiation on *Parabroteas sarsi* (Copepoda, Calanoida). *Biologia, Bratislava*, 62: 210–213.

Diaz, S., Camillón, C., Deferrari, G., Fuenzalida, H., Armstrong, R., Booth, C., Paladani, A., Cabrera, S., Casiccia, C., Lovengreen, C., Pedroni, J., Rosales, A., Zagarese, H. and Vernet, M., 2006. Ozone and UV radiation over southern South America: Climatology and anomalies. *Photochem. Photobiol.*, 82: 834–843.

Gao, K., Li, G., Helbling, W. and Villafañe, V.E., 2008. Variability of UVR effects on photosynthesis of summer phytoplankton assemblages from a tropical coastal area of the South China Sea. *Photochem. Photobiol.*, 83: 802–809.

Helbling, W.E., Villafañe, V. and Barbieri, E., 2001. Sensibilidad de comunidades fitoplanctónicas invernales de lagos andinos a la radiación ultravioleta artificial. *Rev. Chilena Hist. Nat.*, 74: 273–282.

Helbling, W.E., Buma, A.G.J., Van Del Poll, W., Ferández Zenoff, M.V. and Villafañe, V.E., 2008. UVR–induced photsynthetic inhibition dominates over DNA damage in marine dinoflagellates exposed to fluctuating solar radiation regimes. *J. Exp. Mar. Biol. Ecol.*, 365: 96–102.

Helbling, W.E., Farías, M.E., Fernández Zenoff, V. and Villafañe, V.E., 2006. *In situ* responses of phytoplankton from the subtropical Lake La Angostura (Tucuman, Argentina) in relation to solar ultraviolet radiation exposure and mixing conditions. *Hydrobiologia*, 559: 123–134.

Laurion, I. and Roy, S., 2009. Growth and photoprotection in three dinoflagellates (including two strains of *Alexandrium tamarense*) and one diatom exposed to four weeks of natural and enhanced ultraviolet-B radiation. *J. Phycol.*, 45: 16–33.

Llabres, M. and Agusti, S., 2006. Picoplankton cell death induced by UV radiation: Evidence for oceanic Atlantic communities. *Limnol. Ocean.*, 51: 21–29.

Jaramillo, J., 2004. Evaluación del efecto de la radiación ultravioleta y nutrientes en la abundancia de cianófitas en los lagos temperados del sur de Chile. *Ph.D. Thesis*, Sciences Faculty, Austral University of Chile, 104 p.

Marinone, M.C., Menu-Marque, S., Añón Suárez, D., Dieguez, M.C., Pérez, A.P., De los Ríos, P., Soto, D., and Zagarese, H.E., 2006. UV radiation as a potential driving force for zooplankton community structure in Patagonian lakes. *Photochem. Photobiol.*, 82: 962–971.

Morris, D.P., Zagarese, H.E., Williamson, C.E., Balseiro, E.G., Hargreaves, B.R., Modenutti, B.E., Moeller, R.E. and Queimaliños, C.P., 1995. The attenuation of solar UV radiation in lakes and the role of dissolved organic carbon. *Limnol. Ocean.*, 40: 1381–1391.

Rivera, R., De los Ríos, P. and Contreras, A., 2010. Indicadores de contaminación fecal en cuerpos de agua rural y urbano de la ciudad de Temuco. *Cienc. Inv. Agr.*, 37: 141–149.

Soto, D., 2002. Oligotrophic patterns in southern Chile lakes: the relevance of nutrients and mixing depth. *Rev. Chilena Hist. Nat.*, 75: 377–393.

Stein, J.R., 1973. *Handbook of Phycological Methods: Culture Methods and Growth Measurements*. Cambridge University Press, New York, USA, 448 p.

Steinhart, G.S., Likens, G.E., and Soto, D., 1999. Nutrient limitation in Lago Chaiquenes (Parque Nacional Alerce Andino, Chile): Evidence from nutrient experiments and physiological assays. *Rev. Chilena Hist. Nat.*, 72: 559–568.

Villafañe, V.E., Helbling, E.W., and Zagarese, H.E., 2001. Solar ultraviolet radiation and its impact on aquatic ecosystems of Patagonia, South America. *Ambio*, 30: 174–180.

Wangberg, S-A., Andreasson, K.I.M., Gustavson, K., Reinthaler, T. and Henriksen, P., 2008. UV-B effects on microplankton communities in Kongsfjord. *J. Exp. Mar. Biol. Ecol.*, 365: 156–163.

Zagarese, H.E., Williamson, C.E., Vail, T.L., Olsen, O. and Queimaliños, C.P., 1997a. Long-term exposure of *Boeckella gibbosa* (Copepoda, Calanoida) to *in situ* levels of solar UVB radiation. *Freshwat. Biol.*, 37: 99–106.

Zagarese, H.E., Feldman, M. and Williamson, C.E., 1997b. UV-B induced damage and photo-reactivation in three species of *Boeckella* (Copepoda, Calanoida). *J. Plank. Res.*, 19: 357–367.

Zagarese, H.E., Tartarotti, B., Cravero, W. and González, P., 1998a. UV damage in shallow lakes: The implications of water mixing. *J. Plank. Res.*, 20: 1423–1433.

Zagarese, H.E., Cravero, W., González, P. and Pedrozo, F., 1998b. Copepod mortality induced by fluctuating levels of natural ultraviolet radiation simulating vertical water mixing. *Limnol. Ocean.*, 43: 169–174.

Chapter 33

Role of Natural Ultraviolet Radiation (UVR) and Nutrient Addition on Phytoplankton Growth on Urban Wetland (Temuco, Chile)

☆ *Patricio De los Ríos-Escalante, Patricio Acevedo, Jacqueline Acuna, and Aracelly Ulloa*

ABSTRACT

Currently it was observed an increase of penetration of natural ultraviolet radiation due ozone depletion in polar and subpolar latitudes, with respective effects on terrestrial and aquatic ecosystems. In aquatic ecosystems, the natural ultraviolet radiation can penetrate in to surface when the water column is transparent, and this condition generates effects on biotic components, that develop photoprotective strategies. The aim of the present study is determine the effects of natural ultraviolet radiation exposure and nutrient addition on phytoplankton growth in an urban wetland. For this purpose it was developed four factorial experiments in limnocorrals, during southern summer, with four different treatments, exposed and protected against natural ultraviolet radiation and with and without nutrient addition. The results of ANOVA revealed the absence of significant effects with exception to the first experiment, where it was observed the significant effects in combined effect of ultraviolet radiation and nutrient addition. The results would be explained to the presence of photoprotective strategies on phytoplankton assemblages or the presence of dissolved organic carbon that would have an screen effect against natural ultraviolet radiation damage and generate an kind of nutrient recycling that would enhance the phytoplankton activity.

Keywords: *Ultraviolet radiation, Nutrients, Doses, Phytoplankton, Growth rate, Density.*

Introduction

Currently it was observed an increase of natural ultraviolet radiation penetration mainly in polar and subpolar latitudes due to ozone depletion (Díaz *et al.*, 2006; Marinone *et al.*, 2006). The natural ultraviolet radiation affects the aquatic ecosystems, because this can penetrate into water column when this is transparent due low concentration of dissolved organic carbon that is a natural screen against ultraviolet radiation (Díaz *et al.*, 2006; Marinone *et al.*, 2006). Under this scenario the biotic components when are exposed to natural ultraviolet radiation develops a photoprotective strategies such as vertical migration in a negative phototactism for avoid ultraviolet radiation exposure or synthesis of photoprotective substances for avoid ultraviolet radiation damage (Morris *et al.*, 1995; Villafañe *et al.*, 2001).

In Patagonia (38-54° S) the increase of ultraviolet radiation and its effects in inland water ecosystems were reported mainly for zooplankton (Zagarese *et al.*, 1997a,b, 1998a,b; De los Ríos and Soto 2005; De los Ríos, 2005, 2007), fishes (Battini *et al.*, 2000) and amphibians (Barriga *et al.*, 2006). Whereas the effects on phytoplankton were documented mainly for marine environments (Villafañe *et al.*, 2001; Helbling *et al.*, 2008). The aim of the present study is determine the effects of natural ultraviolet radiation and nutrient addition on phytoplankton assemblage in a shallow urban pond close to Temuco (38° S, northern Patagonia).

Materials and Methods

The study site is a artificial wetland in the surrounding of Temuco (Catholic University of Temuco, 38° 42' 11" S; 72° 32' 50" W; Cf. Rivera *et al.*, 2010), the experiment consisted in limnocorrals in a factorial design with four treatments with three replicated each one, the first treatment, was exposed to natural ultraviolet radiation, and the remaining three were covered monolithic polycarbonate that has 0.9 per cent transmissibility of total ultraviolet radiation (Table 33.1). The experiments were done three times in January and March 2008 (Table 33.2). The natural ultraviolet radiation was measured in according to the descriptions from literature, considering total ultraviolet radiation (UV) and ultraviolet B radiation (UVB) doses (De los Ríos *et al.*, 2007a,b; Table 33.2), the nutrient addition consisted in addition of nitrogen and phosphorus in according to the descriptions of Soto (2002). The phytoplankton activity was measured considering growth rate and total cells (Stein, 1973), the growth rate and total

Table 33.1: Transmittance for Total UVR and UVB of Plastic screen Used during the Present

Plastic screen	Transmittance	
	UV total	*UVB*
Monolytic policarbonate	0.003	0.009

Table 33.2:Total UVR and UVB Doses Measured during the Experimental Period

Experiment	Date	Dose UV (280-400) (kJ*m⁻²)	Dose UVB (280-320) (kJ*m⁻²)
1	14-18th January 2008	6300.00	370.30
2	18-22th January 2008	6407.00	365.00
3	22-25th January 2008	4777.00	271.00
4	03-07th March 2008	4604.00	261.00

Advances in Aquatic Ecology Volume 7

cells were measured by cellular counting using an Newbauer counting chamera. For statistical analysis was applied a two way ANOVA previous verification of normality and variance homogeneity (Zar, 1999), and it was applied a Tukey multiple comparison test for determine potential significant differences between treatments. All statistical analysis was applied using the software Xlstat 5.0.

Results and Discussion

The results revealed the absence of significant effects for both treatments separated and combined (Tables 33.3 and 33.4), with exception to experiment 1, where was observed a significant effect on combined effects (Tables 33.3 and 33.4) although there are not significant effects, probably the exposure to natural ultraviolet radiation would have a weak non significant effect on phytoplankton cell abundances and growth (Tables 33.3 and 33.4). These results revealed that the ultraviolet radiation exposure can generate effects such as mortality and low photosynthetic activity (Helbling *et al.*, 2001, 2008; Villafañe *et al.*, 2001), that is due to cellular damage due exposure to UVR-B (Villafañe *et al.*, 2001; Helbling *et al.*, 2006). Although in this present study was considered the whole phytoplankton community, that was a similar procedure using for phytoplankton and nutrient limitation in Chilean lakes (Steinhart *et al.*, 2002), the literature agree with the obtained results about phytoplankton mortality, nevertheless the responses are different for each phytoplankton group (Warnberg *et al.*, 2008). On a community ecology view point, the responses of phytoplankton assemblages are different under UVR

Table 33.3: Phytoplankton Cell Abundances (cell*ml^{-1}) and Growth Rates (cell * day^{-1}) for Experiments Carried Out in the Present Study

Treatment		Cell Abundance	Growth Rate
Experiment 1: 14th-18th January, 2008. Initial cell number 32500 cell*ml^{-1}			
Nutrient addition	Exposed	158333.3 ± 25426.9	0.388 ± 0.34
Nutrient addition	Protected	136666.7 ± 2204.7	0.359 ± 0.36
Without nutrient addition	Exposed	135833.3 ± 8207.4	0.356 ± 0.36
Without nutrient addition	Protected	203333.3 ± 4409.6	0.458 ± 0.46
Experiment 2: 18th-22th January, 2008. Initial cell number 70000 cell*ml^{-1}			
Nutrient addition	Exposed	533333.3 ± 2204.8	-0.06 ± 0.01
Nutrient addition	Protected	641666.7 ± 10833.3	-0.03 ± 0.04
Without nutrient addition	Exposed	608333.3 ± 10137.9	-0.04 ± 0.04
Without nutrient addition	Protected	108333.3 ± 46240.6	0.06 ± 0.01
Experiment 3: 22th-25th January, 2008. Initial cell number 80000 cell*ml^{-1}			
Nutrient addition	Exposed	73333.3 ± 24678.4	-0.05 ± 0.01
Nutrient addition	Protected	1675000 ± 29825.8	0.18 ± 0.04
Without nutrient addition	Exposed	166666.6 ± 44775.6	0.16 ± 0.07
Without nutrient addition	Protected	198333.3 ± 34681.1	0.21 ± 0.04
Experiment 4: 03th-07th March, 2008. Initial cell number 12500 cell*ml^{-1}			
Nutrient addition	Exposed	60000.0 ± 18027.8	0.37 ± 0.07
Nutrient addition	Protected	49166.7 ± 8457.4	0.33 ± 0.05
Without nutrient addition	Exposed	675000.0 ± 33757.7	0.36 ± 0.11
Without nutrient addition	Protected	283333.3 ± 5833.3	0.19 ± 0.04

exposure, because the phytoplankton with xantophyll pigments such as diatoms are more tolerant to UVR exposure in comparison to other phytoplankton groups without xanthophyll pigments such as chlorophytes (Gao *et al.*, 2008).

Table 33.4: Results of Two-Way ANOVA for the Experiments Carried Out in the Present Study.
"P" values lower than 0.05 denotes significant effects

Effect	Cell Abundance	Growth Rate
Experiment 1: 14th-18th January, 2008		
Nutrients	F = 2.643; P = 0.143	F = 0.828; P = 0.389
Ultraviolet Radiation	F = 2.846; P = 0.130	F = 1.248; P = 0.296
Nutrients and Ultraviolet Radiation	F= 10.770; P = 0.010*	F = 0.255; P = 0.627
Experiment 2: 18th-22th January, 2008		
Nutrients	F = 1.130; P = 0.319	F = 0.828; P = 0.389
Ultraviolet Radiation	F = 1.440; P = 0.264	F = 1.248; P = 0.296
Nutrients and Ultraviolet Radiation	F = 0.569; P = 0.472	F = 0.255; P = 0.627
Experiment 3: 22th-25th January, 2008		
Nutrients	F = 3.276; P = 0.108	F = 4.010; P = 0.080
Ultraviolet Radiation	F = 3.364; P = 0.104	F = 4.874; P = 0.058
Nutrients and Ultraviolet Radiation	F = 0.830; P = 0.389	F = 1.735; P = 0.224
Experiment 4: 03th-07th March, 2008		
Nutrients	F = 0.113; P = 0.745	F = 0.908; P = 0.368
Ultraviolet Radiation	F = 1.592; P = 0.243	F = 1.809; P = 0.216
Nutrients and Ultraviolet Radiation	F = 0.511; P = 0.495	F = 0.751; P = 0.411

The non significant effect observed in the present study would be explained probably due the presence of interaction between bacteria and dissolved organic carbon, under ultraviolet radiation exposure, because in these scenario, it happens a photopleaching of dissolved organic carbon that generate basic organic substances that sustain abundant bacterial biomass that generate nutrient recycling that can be used by phytoplankton communities (Reche *et al.*, 1998; Cole, 2000; Mc Callister *et al.*, 2005). If we considered the presence of coliform fecal bacteria in the site where was carried out the present experiment (Rivera *et al.*, 2010), it would be probably that the we have a complex scenario where it is possible that the bacteria would have an important role in nutrient availability for phytoplankton, and the potential presence of dissolved organic carbon would have a possible photoprotection effect against natural ultraviolet radiation that would be an important advantage for phytoplankton activity.

Acknowledgements

The present study was founding by Research Direction of the Catholic University of Temuco (Project DGI-CDA-UCT-01) and projects DI-08-0040 of the Universidad de la Frontera.

References

Barriga, J.P., Battini, M.A., Macchi, P.J., Milano, D. and Cussac, V.E., 2002. Spatial and temporal distribution of landlocked *Galaxias maculatus* and *Galaxias platei* (Pisces: Galaxidae) in a lake in the South American Andes. *New Zealand J. Mar. Freshwat. Res.*, 36: 349–363.

Battini, M., Rocco, V., Lozada, M., Tartarotti, B. and Zagarese, H.E., 2000. Effects of ultraviolet radiation on the eggs of landlocked *Galaxias maculatus* in northwestern Patagonia. *Freshwat. Biol.*, 44: 547–552.

Cole, J.J., 2000. Aquatic microbiology for ecosystem scientist: new and recycled paradigms in ecological microbiology. *Ecosystems,* 2: 215–225.

De los Ríos, P. and Soto, D., 2005. Survival of two species of crustacean zooplankton under to two chlorophyll concentrations and protection or exposure to natural ultraviolet radiation. *Crustaceana,* 78: 163–169.

De los Ríos, P., Hauenstein, E., Acevedo, P. and Jaque, X., 2007a. Littoral crustaceans in mountain lakes of Huerquehue National Park (38°S, Araucania Region, Chile). *Crustaceana,* 80: 401–410.

De los Ríos, P., Acevedo, P. and Verdugo, K., 2007b. Survival of *Ceriodaphnia dubia* (Crustacea, Cladocera) exposed to different screens against natural ultraviolet radiation. *Pol. J. Env. Stud.*, 16: 481–485.

De los Ríos, P., 2005. Survival of pigmented freshwater zooplankton exposed to artificial ultraviolet radiation and two levels of dissolved organic carbon. *Pol. J. Ecol.*, 53: 113–116.

De los Ríos, P., 2007. Short term effects of exposition to natural ultraviolet radiation on *Parabroteas sarsi* (Copepoda, Calanoida). *Biologia, Bratislava,* 62: 210–213.

Diaz, S., Camillón, C., Deferrari, G., Fuenzalida, H., Armstrong, R., Booth, C., Paladani, A., Cabrera, S., Casiccia, C., Lovengreen, C., Pedroni, J., Rosales, A., Zagarese, H. and Vernet, M., 2006. Ozone and UV radiation over southern South America: Climatology and anomalies. *Photochem. Photobiol.,* 82: 834–843.

Gao, K., Li, G., Helbling, W. and Villafañe, V.E., 2008. Variability of UVR effects on photosynthesis of summer phytoplankton assemblages from a tropical coastal area of the South China Sea. *Photochem. Photobiol.*, 83: 802–809.

Helbling, W.E., Villafañe, V. and Barbieri, E., 2001. Sensibilidad de comunidades fitoplanctónicas invernales de lagos andinos a la radiación ultravioleta artificial. *Rev. Chilena Hist. Nat.*, 74: 273–282.

Helbling, W.E., Buma, A.G.J., Van Del Poll, W., Ferández Zenoff, M.V. and Villafañe, V.E., 2008. UVR–induced photsynthetic inhibition dominates over DNA damage in marine dinoflagellates exposed to fluctuating solar radiation regimes. *J. Exp. Mar. Biol. Ecol.*, 365: 96–102.

Helbling, W.E., Farías, M.E., Fernández Zenoff, V. and Villafañe, V.E., 2006. *In situ* responses of phytoplankton from the subtropical Lake La Angostura (Tucuman, Argentina) in relation to solar ultraviolet radiation exposure and mixing conditions. *Hydrobiologia,* 559: 123–134.

Jaramillo, J., 2004. Evaluación del efecto de la radiación ultravioleta y nutrientes en la abundancia de cianófitas en los lagos temperados del sur de Chile. *Ph.D. Thesis,* Sciences Faculty, Austral University of Chile, 104 p.

Marinone, M.C., Menu-Marque, S., Añón Suárez, D., Dieguez, M.C., Pérez, A.P., De los Ríos, P., Soto, D., and Zagarese, H.E., 2006. UV radiation as a potential driving force for zooplankton community structure in Patagonian lakes. *Photochem. Photobiol.*, 82: 962–971.

Mc Callister, S.L., Bauer, J.E., Kelly, J. and Ducklow, H.W., 2005. Effects of sunlight on decomposition of estuarine dissolved organic C, N and P and bacterial metabolism. *Aq. Microb. Ecol.*, 40: 25–35.

Morris, D.P., Zagarese, H.E., Williamson, C.E., Balseiro, E.G., Hargreaves, B.R., Modenutti, B.E., Moeller, R.E. and Queimaliños, C.P., 1995. The attenuation of solar UV radiation in lakes and the role of dissolved organic carbon. *Limnol. Ocean.*, 40: 1381–1391.

Reche, I., Pace, M.L. and Cole, J.J., 1998. Interactions of photobleaching and inorganic nutrients in determining bacterial growth on colored dissolved organic carbon. *Microb. Ecol.*, 36: 270–280.

Rivera, R., De los Ríos, P. and Contreras, A., 2010. Indicadores de contaminación fecal en cuerpos de agua rural y urbano de la ciudad de Temuco. *Cienc. Inv. Agr.*, 37: 141–149.

Soto, D., 2002. Oligotrophic patterns in southern Chile lakes: the relevance of nutrients and mixing depth. *Rev. Chilena Hist. Nat.*, 75: 377–393.

Stein, J.R., 1973. *Handbook of Phycological Methods: Culture Methods and Growth Measurements*. Cambridge University Press, New York, USA, 448 p.

Steinhart, G.S., Likens, G.E., and Soto, D., 1999. Nutrient limitation in Lago Chaiquenes (Parque Nacional Alerce Andino, Chile): Evidence from nutrient experiments and physiological assays. *Rev. Chilena Hist. Nat.*, 72: 559–568.

Villafañe, V.E., Helbling, E.W., and Zagarese, H.E., 2001. Solar ultraviolet radiation and its impact on aquatic ecosystems of Patagonia, South America. *Ambio*, 30: 174–180.

Wangberg, S.A., Andreasson, K.I.M., Gustavson, K., Reinthaler, T. and Henriksen, P., 2008. UV-B effects on microplankton communities in Kongsfjord. *J. Exp. Mar. Biol. Ecol.*, 365: 156–163.

Zagarese, H.E., Williamson, C.E., Vail, T.L., Olsen, O. and Queimaliños, C.P., 1997a. Long-term exposure of *Boeckella gibbosa* (Copepoda, Calanoida) to *in situ* levels of solar UVB radiation. *Freshwat. Biol.*, 37: 99–106.

Zagarese, H.E., Feldman, M. and Williamson, C.E., 1997b. UV-B induced damage and photo-reactivation in three species of *Boeckella* (Copepoda, Calanoida). *J. Plank. Res.*, 19: 357–367.

Zagarese, H.E., Tartarotti, B., Cravero, W. and González, P., 1998a. UV damage in shallow lakes: The implications of water mixing. *J. Plank. Res.*, 20: 1423–1433.

Zagarese, H.E., Cravero, W., González, P. and Pedrozo, F., 1998b. Copepod mortality induced by fluctuating levels of natural ultraviolet radiation simulating vertical water mixing. *Limnol. Ocean.*, 43: 169–174.

Chapter 34

Beneficial Effects of *Aegle marmelos* Leaves on Blood Glucose Levels and Body Weight Changes in *Alloxan-*Induced Diabetic Rats

☆ *Leena Muralidharan*

ABSTRACT

Diabetes mellitus is one of the most recognized and clinically significant disorders of the endocrine system. It is characterized by the disturbances of carbohydrate, lipid and protein metabolism and an abnormal response to the glucose load. Feeding with aqueous extract of leaves of *Aegle marmelos* commonly called as *bael* in *alloxan*- induced diabetic rats significantly (P< 0.001) decreased blood glucose levels and significant (P< 0.01) increase in body weight changes were observed. In non-diabetic rats, the experimental *bael* leaves did not cause any hypoglycaemia effect and no significant body weight changes were found indicating that *Aegle marmelos* has anti- diabetic activity.

Keywords: Aegle marmelos, Diabetis mellitus, Alloxan, Blood glucose level, Body weight.

Introduction

All over the world, *Diabetes mellitus* is increasing. India is predicted to have the largest number of people with *Diabetes mellitus* (Shaw *et al.*, 1999). Though many new oral hypoglycaemia agents are now available, there is a great difficulty in choosing the right medication for longer period either because of their side effects or due to the poor response. Herbal drugs are traditionally used in various parts of the world to cure different diseases. The trend of using natural products has increased and the active plant extracts are frequently screened for new drug discoveries (Shridhar, 2002). Growing demand for herbal medicines is due to their effectiveness, minimal side effects and economical aspects.

The plant *Aegle marmelos* (*Roxb*) is popularly known as *bael* in India. It is a spine tree belonging to the family *Rutaceae*. It grows up to a height of 3- 6 metres. The leaves of this tree are oval in shape and the flowers have pleasant fragrance. The medicinal properties of this plant have been described in the *Ayurveda*. In fact, as per Charaka (1500 B.C.), no drug has been longer or better known by the inhabitants of India than *bael*.

Aegle marmelos leaf extract has been reported to regenerate the damaged pancreatic β cells in the diabetic rats (Shukla, 2000). Fresh aqueous and alcoholic leaf extracts of *Aegle marmelos* were reported to have cardio tonic effects in mammals (Grover *et al.*, 2002). The ethanol extract of *Aegle marmelos* leaf possesses anti- spermatogenic (Babu *et al.*, 1988).

Considering the importance of *bael*, the present study was undertaken to evaluate the anti- diabetic effect of *Aegle marmelos* on blood glucose level, body weight and behavioural changes in *alloxan-*induced diabetic rats to provide the scientific evidences on modern line.

Materials and Methods

Adult albino rats (Winstar strain) aged about 3 months weighing between 400- 420 g and free from any kind of infections were used. The animals were maintained as per the guidelines. For experimental purposes, the rats were kept fasting overnight but were allowed free access to water. The standard protocol for laboratory animal care was followed. Animal experimental design for the present study is shown in Figure 34.1.

Twenty male albino rats (Winstar strain) were divided into four groups of five rats each by random block design and were housed individually in wire mesh cages. Animals were maintained under the laboratory conditions with standard stock diet and water adlibitum.

Diabetes was introduced by intra- peritoneal administration of 150 mg/kg body weight of ice cold aqueous alloxan monohydrate (Kameshwar *et al.*, 1997) to two groups of rats served on diabetic control and diabetic experimental, respectively. After a fortnight, hyperglycemia was observed in both the groups of rats. The other two groups were kept as non- diabetic control and non-diabetic experimental, respectively. The rats were given high fibre and high protein diet. 20 g diets were fed and distilled water was provided adlibitum. The leftover food residues were collected to calculate the actual food intake.

The rats were weighed every week up to 4 weeks of experimental period to record the body weight changes. The initial and final blood glucose levels were measured from the tail veins with the help of glucometer using dextrostix. The results were compared with the control groups of non- diabetic and diabetic rats with the initial values of the same groups.

Experimental leaves were collected from the garden and washed well with the distilled water. 50 g of air- dried leaves were extracted in one litre of boiling water for 2 hours and were concentrated to half the volume. The resulting brown extract was cooled and filtered using Whatmann filter paper. Leaf extract was introduced by intra- peritoneal administration of a single dose of 10 mg/100 g every-day morning for a period of 30 days.

Urine sugar was checked by uristrix strips from Bayer. The rats were observed continuously for gross behavioural changes. Blood glucose was measured by One Touch Glucometer. The data was analysed statistically using variance 't' test.

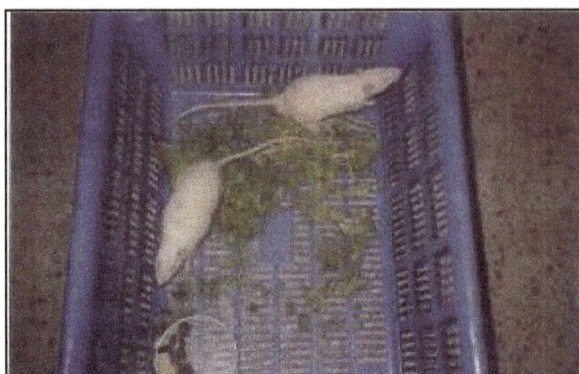

Figure 34.1: Experimental Animals Winstar Rats (200g) (Control)

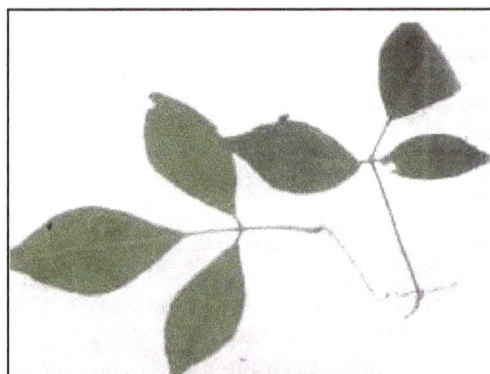

Figure 34.2: Experimental Plant (*Aegle marmelos*)

Results and Discussion

Experimental observation and preliminary data about the mechanism of action of *Aegle marmelos* is reported to offer scientific explanation towards the potential use of such plants for the treatment of *Diabetes mellitus*.

The study on LD_{50} of the leaf extract was observed to be appreciably high. No death occurred during the 30 days of experimental period. No mortality was seen even with the 20 times high dose feed of the leaf extracts. This further indicates high margin of safety. The leaf extracts of *Aegle marmelos* appears to be useful in inhibiting glucose-6-phosphate dehydrogenase, hepatic glucose output and controlling the elevated blood glucose levels. *Aegle marmelos* changed the insulin action in tissues. It is an insulin sensitizer which can be used in the treatment of diabetes. It improves the glycemia control by enhancing the insulin sensitivity in liver and muscle. Improved metabolic control with *Aegle marmelos* did not cause weight gain.

The behaviour of diabetic rats appeared sluggish and abnormally active initially but returned to normalcy after a week of treatment. The consumption of food increased initially which became normal in the treated rats. Fluid intake increased six times in the diabetic untreated rats while the intake of water was twice in *Aegle marmelos* treated rats.

Table 34.1: Effect on PPG and FBG Levels in Blood Glucose of Diabetic Rats after 30 Days Treatment with *Aegle marmelos*

Sl.No.	Group	FBG (Initial) mg/dl	FBG (Final) mg/dl	PPG (Initial) mg/dl	PPG (Final) mg/dl
1.	Control (Treated)	69 ± 8.6	74 ± 5.4	100 ± 9.8	102 ± 7.6
2.	Control (Untreated)	68 ± 7.8	79 ± 5.8	102 ± 6.7	104 ± 9.8
3.	Experimental (Treated)	275 ± 9.5	150 ± 3.8	278 ± 4.6	202 ± 6.2
4.	Experimental (Untreated)	279 ± 9.4	438 ± 4.6	287 ± 5.7	523 ± 7.8

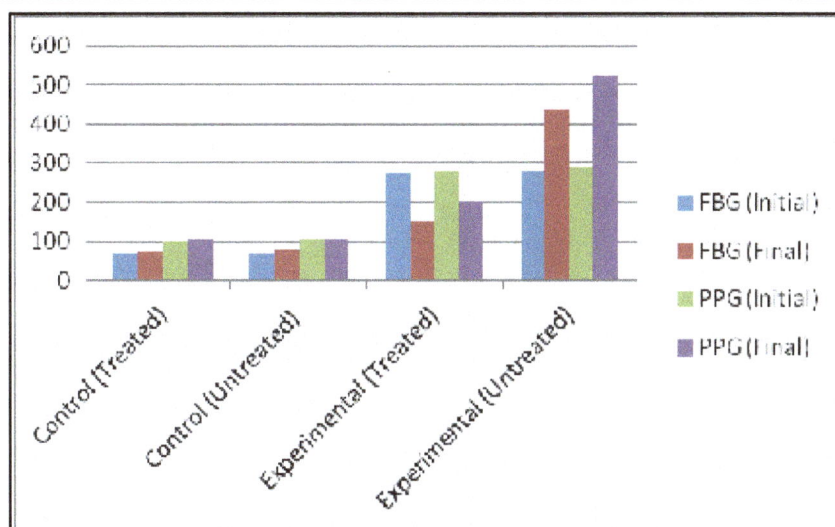

Table 34.2: Effect on Urine Sugar Levels in Diabetic Rats after 30 Day Treatment with
Aegle marmelos

Sl.No.	Group	Urine Sugar (Initial)	Urine Sugar (Final)
1.	Control (Treated)	-ve	-ve
2.	Control (Untreated)	-ve	-ve
3.	Experimental (Treated)	+4	+1
4.	Experimental (Untreated)	+4	+5

Table 34.3: Effect on Urine Sugar Levels in Diabetic Rats after 30 Day Treatment with
Aegle marmelos

Sl.No.	Group	Urine Sugar (Initial)	Urine Sugar (Final)
1.	Control (Treated)	-ve	-ve
2.	Control (Untreated)	-ve	-ve
3.	Experimental (Treated)	+4	+1
4.	Experimental (Untreated)	+4	+5

Table 34.4: Effect on Body Weight (B.W.) Changes in Diabetic Rats after 30 Day Treatment with
Aegle marmelos

Sl.No.	Group	Initial (g)	Final (g)
1.	Control (Treated)	200 ± 5	215 ± 9
2.	Control (Untreated)	195 ± 6	170 ± 4
3.	Experimental (Treated)	198 ± 9	189 ± 10
4.	Experimental (Untreated)	200 ± 6	178 ± 8

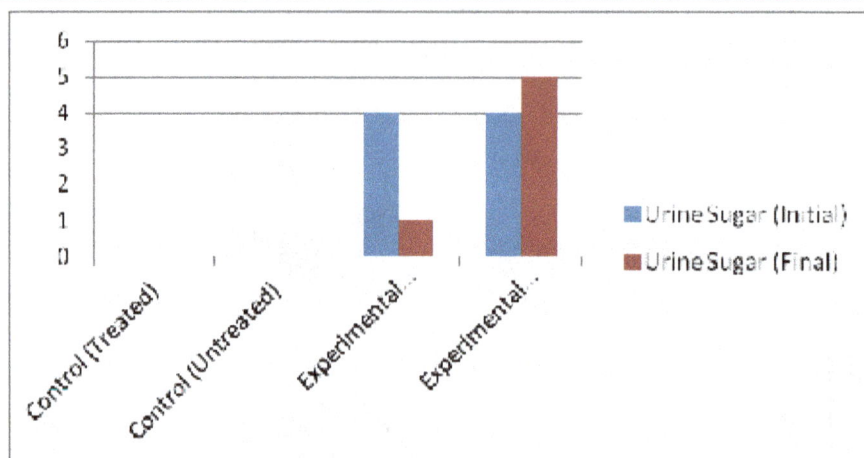

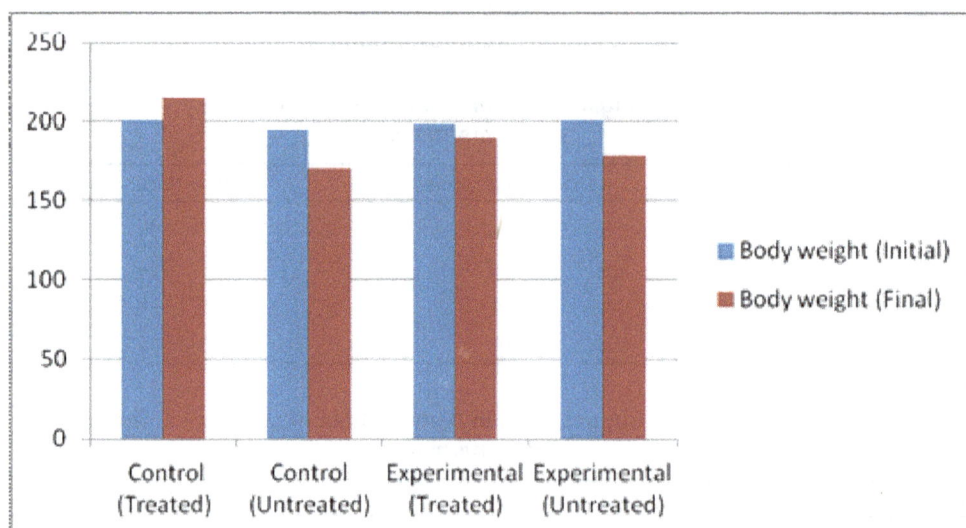

Conclusion

The present study concluded that the leaf extracts of experimental plant taken for the study *i.e. Aegle marmelos*, helps in regulating and maintaining the homeostatic metabolism in the body. Extracts were found to be effective as an antidiabetic agent. A detailed study on the metabolites of plant extracts of *Aegle marmelos* on release of insulin, release of glucose and uptake of glucose is very essential to throw light on its anti diabetic activities.

Acknowledgements

The authors wish to thank University of Mumbai for accepting the project proposal and providing financial assistance for carrying out the experiment.

References

Babu, B.V., Murthy, R., Prabhu, K.M. and Murthy, P.S., 1988. *Indian J. Biochem. Biophysics*, 25: 714–718.

Grover, J.K., Yadav, S. and Vats, V., 2002. *J. Ethnopharmacol.*, 81: 81–100.

Kameshwar, Rao, Giri, R., Kesavulu, M.M. and Apparao, C., 1997. *Manphar Vaidhya Patrika*, p. 33–35.

Shaw, J.E., De Courten, M.P. and Zimma, P.Z., 1999. *Diabetes in new Millennium*. The Endocrinology and Diabetic Research Foundation of University of Sydney, p. 1–9.

Shukla, R., Sharma, S.B., Puri, D. and Muthu, 2000. *Ind. J. Clinical Biochem.*, 15: 169–171.

Shridhar, G.R., 2002. *Current Science*, p. 83–91.

Previous Volumes–Contents

— Volume 1 —

2007, xvi+194p., figs., tabls., ind., 25 cm   Rs. 950

ISBN 81-7035-483-8

Preface   *vii*

List of Contributors   *xiii*

1. Ecology of Sahastradhara Hill-Stream at Dehradun, Uttaranchal   1
   D.S. Malik and P.K. Bharti

2. Effects of Habitats on the Occurrence and Distribution of Blue Green Algae
   in North Maharashtra, India   12
   S.R. Mahajan and S.N. Nandan

3. Zooplankton Composition in Vellar Estuary in Relation to Shrimp Farming   17
   M. Rajasegar, M. Srinivasan and S. Ajmal Khan

4. Kinetic of Planktons in Integrated Fish-Livestock Ponds Situated at Mid-Hill
   Altitude of Meghalaya, North Eastern India   24
   Sullip K. Majhi

5. Studies on the Ecology and Fish Fauna of Makroda Reservoir of Guna District
   (Madhya Pradesh)   31
   Dushyant Kumar Sharma

6. Marine Finfish Resources of Nagapattinam Coast, Tamil Nadu, South India   35
    M. Rajasegar, J. Gopal Samy and B. Bama Devi

7. Microbial Ecology of Groundwater at Lucknow City   38
    D.K. Singh, P.K. Bharti and D.S. Malik

8. Physico-chemical Parameters of the Benthic Environment of the Continental
    Slope of Bay of Bengal   52
    Surajit Das, P.S. Lyla and S. Ajmal Khan

9. Role of Limno-chemical Factors on Phytoplankton Productivity   64
    N. Shiddamallayya and M. Pratima

10. Studies on Certain Physico-chemical Parameters of Hataikheda Reservoir
    Near Bhopal, India   75
    Vikas Salgotra, S.A. Mastan, Adarsh Kumar and T.A. Qureshi

11. Water Quality of Bastawade Pond from Tasgaon Tahsil of Sangli District
    (Maharashtra) and its Significance to Prawn Farming   79
    S.A. Khabade and M.B. Mule

12. Bioremediation to Restore the Health of Aquaculture Pond Ecosystem   85
    A. Venkateswara Rao

13. Development of Fisheries through Biotechnology: An Appeal   95
    Indranil Ghosh

14. Optimization of Resources for Food and Income Security   100
    Salim Sultan

15. Development and Management of Reservoirs in India through Morpho-edaphic
    Index and Stocking Density: A Review   110
    B.R. Kiran, E.T. Puttaiah and K. Harish Kumar

16. Preliminary Study on Status of Fisheries in Salal Reservoir and Suggestive
    Measures for Development   124
    K.S. Charak and F.A. Fayaz

17. Effect of Slaughter House Waste Based Diets on Growth, Feed Utilization
    and Body Carcass Composition of *Labeo rohita* (Hamilton) Fingerlings   132
    Pradeep Kumar Singh and Sandhya Rani Gaur

18. Quantitative Protein Requirement of the Fingerlings of Indian Major Carp Rohu,
    Labeo rohita (Hamilton)   141
    Pradeep Kumar Singh, Sandhya Rani Gaur and M.S. Chari

19. Comparative Study of Effect of Three Different Types of Food on
Population Density of Freshwater Rotifer *Monostyla* sp. 148

Sampada Ketkar and Madhuri Pejaver

20. Structure Immune Organ (Spleen) in *Mystus gulio* (Hamilton) 157

B. Deivasigamani

21. Effects on Nucleic Acid Content of Muscles in the Starved Mosquito
Fish *Gambusia affinis* 163

I.A. Raja and K.M. Kulkarni

22. Toxicity Effect of Zinc on the Rice Field Crab *Spiralothelphusa hydrodroma* 166

S. Jayakumar, M. Deecaraman and S. Karpagam

23. Impact of Zinc and Cadmium on Hepatopancreas and Gonad of Estuarine
Oyster *Crassostrea cattuckensis* 173

G.D. Suryawanshi

24. The Pathobiology of Intestinal Neoechinorhynchosis, Caused by
Neoechinorhynchus rutili in Freshwater Fish 184

G. Benarjee and B. Laxma Reddy

Index 193

— Volume 2 —

2008, xvi+143p., col. plts., figs., tabls., ind., 25 cm Rs. 750

ISBN 81-7035-559-5

Preface vii

List of Contributors xiii

1. Biochemical Composition of Wild Copepods, *Acartia erythraea* Giesbrecht
and *Oithona brevicornis* Giesbrecht from Coleroon Coastal Waters,
Southeast Coast of India 1

M. Rajkumar, K.P. Kumaraguru Vasagam and P. Perumal

2. Studies on the Toxic Effects of Arsenic on Glucose Content in Freshwater Fish,
Labeo rohita (Hamilton) 21

K. Pazhanisamy

3. Microbial Resources of the Mangrove and Vellar Estuary, Southeast Coast of India 26

Surajit Das, P.S. Lyla and S. Ajmal Khan

4. A Study on the Physico-chemical Characteristics of Water in Vallur Village of Tiruvallur District, Tamil Nadu, India — 34
   M. Sangeetha, P. Lakshmi Devi, P. Indumathy, J. Jayanthi and M.G. Ragunathan

5. Zooplankton Diversity of Goutami Godavari Estuary, Yanam, Pondicherry (U.T.) — 38
   K. Sree Latha and S. Rajalakshmi

6. Research on Bioactive Compounds Producing Marine Actinobacteria — 43
   K. Sivakumar, Maloy Kumar Sahu, J. Rajkumar, T. Thangaradjou and L. Kanna

7. Physico-chemical Factors of the Waters of Vattambakkam Lake in Kanchipuram District, Tamil Nadu, India — 65
   R. Eswaralakshmi, J. Jayanthi and M.G. Ragunathan

8. Water Quality Index (WQI) of Shetty Lake and Hadhinaru Lake of Mysore District, Karnataka, India — 69
   B.M. Sudeep, S. Srikantswamy and S.P. Hosmani

9. Preliminary Observations on Physico-chemical Characteristics of Water and Soil of Domestic Sewage Water and It's Suitability for Aquaculture — 80
   S.S. Dhane, S.J. Meshram and P.E. Shinagre

10. Assessment of Microbial Load Along with Limnological Analysis in Flood Affected Areas of Thar Desert of Rajasthan, India — 85
    Mamta Rawat, Sumit Dookia and G.R. Jakher

11. Diversity of Fish Fauna in Nagathibelagulu Pond, Shimoga, Karnataka — 95
    H.M. Ashashree, M. Venkateshwarlu and H.M. Renuka Swamy

12. Culture Potential of Clams Along the Coast of India — 98
    Milind M. Girkar, Sachin B. Satam and Prakash Shingare

13. Vibriosis and Parasitic Isopod Infections in the Black Fin Sea Catfish, *Arius jella* — 102
    M. Rajkumar, R. Thavasi, J.P. Trilles, and P. Perumal

14. Assessment of Polyculture Trial of Giant Freshwater Prawn (*M. rosenbergii*) with *O. mossambicus* (Tilapia) and Carps in the Ponds of Aquafarmer — 110
    P.E. Shingare, G.A. Shirgur and V.B. Mehta

15. Role of Women in Aquaculture — 114
    K. Sree Latha

16. Polyculture Trials of Lates calcarifer (Jitada) with *O. mossambicus* (Tilapia) and Carps in the Station Pond and in Private Aquafarmer — 116
    P.E. Shingare, G.A. Shirgur and V.B. Mehta

17. Seasonal Variations in the Haematological Parameters of Three Species of the
Genus *Channa* from Khandala Reservoir, Dist. Osmanabad (MS) 120

K.R. Reddy, M.G. Babare and M.V. Mote

18. The Saline Soil Resources of Maharashtra and its Suitability for Aquaculture 123

S.S. Dhane, P.E. Shingare and S.J. Meshram

19. Effect of Water Probiotic on Water Quality and Growth of Catla Seed in Rain Fed Ponds 130

P.E. Shinagre, S.J. Meshram and S.S. Dhane

20. Biology of Anostracan Zooplankton 134

S.P. Chavan and M.S. Kadam

Index 141

— Volume 3 —

2010, xiv+176p., col. plts., figs., tabls., ind., 25 cm   Rs. 800

ISBN 978-81-7035-633-2

Preface *v*

List of Contributors *xi*

1. Status of Biodiversity in Inland Wetlands of Gwalior-Chambal Region in
Madhya Pradesh 1–11

R.J. Rao, R.K. Garg, S. Taigore, M. Arya, H. Singh, Bidyalakshmi and K. Kushwah

2. Marine Ornamental Fishes in the Little Andaman Island 12–18

*M. Murugan, Maloy Kumar Sahu, M. Srinivasan, Kamala Devi, S. Ajmal Khan
and L. Kannan*

3. Influence of Supplementary Feeds on the Growth and Excretory Metabolite
Levels in *Heteropneustes fossilis* 19–31

Meenakshi Jindal

4. Bactericides from Actinobacteria Isolated from the Sediments of Shrimp Pond 32–38

*K. Sivakumar, Maloy Kumar Sahu, V. Arul, Prashant Kumar, S. Raja,
T. Thangaradjou and L. Kannan*

5. The Dynamics of Gonad Growth and Ascorbate Status in Certain Commercially
Valued Marine and Freshwater Fishes of Orissa 39–48

A.K. Patra

6. Effect of Feeding Dietary Protein Sources on Daily Excretion in *Channa punctatus*
for Sustainable Aquaculture 49–60

Meenakshi Jindal, S.K. Garg and N.K. Yadava

7. Ornamental Fish Packing and Health Management   61–70
Sachin Satam and Balaji Chaudhari

8. Brachyuran Crab Resources of the Little Andaman Islands, India   71–77
Maloy Kumar Sahu, M. Murugan, R. Balasubramanian, S. Ajmal Khan and L. Kannan

9. Investigation on Tourism Effects of Macro Pollutants in the Beaches and
Mangrove Environment, Southeast Coast of India   78–87
K. Balaji, S. Sudhakar, P. Raja, G. Thirumaran and P. Anantharaman

10. A Comparison of Live Feed and Supplementary Feed for the Growth of Catfish Fry,
Clarias batrachus (Linn)   88–96
Meenakshi Jindal, N.K. Yadava and Manju Muwal

11. Culture of Indian Magur   97–106
M.M. Girkar, S.B. Satam and S.S. Todkari

12. Influence of Heavy Metals on Abundance of Cyanophyceae Members in Three
Spring-Fed Lake in Kempty, Dehradun   107–111
P.K. Bharti, D.S. Malik and Rashmi Yadav

13. Aquarium Keeping and Maintenance of Marine Ornamental Invertebrates   112–129
*K. Balaji, G. Thirumaran, R. Arumugam, K.P. Kumaraguruvasagam
and P. Anantharaman*

14. A Note on Water Quality in Aquarium   130–131
Indranil Ghosh

15. Water Borne Diseases   132–140
Deepa Dev

16. Studies on Aquatic Insects in Relation to Physico-chemical Parameters of
Anjani Reservoir in Sangli District of Maharashtra   141–147
S.A. Khabade and M.B. Mule

17. Studies on Groundwater Quality of Latur City in Maharashtra   148–151
M.V. Lokhande, K.G. Dande, S.V. Karadkhele, D.S. Rathod and V.S. Shembekar

18. Studies on Oxygen Levels and Temperature Fluctuation in Dhanegaon Reservoir
in Osmanabad District of Maharashtra   150–157
M.V. Lokhande, D.S. Rathod, V.S. Shembekar and K.G. Dande

19. Diurnal Changes of Some Physico-chemical Factors in Thodga Reservoir of Latur
District in Maharashtra   158–162
P.V. Patil and A.N. Kulkarni

20. **Impact of Heavy Metals on Aquaculture and Fisheries: Its Determination in Water, Sediment, Fish and Feed Samples** 163–169

 P.H. Sapkale, V.B. Mulye and R.K. Sadawarte

21. **Effect of Dimethoate on Blood Sugar Level of Freshwater Fish,** *Macronus vittatus* 170–173

 D.S. Rathod, M.V. Lokhande and V.S. Shembekar

 Index 175–176

— Volume 4 —

2010, xvii+182p., figs., tabls., ind., 25 cm Rs. 750

ISBN 978-81-7035-657-8

 Preface *vii*

 List of Contributors *xiii*

1. **Effect of Fluoride on Tissue Glycogen Levels in Freshwater Catfish,** *Clarias batrachus* (Linn) 1

 Achyutha Devi. J. and Ravi Shankar Piska

2. **Effect of Three Different Doses of Dried Water Hyacinth and Poultry Manure in the Production of Cladocerans** 6

 P.K. Srivastava and D. Roy

3. **Effet of Photosensitizer on Melanophore Responses of Blue Gouramei** (*Trichogaster trichopterus*) 14

 O.A. Asimi, P.P. Srivastava and T.H. Shah

4. **Dietary Effect of Chitin, Chitosan and Levamisole on Survival, Growth and Production of** *Cyprinus carpio* **Against the Challenge of** *Aeromonas hydrophila* **Under Temperate Climatic Conditions of Kashmir Valley** 22

 Sajid Maqsood, Samoon, M.H. and Prabjeet Singh

5. **Estimation of Activity of Some Metabolic Enzymes in the Fish** *Cyprinus carpio* **as a Function of their Age in Juveniles** 33

 Meenakshi Jindal, K.L. Jain, R.K. Verma and Simmi

6. **Studies on Oxygen Levels and Temperature Fluctuation in Dhanegaon Reservoir of Maharashtra** 37

 M.V. Lokhande, D.S. Rathod, V.S. Shembekar and K.G. Dande

7. **Seasonal Variation in Turbidity, Total Solids, Total Dissolved Solids and Total Suspended Solids of Dhanegaon Reservoir in Maharashtra** 42

 M.V. Lokhande, D.S. Rathod, V.S. Shembekar and S.V. Karadkhele

8.   Study of Physical Parameters of Ujani Reservoir in Solapur District of Maharashtra   48
     A.K. Kumbhar, D.A. Kulkarni, P.S. Salunke and B.N. Ghorpade

9.   Studies on Water Quality Parameters and Prawn Farming of Bastawade Pond
     from Tasgaon Tahsil of Sangli District, Maharashtra   54
     S.A. Khabade and M.B. Mule

10. Water Quality Management in Freshwater Fish Ponds   60
     M.M. Girkar, S.S. Todkari, A.T. Tandale and B.S. Chaudhari

11. Studies on Phytoplankton Diversity of Vinjasan Lake in Bhadrawati Town
     of Chandrapur District, Maharashtra   66
     P.N. Nasare, N.S. Wadhave, N.V. Harney and S.R. Sitre

12. Aquatic Protected Areas in River Narmada Around Hoshangabad   71
     Vipin Vyas

13. Selective Study on the Availability of Indigenous Fish Species having
     Ornamental Value in Some Districts of West Bengal   77
     A.K Panigrahi, Sarbani Dutta (Roy) and Indranil Ghosh

14. Utilization of Seaweeds in India and Global Scenario   82
     G. Thirumaran, R. Arumugam and P. Anantharaman

15. Helminthic Dynamics of Freshwater Clupeid, *Notopterurs notopterus* (Pallas)
     of Nizamabad District, Andhra Pradesh   106
     G.S. Jyothirmai, K. Geetha, D. Suneetha Devi, P. Manjusha and Ravi Shankar Piska

16. Role of Probiotics in Aquaculture   113
     M.M. Girkar, A.T. Tandale, S.S. Todkari and B.S. Chaudhari

17. Role of Organics and Probiotics in Shrimp Culture   116
     R. Saravanan, S. Rajagopal and P. Vijayanand

18. Prospectus of Magur Culture   123
     M.M. Girkar, S.S. Todkari, S.B. Satam and A.T. Tandale

19. Cage Culture: Future Potential Culture System   132
     Mangesh Gawde, Abhijeet Thakare and Gauri Sawant

20. Crafts and Gears Used in Yeldari Reservoir, Maharashtra   136
     S.D. Niture and S.P. Chavan

21. Socio-economic Status of Fishermen Community Around Bori Tank Near
     Naldurg, Maharashtra   143
     M.G. Babare and M.V. Mote

22. Success Rate of Fishing in Potential Fishing Zone (PFZ) Grounds Over
    Non-Potential Fishing Zone (Non-PFZ) Grounds   147
    A.U. Thakare and M.M. Shirdhankar

23. Fisheries Management of Yeldari Reservoir, Maharashtra   152
    S.D. Niture and S.P. Chavan

    Previous Volumes   173

    Index   181

— Volume 5 —

2011, xviii+231p., col. plts., tabls., figs., ind., 25 cm   Rs. 1200

ISBN 978-81-7035-697-4

1. Agro-based Material for Purifying Turbid Water   1
   Syeda Azeem Unnisa, P. Deepthi and K. Mukkanti

2. Gambusia Fish for Controlling of Pathogenic Water Microorganisms   7
   Chand Pasha, Vinod Kumar Chaouhan, M.D. Hakeem, B. Sri Divya and J. Venkateswar Rao

3. Ultraviolet Radiation Tolerance in Zooplankton Species of Tinquilco Lake
   (38°S Araucania Region, Chile): Experimental and Field Observations   14
   Patricio De los Rios, Enrique Hauenstein, Patricio Acevedo and Ximena Jaque

4. Phytoplankton Diversity in Ramdara Reservoir Near Tuljapur, Maharashtra   22
   J.S. Mohite and P.K. Joshi

5. Incidence of Vibriosis in the Indian Magur (*Clarius batracus* L.) in Saline
   Water Ponds of Haryana: A New Report from India   24
   T.P. Dahiya and R.C. Sihag

6. Breakthrough in Breeding and Rearing of Gold Fish (*Carassius auratus*)
   in Temperate Climatic Conditions of Kashmir Valley   32
   Sajid Maqsood, Prabjeet Singh and M.H. Samoon

7. Effect of Papain on Growth Rate and Feed Conversion Ratio in Fingerlings
   of *Cyprinus carpio* Under Temperate Climatic Conditions of Kashmir Valley   36
   Prabjeet Singh, M.H. Balkhi, Sajid Maqsood and M.H. Samoon

8. On the Aspect of Seed Production and Prospects for Brackishwater Culture
   of Pearl Spot, *Etroplus suratensis*   42
   S.D. Naik, S.T. Sharangdher, H.B. Dhamagaye and R.K. Sadawarte

9. Food and Feeding Habits of Mudcrab, *Scylla* spp. of Ratnagiri Coast, Maharashtra   47
    A.B. Funde, S.D. Naik, S.A. Mohite, G.N. Kulkarni and A.V. Deshmukh

10. Potential for Ammonia and Nitrite Reducing Products for Shrimp Farms
    in Andhra Pradesh   53
    N.A. Sadafule, S.S. Salim and A.D. Nakhawa

11. Evaluation of Effect of Different Salinities on Growth and Maturity of
    Indian Medaka, *Oryzias melastigma* (Mcclelland, 1839)   61
    A.S. Pawar, S.T. Indulkar and B.R. Chavan

12. Acute Toxicity of Dimethoate on Freshwater Fish, *Channa gachua*   70
    R.M. Reddy and V.S. Shembekar

13. Impact of Copper Sulphate on the Oxygen Consumption in the
    Freshwater Fish, *Catla catla*   75
    C.M. Bharambe

14. Effect of Temperature on Biochemical Composition of *Lamellidens marginalis*   79
    S.S. Surwase, D.A. Kulkarni and R.S. Chati

15. Water Quality Management for Sustainable Aquaculture   88
    Meenakshi Jindal

16. Modified Atmosphere Packaging (MAP) in Fisheries   93
    S.S. Todkari and M.M. Girkar

17. Piscivorous Birds of Dhanora Tank in Bhokar Tahasil of
    Nanded District, Maharashtra   96
    V.S. Kanwate and V.S. Jadhav

18. Chloride Content in Water from Shikara Dam Near Mukhed in Nanded District   99
    M.S. Pentewar, V.S. Kanwate and V.R. Madlapure

19. Studies on Aquatic Insects in Relation to Physico-chemical Parameters
    of Anjani Reservoir from Tasgaon Tahsil of Sangli District, Maharashtra   101
    S.A. Khabade and M.B. Mule

20. Physico-chemical Analysis of Small Reservoir Budha Talab in Raipur   111
    P. Biswas, H.K. Vardia and A. Ghosh

21. Studies on Algal Flora and Physico-chemical Characteristics of Shikara Reservoir
    in Nanded District, Maharashtra   122
    S.D. Dhavle, H.M. Lakde and S.D. Lohare

22. On a New *Senga waranensis* (Cestoda : Ptychobothridae) from
    Mastacembellus aramatus at Warnanagar in Maharashtra — 125
    L.P. Lanka, S.R. Patil, A.D. Mohekar and B.V. Jadhav

23. On a New *Hexacanalis trygoni* (Cestoda : Lecanisephalidae) from
    Trygon zugei at Malvan in Sindhudurg District of Maharashtra — 131
    L.P. Lanka and B.V. Jadhav

24. Length-Weight Relationship in *Mystus seenghala* — 135
    M.S. Pentewar and V.S. Kanwate

25. Costs and Earnings Analysis of Gill Net and Trawl Net Operation
    along the Ratnagiri Coast — 138
    S.K. Barve, P.C. Raje, M.M. Shirdhankar, K.J. Chaudhari and M.M. Gawde

26. Study of Protein Metabolism in Hepatopancreas and Muscle of Prawn
    Penaeus monodon on Exposure to Altered pH Media — 143
    V. Sailaja, E. Madhuri, K. Ramesh Babu, S. Rama Krishna and M. Bhaskar

27. Reproductive Biology of *Xancus pyrum* Varieties Acuta and Obtusa from
    Thondi Coastal Waters-Palk Strait (South East Coasts of India) — 150
    C. Stella and J. Siva

28. Protective Role of Ascorbic Acid on the Lead and Cadmium Induced
    Alteration in Total RBC of the Freshwater Fish, *Channa orientalis* (Schneider) — 156
    V.R. Borane, B.R. Shinde and R.D. Patil

29. Seasonal Variation in Faecal Coliform and Total Coliform Bacterial
    Density at Pamba River in Comparison with River Achencovil — 160
    Firozia Naseema Jalal and M.G. Sanal Kumar

30. Checklist of Birds of Ghodpeth Reservoir of Bhadrawati Tahsil in
    Chandrapur District of Maharashtra State — 165
    N.V. Harney, S.R. Sitre, N.S. Wadhave and P.N. Nasare

31. Impact of Sugar Mill Wastewater on Groundwater Quality — 170
    R.D. Joshi and S.S. Patil

32. Physico-chemical Properties of Groundwater of Jintur Taluka in
    Parbhani District of Maharashtra — 177
    V.B. Pawar and Kshama Khobragade

33. Marine Algae Hemagglutinins from the Coast of Goa — 185
    Kumar Sudhir, Tiwary Mukesh, Kumari Switi and Barros Urmila

34. Two New Distributional Records of Bivalve Species of Family Spondylidae
    from Mandapam Area, South-East Coast of India   190
    C. Stella, S. Vijayalakshmi and A. Murugan

35. Protein Content Variation in Some Body Components of *Barytelphusa guerini*
    After Exposure to Zinc Sulphate   196
    R.P. Mali and Shaikh Afsar

36. Gonadosomatic Index and Spawning Season of Snow Trout
    Schizopyge esocinus (Heckel, 1838)   199
    Shabir Ahmad Dar, A.M. Najar, M.H. Balkhi, Mohd. Ashraf Rather and Rupam Sharma

37. Constraints in Shrimp Farming in the North Konkan Region of Maharashtra State   203
    A.R. Sathe, R. Pai, M.M. Shirdhankar and M.M. Gawde

38. Studies on pH and Total Dissolved Solids Fluctuations in Kaij City, Maharashtra   208
    A.D. Chalak

39. Effect of *Piper longum* Extract on the Gross Primary Productivity of an
    Angel Fish Mass Culture System   211
    K. Kannan and G. Rajasekaran

40. A Study on the Development of Immunity in the Common Carp,
    Cyprinus carpio by Neem Azal Formulation   216
    N. Muthumurugan, M. Pavaraj and V. Balasubramanian

41. Toxic Effect of Imidacloprid on the Lipid Metabolism of the Estuarine Clam,
    Katelysia opima (Gmelin)   220
    V.B. Suvare, A.S. Kulkarni and M.V. Tendulkar

42. Sodium Fluoride Induced Protein Alterations in Freshwater Fish, *Labeo rohita*   224
    M.D. Kale, S.A. Vhanalakar, S.S. Waghmode and D.V. Muley

    Index   229

— Volume 6 —

2012, xv+345p., col. plts., figs., tabls., ind., 25 cm   Rs. 1800

ISBN 978-81-7035-782-7

1. A Preliminary Study on Biotic Community and Water Quality of Umaim
   River in East Khasi Hills, Meghalaya   1
   Umesh Bharti, P.K. Bharti, D.S. Malik, Pawan Kumar and Vijender Singh

2. Biodiversity of Yashwant Lake of Toranmal: Seasonal Variation in
   Molluscs Density and Species Richness   9
   A.P. Ekhande, J.V. Patil and G.S. Padate

3. Study on Ichthyofauna of Waterbodies Around Ambajogai, Maharashtra 22
G.B. Sanap, R.B. Haregaokar and V.B. Sakhare

4. Seasonal Fluctuation of Phytoplankton in Nagzari Tank Near Ambajogai, Maharashtra 30
S.W. Bhivgade, A.S. Taware, U.S. Salve and N.B. Pandure

5. Total Zooplankton Abundance and Biomass in Three Contrasting Lentic
Ecosystems of Mysore, Karnataka State 34
Koorosh Jalilzadeh and Sadanand M. Yamakanamardi

6. Water Quality Status of Kirung Ri River at Nganglam, Pemagatshel (Bhutan) 50
Umesh Bharti, P.K. Bharti, D.S. Malik, Pawan Kumar and Gaurav

7. Food and Feeding Habits of the Mudskipper, *Boleophthalmus boddarti* (Pallas)
from Vellar Estuary, Southeast Coast of India 57
V. Ravi

8. Limnochemistry of Bennithora Dam Near Gulbarga, Karnataka 67
B. Vasanthkumar and K. Vijaykumar

9. A Study on Physico-chemical Parameters of Perennial Tank, Laxmiwadi,
Kolhapur District, Maharashtra 72
S.A. Manjare, S.A. Vhanalakar and D.V. Muley

10. Monthly Variations in Physico-Chemical Parameters of River Tunga,
Shivamogga, Karnataka 78
H.A. Sayeswara, K.L. Naik and Mahesh Anand Goudar

11. Studies on Surface Water Quality Evaluation of Gowdana Pond,
Shivamogga, Karnataka 83
H.A. Sayeswara, Nafeesa Begum and Mahesh Anand Goudar

12. Heavy Metals in Water and Sediment of Ichchamati River in
East Khasi Hills, Meghalaya 88
P.K. Bharti, D.S. Malik, Pawan Kumar and Umesh Bharti

13. Assessment of Heavy Metals in Water and Sediment of Hatmawdon River
near Bangladesh Border in District East Khasi Hills, Meghalaya 95
Umesh Bharti, P.K. Bharti, D.S. Malik, Pawan Kumar, Gaurav and Vijender Singh

14. Benthic Biodiversity in the Marine Zone of Vellar Estuary, Southeast Coast of India 102
P. Murugesan and S. Muthuvelu

15. Mangrove Biodiversity and its Conservation at Ratnagiri 124
A.S. Kulkarni, M.V. Tendulkar, S.M. Nikam and A.S. Injal

16. Status of Fishes in Mangroves of Kali Estuary, Karwar, Central West Coast of India   136
    S.V. Roopa, J.L. Rathod and B. Vasanth Kumar

17. Marine Organisms as a Source of Drug: An Overview   141
    J.L. Rathod and P.V. Khajure

18. Haematological Responses of *Anabas testudineus* (Bloch) exposed to
    Sublethal Concentrations of Plant Extract of *Calotropis gigantea* (L.) R.Br.   164
    K. Sree Latha

19. Adaptive Changes in Respiratory Movements of Air-breathing Fish, *Anabas testudineus*
    (Bloch) Exposed to Latex and Plant extract of *Calotropis gigantea* (L.) R.Br. and
    Recovery of Toxicity with Additive Nutrients   171
    K. Sree Latha

20. Role of Purna Fish Co-operative Society to Rejuvinate Yeldari Reservoir Fishery
    in Parbhani District of Maharashtra   176
    S.D. Niture and S.P. Chavan

21. Distribution of Bacterial Flora in Pearl Spot *Etroplus suratensis* (Bloch)
    during Captivity   188
    J.L. Rathod and U.G. Naik

22. Effect of Climatic Changes on Prawn Biodiversity at Mochemad Estuary of
    Vengurla, South Konkan, Maharashtra   200
    V.M. Patole, S.G. Yeragi and S.S. Yeragi

23. Impact of Climate Change on the Fish Farming   204
    Meenakshi Jindal and Kavita Sharma

24. Standardisation of Natural Colour in Surimi made Shrimp Analogs   214
    Y.T. Patil, A.T. Tandale, S.S. Todkari, M.M. Girkar and S.B. Gore

25. Antimicrobial Activity of *Acanthus ilicifolius* Extracted from the Mangroves
    Forest of Karwar Coast   220
    P.V. Khajure and J.L. Rathod

26. Mangroves Diversity in Kali Estuary, Karwar   224
    Pradeep V. Khajure and J.L. Rathod

27. Assessment of Water Quality of Byramangala Lake in Bangalore City   234
    H. Krishnaram, M. Ramachandra Mohan Shivabasavaiah and Rmanjnath

28. Biological Index of Pollution in Ponds and Lakes of Mysore District, Karnataka   244
    T.B. Mruthunjaya and S.P. Hosmuni

29. Rotifer Biodiversity of Phutala Lake of Nagpur City, Maharashtra 248
S.R. Sitre and S.B. Zade

30. Phytoplankton Diversity in Wet Soil of Wetlands from Tasgaon Tahsil, Maharashtra 253
S.A. Khabade, M.B. Mule and S.S. Sathe

31. The Bray-Curtis Similarity Index for Assessment of Euglenaceae in Lakes of Mysore 257
T.B. Mruthunjaya

32. Study of Hydrophytes and Amphibious Plants Occurred in Panchanganga River,
Maharashtra 261
V.G. Patil, S.A. Khabade and S.K. Khade

33. An Overview of Nile Tilapia (*Oreochromis niloticus*) and Low Cost Feed Formulation
Technique for its Culture 264
B.R. Chavan and A. Yakupitiyage

34. Impact of Probiotic Supplementation in Freshwater Fish *Cyprinus carpio* 285
R. Thamarai Selva, V. Vignesh, S. Jothi Lakshmi and R. Thirumurugan

35. Host-Parasite Relationship with Reference to Histochemical Changes in
an Acanthocephalan Infected Fish 295
B. Laxma Reddy and G. Benarjee

36. Seasonal Occurrence of *Genarchopsis goppo* in Freshwater Teleost,
Clarias batrachus 301
B. Laxma Reddy, P. Gowri and G. Benarjee

37. Modified Chinese Circular Hatchery of Marathwada, India 308
S.D. Niture and S.P. Chavan

38. Variations in Growth of *Perionyx excavatus* Fed *Ad libitum* on Different Feed during
Different Seasons from Hatchlings to Post Reproductive Statges 324
Manjunathr, Krishna Ram H. and Shivabasavaiah

Previous Volumes 331

Index 343

Index

A

Acidity 28, 30, 33, 34, 35, 36, 37

Aegle marmelos 214-219

Aeromonas hydrophila 135-144

Air temperature 28, 30, 31, 35, 36, 37, 41, 52, 53, 184, 185

Alkalinity 28, 30, 32, 33, 34, 30, 32, 34, 35, 36, 37, 40, 41, 42, 45, 48, 62, 64, 151, 152, 153, 156, 157, 158, 159, 183, 184, 185, 193

Ammonia 4, 6, 22-26, 40, 42, 45, 46, 47, 48, 62, 64,

Anabas testudineus 144-149, 162, 164, 197

Anguilla Anguilla 147

Anthropogenic activities 58, 63

Aquaculture 1, 2, 3, 7, 28

Arboreal birds 116-117

Avifauna 108-118

B

Bamni 89

Bhadra wildlife Sanctuary 27-39

Biochemical oxygen demand 28, 30, 32, 34, 35, 36, 37, 40, 42, 44, 45, 46, 47, 48, 97-103

Brackish water 5

C

Calcium 33, 35, 36, 37, 46, 61, 183, 184, 186

Carassius auratus 135-144

Carbon dioxide 28, 30, 33, 35, 36, 37, 40, 45, 48, 55, 59, 144, 150, 151, 152, 153, 156, 157, 158, 159, 183, 184, 185, 186

Catla catla 2, 10, 11, 13, 131-134, 162, 163, 164, 196, 197

Chemical Oxygen Demand 28, 30, 33, 34, 35, 36, 37, 97-103

Chloride 28, 30, 33, 34, 35, 36, 37, 40, 42, 46, 48, 61, 183, 184, 185, 191, 192, 193

Cirrhinus mrigala 2, 10, 11, 13, 71, 162, 163, 197

Cladocera 153, 154, 156, 159, 160, 166, 167, 169, 170, 171, 172, 173, 175, 176, 178

Clarias batrachus 4-9, 162

Conductivity 40, 44

Copepods 153, 154, 156, 159, 160, 166, 167, 168, 169, 171, 172, 174, 175, 176, 178, 179,

Coracle 189

Ctenopharyngodon idella 11, 162, 163, 164

Cypermethrin 17, 18, 19, 20, 76

Cyprinus carpio 10, 11, 71, 124-130, 162, 163, 197

D

Diabetes mellitus 214, 216

Diatoms 59

Dissolved oxygen 4, 6, 18, 19, 28, 32, 34, 35, 36, 37, 40, 41, 42, 44-44, 47, 48, 51, 52, 54-55, 56, 59, 60, 61, 62, 64, 97-103, 150, 151, 152, 156, 157, 158, 167, 183, 185

E

Electrical conductivity 28, 31, 32, 34, 35, 36, 37

Eutrophication 63, 178

F

Faecal coliforms 65, 66

Faecal streptococci 65, 66

Feed conversion efficiency 68

Feed Conversion ratio 4, 6, 69, 72

Fenthion 124-130

Fenneropenaeus merguiensis 68-74

Fish culture 7, 11, 45, 119

Fish farming 5

Fishing crafts 188-190

Fishing gears 188-190

Floating and diving water birds 115-116

G

Gadus morhua 147

Glossogobius giuries 80-88, 162

Grass carp 10, 12, 13, 14

Groundwater quality 191-194

H

Hardness 28, 34, 35, 36, 37, 40, 41, 42, 45, 46, 151, 153, 156, 157, 158, 159, 183, 184, 185, 186, 191, 192, 193

I

Ichthyofauna 162-166

Imidacloprid 75-79

Industrial effluents 18

Insect larvae 1

K

Katelysia opima 75-79

L

Labeo calbasu 2, 14

Labeo rohita 2, 10, 11, 13, 80, 82, 83, 84, 85, 86, 87, 162, 163, 164

Lactobacillus sp. 80-88

Laxmiwadi tank 156-161

Lepidocephalichthys guntea 17-21

Limnology 41

M

Macrobrachium rosenbergi 2

Magnesium 33, 35, 36, 37, 46

Malathion 17, 18, 19

Mankeshwar beach 58-59

Mosquito control 17

N

Nemipterus japonicus 89-96

Nitrates 40, 42, 44, 47, 61, 151, 152, 153, 157, 158, 159

Nitrites 40, 42, 44, 47

Notopterus notopterus 145, 147, 162, 163, 164, 195-199

O

Omnivorous 5, 132

Oreochromis mossambicus 22-26, 162, 163, 164

Oreocromis niloticus 71

Oxygen consumption 17-21

Organochlorine pesticide 131-134

P

Pathogenic bacteria 62

Penaeus monodon 73

Pesticides 22, 23

pH 4, 6, 19, 23, 28, 34, 35, 36, 37, 40, 41, 42, 44, 47, 48, 51, 52, 55, 56, 59, 61, 62, 64, 121, 150, 151, 156, 157, 167, 177, 183, 184, 185, 191, 192

Phenology 105, 106

Phosphate 34, 35, 36, 37, 42, 45, 47, 62, 152, 153, 156, 158, 159

Phytoplankton 28, 55, 58, 62, 63, 64, 167-170, 173, 174, 175, 200-207, 208-213, 159, 166, 167

Polychaetes 92

Protein efficiency ratio 4, 6, 7

Poultry manure 5

R

Rani masa 89

Rankala Lake 60-61

Raviwar Peth Lake 62-67

Rotifers 1, 150, 153, 154, 156, 159, 160, 166, 167, 168, 169, 170, 171, 172, 173, 175, 176, 177, 178

S

Salinity 51, 52, 55-56, 59

Sedgwick rafter Cell 157

Serum Acetylcholinesterase levels 195-199

Shannon-Weiner Index 178

Siddeshwar dam 183-187

Silicates 40, 41, 46

Specific growth rate 4, 6

Sugarcane bagasses 1, 2, 4-9

Sulphates 40, 42, 46, 61, 192, 193

T

Topography 51

Total Dissolved Solids 28, 30, 31, 34, 35, 36, 37, 40, 45, 47, 48, 61, 64, 191, 193

Transparency 64, 167

Turbidity 44, 45, 48, 150, 151, 156, 157

U

Ultraviolet radiation 200-207, 208-213

V

Vibrio harveyi 119-123

W

Water temperature 28, 30, 31, 32, 35, 36, 37, 40, 41, 44, 52, 54, 151, 157, 158, 184, 185

W.H.O. standards 27

Z

Zooplankton 7, 46, 150-155, 156-161, 166, 167, 168, 170-173, 175, 176, 177, 178